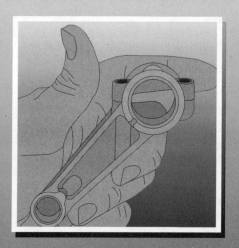

Introduction

Learning Inventor is a text designed with the student as well as the engineering professional in mind. You will find it is presented in a manner designed to facilitate learning—using practical examples and clear instructions. If you are looking for loads of theory, you will not find it in this text. The intention of this text is to recreate the actual workflow experienced by professionals as they use the software. After all, this text was written by those very same professionals. By the time you have finished this text, you will have a keen understanding of the methods used to produce a viable solid model part or assembly in Inventor.

Parametric design is very important in Inventor. You will encounter this a great deal. Throughout this text you will find example after example of parametric design principles. This text cannot stress this enough. The fact that your solid model should not be a static part but a dynamic part, able to withstand revision after revision is of great importance. You will hear this principle echoed throughout this text. The software was designed for this capability, so take advantage of it.

The goal of *Learning Inventor* is to present a process-based approach to the Inventor tools, options, and techniques. Each topic is presented in a logical sequence where they naturally fit in the design process of real-world products. In addition, this text offers the following features:

- Inventor tools are introduced in a step-by-step manner.
- Easily understandable explanations of how and why the tools function as they do.
- Numerous examples and illustrations to reinforce concepts.
- Professional tips explaining how to use Inventor effectively and efficiently.
- Practices involving tasks to reinforce chapter topics.
- Chapter tests for reviewing tools and key Inventor concepts.
- Chapter exercises to supplement each chapter.

Fonts Used in This Text

Different typefaces are used throughout the chapters to define terms and identify Inventor tools. Important terms always appear in bold-italic face, serif type. Inventor menus, tools, variables, dialog box names, and button names are printed in bold-face, sans serif type. File names, directory names, paths, and keyboard-entry items appear in the text in Roman, sans serif type. Keyboard keys are shown inside brackets [ ] and appear in Roman, sans serif type. For example, [Enter] means to press the enter (return) key.

Flexibility in Design

Flexibility is the key word when using *Learning Inventor*. This text is an excellent training aid for individual as well as classroom instruction. *Learning Inventor* teaches you Inventor and its applications to real-world problems. It is also a useful resource for professionals using Inventor in the work environment.

Notices

There are a variety of notices you will see throughout the text. These notices consist of technical information, hints, and cautions that will help you develop your Inventor skills. The notices that appear in the text are identified by icons and rules around the text. The notices are as follows:

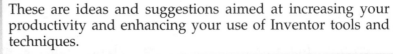

PROFESSIONAL TIP These are ideas and suggestions aimed at increasing your productivity and enhancing your use of Inventor tools and techniques.

NOTE A note alerts you to important aspects of the tool or activity that is being discussed.

CAUTION A caution alerts you to potential problems if instructions or tools are used incorrectly, or if an action could corrupt or alter files, folders, or disks. If you are in doubt after reading a caution, always consult your instructor or supervisor.

Reinforcement and Evaluation

The chapter examples, practices, tests, and exercises are set up to allow you to select individual or group learning goals. Thus, the structure of *Learning Inventor* lends itself to the development of a course devoted entirely to Inventor training. *Learning Inventor* offers several ways for you to evaluate your performance. Included are:

- **Examples.** The chapters include mini-tutorial examples that offer step-by-step instructions for producing Inventor drawings, parts, and/or assemblies. The examples not only introduce topics, but also serve to help reinforce and illustrate principles, concepts, techniques, and tools.
- **Practices.** Chapters have short sections covering various aspects of Inventor. A practice composed of several instructions is found at the end of most sections. These practices help you become acquainted with the tools and techniques just introduced. They emphasize a specific point.
- **Chapter Tests.** Each chapter also includes a written test. Questions may require you to provide the proper tool, option, or response to perform a certain task.

- **Chapter Exercises.** A variety of drawing exercises follow each chapter. The exercises are designed to make you think and solve problems. They are used to reinforce the chapter concepts and develop your skills.

The *User's Files* CD

Each chapter consists of examples, practices, and exercises. Most of these activities require a file that has been created and supplied to you on the *User's Files* CD. This CD is packaged with the text. At various points throughout this text, you will be instructed to open or access a file. The file will be used to develop, emphasize, and reinforce Inventor concepts and techniques.

The CD is set up so each chapter has a file folder with specific chapter-related files. You may install these files on your hard drive or run them from the CD. Before installing any file to your hard drive, be sure to check with your instructor or network administrator.

About the Authors

The authors have been using the software since its infancy. They can safely say that you have the best of all releases in Inventor 9.

Thomas Short

Thomas Short is a nationally recognized expert in Inventor and 3D solid/surface modeling. He is a registered mechanical engineer in Michigan and has his B.S. and M.S. in Mechanical Engineering. He was a faculty member in the Mechanical Engineering Department at Kettering University (formerly General Motors Institute). He is also a member of the Society of Manufacturing Engineers and the Society of Automotive Engineers.

Thomas has been using, teaching, and consulting in CAD since 1975 and AutoCAD since 1983. In 1984, he founded CommandTrain, Inc., which was a Premier Authorized AutoCAD Training Center. He is now a consulting engineer for Munro & Associates in Troy, Michigan. He is a Certified AutoCAD Instructor and Certified Technical Instructor. He has written for and taught many courses in Autodesk software including AutoCAD, AutoCAD 3D, AutoLISP, and Inventor. He has taught at every Autodesk University. He has also taught AutoCAD classes in the United States, Mexico, Brazil, Canada, and England.

As a CAD consultant and trainer he has worked for many companies, including Ford Motor Company, General Motors, Visteon, 3M, and McDonnell Douglas. He has helped several tooling and manufacturing companies in the Detroit area implement successful strategies for using Inventor as their solid modeling system.

Anthony Dudek

Anthony Dudek has been working with AutoCAD and Inventor in the mechanical engineering industry since 1985. He is now a nationally recognized expert in Inventor and 3D solid/surface modeling. Anthony is a Certified Autodesk Instructor and Autodesk Training Specialist. For the past several years, he has been a speaker at Autodesk University.

Anthony has been a Mechanical Designer, 3D Modeler, CAD Manager, programmer, and management/networking consultant. For the past 10 years, he has run A.F. Dudek & Associates, which is a CAD consulting firm in the Chicago area. He is also an Inventor instructor at Moraine Valley Community College's Premier Authorized AutoCAD Training Center, as well as other venues. He is the author of numerous articles and training materials. He is currently on permanent contract with BP as CAD and Document Management Project Manager.

Bill Kramer

Bill Kramer is a computer scientist who first began popularizing programming languages in 1982. Since then, Bill has been writing about the customization of CAD/CAM systems covering a wide range of languages and platforms in both books and magazines. His article credits include the longest running series in *CADENCE*, *Programmer's Toolbox*, and currently he is writing "Hot Tip Harry" for *CADALYST* magazine. Bill has written in areas that include AutoLISP macros, LISP tutorials, VBA tutorials, and ObjectARX programming with C++.

Bill first started working with engineering system graphics in the late 1970s. His expertise in building translation systems between different graphic databases landed him a position in the special products group at AutoTrol Technology. From there he began working with end users addressing their special requirements. In 1985, Bill and his wife Denise formed Kramer Consulting, Inc. to provide custom programming services to a wide variety of clients ranging from smaller shops to the world's largest companies.

Bill has developed or been involved in the development of numerous systems that augment off-the-shelf CAD solutions. These improvements are largely related to the manufacturing industry and NC/CNC program generation. Bill is part owner of a patent "Expert Manufacturing System" where engineering designs and custom manufacturing instructions are created from rules and parameters. These principles are used in many of the interfaces Bill creates using the AUTO-CODE software he developed and markets through his company AUTO-CODE MECHANICAL.

Notice to the User

This text is designed as a complete entry-level Inventor teaching tool. The authors present a typical point of view. Users are encouraged to explore alternative techniques for using and mastering Inventor. The authors and publisher accept no responsibility for any loss or damage resulting from the contents of information presented in this text. This text contains the most complete and accurate information that could be obtained from various authoritative sources at the time of production. The publisher cannot assume responsibility for any changes, errors, or omissions.

Acknowledgments

The authors and publisher would like to thank the following individuals for their assistance and contributions.

Mike Berna	Jim Irvine	Doug Montgomery
David Boomer	J.C. Malitzke	Rick Oprisu
Mary Dudek	Lawrence Maples	Eliza Perry
Fern Espino	Jerry McNaughton	Del Radloff

Cover Art

The cover image for this textbook was supplied by Mr. Sean Dotson. Sean is a Professional Engineer and an Autodesk Inventor Certified Expert. He provides training and VB and VBA customization of Autodesk Inventor to client companies via his consulting firm, Fenris Consulting.

Sean is perhaps best known for his numerous tutorials on advanced Autodesk Inventor subjects, which he offers free-of-charge at his web site, www.sdotson.com. Sean also coauthored *Animator for Autodesk Inventor*, an animation add-on application, and is the author of *iPropertiesWizard*, a data management tool.

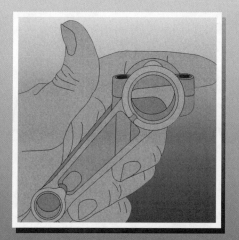

Brief Contents

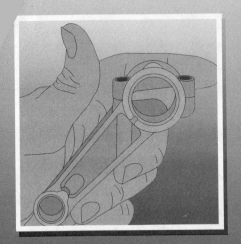

Contents

Chapter 3

Sketching, Constraints, and the Base Feature

Chapter 4

More Complex Sketching, Constraints, Formulas, and the Construction Geometry

Chapter 5

Secondary Sketches and Work Planes

Chapter 6

Adding Features

Chapter 7

Adding More Features

Chapter 8

Creating Drawings

Chapter 9

Dimensioning and Annotating Drawings

Chapter 10

Sweeps and Lofts

Chapter 11

Building Assemblies with Constraints

Chapter 12

Working with Assemblies

Chapter 13

Motion Constraints and Assemblies

Chapter 14

iParts and iFactories

Chapter 15

Parameters in Assemblies

Chapter 16

Surfaces

Chapter 17

Assembly Drawings

Chapter 18

Presentation Files

Chapter 19

Sheet Metal Parts

Chapter 20

Using VBA

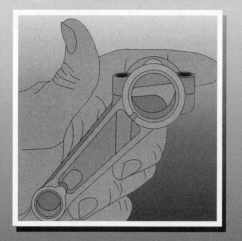

CHAPTER 1

Introduction to Inventor

Objectives

After completing this chapter, you will be able to:

* Define a feature in Inventor.
* List the types of features in Inventor.
* Explain how to edit a part.
* Define an assembly in Inventor.
* Explain how to model motion in Inventor.
* Explain the engineer's notebook.

A Journey Begins

It was a historic decision to leave the AutoCAD DWG format and create a new 3D modeling program from the ground up. While being developed, the Inventor project was code-named *Rubicon,* and with good reason. When Julius Caesar crossed the river Rubicon, it was considered an action that could not be undone. Not easily would Autodesk "cross the Rubicon" and, once crossed, there would indeed be no turning back. As with Julius Caesar's crossing of the river, astounding success has followed Autodesk's decision to develop a new solid modeling format—Inventor.

Inventor builds on a sound foundation of feature-based parametric solid modeling technology that so many other programs share. The term *feature-based, parametric solid modeling* is widely used in the industry and will be explained as you read further. There is no lack of solid modeling programs. Yet Inventor takes the technology further than its competitors, and further than its Autodesk predecessor Mechanical Desktop.

Feature-Based Modeling

To explore the *feature-based* aspect of Inventor, refer to **Figure 1-1.** This is a part tree in the **Browser** for a completed part. The **Browser** is an important part of the user interface. The part tree in the **Browser** lists the pieces and processes that make up the part. Inventor refers to these pieces and processes as *features.* If you think about any "single" part, it is really the end result of several features. The .ipt file extension is used for Inventor part files.

Figure 1-1.
The part tree is displayed in the **Browser**. The tree contains all of the features that make up the part.

The part tree in **Figure 1-1** shows several features that together form the actual solid geometry of the final part. These features are named and have a colored icon next to their name that represents the type of feature. In this example, features include the Hole for Outlet, Cleanup, Bolt Hole, and Bolt Pattern. These descriptive names were entered by the designer. By default, a feature is created with a generic name that represents the feature, such as Extrusion1 for an extrusion or Revolution2 for a revolution.

Work features are used for construction purposes. When created, they have names with the word Work in them, such as Work Plane2 and Work Plane3. The designer may elect to give work features descriptive names, such as Work Plane for Extrusion. Work features are not typically displayed in the graphics window, except when needed. Features that are not visible in the graphics window are grayed out in the **Browser**. The visibility can be turned on if needed.

All the features are listed in the part tree in the order in which they were created, with the first feature at the top. Features may be reordered by simply picking and dragging them to a new position in the part tree. Reordering features may alter the part. In addition, some features cannot be moved above other features on which they are based. For example, if a hole has a fillet applied to it, the fillet feature cannot be moved above the hole feature. The last feature to be listed in any Inventor part tree is the End of Part marker. It has an icon that is a red sphere with a white X. This serves as an "end-of-file" marker for the part.

Features can be edited on an individual basis to change their size, shape, and, in some cases, location on the part. Features can be suppressed. This means that the features are still in the part tree, but its effect on the part is not applied. To permanently remove a feature from the part, it is simply deleted. Features can even be exported for use in other parts.

The part that you are creating is never really edited; its *features* are edited. Since the part is made up of features, as the features are edited, the part is altered. Inventor is truly a feature-based solid modeling program and its features form the heart of the system.

Inventor's features can be broken down into three categories—sketched, placed, and work. These categories are introduced in the next sections. As you work through this book, you will become very familiar with features and the tools used to create and edit them.

Sketched Features

A 2D sketch forms the basis for a sketched feature. This sketch can be drawn or sketched like any 2D geometry from lines, arc, circle, and so on. The sketch is then dimensioned to exact size. Furthermore, the geometric relationship of the sketch geometry is constrained to a final shape. See **Figure 1-2.** A *fully constrained sketch* completely describes the size, shape, and location of the sketched geometry so Inventor cannot inadvertently change the design in future operations. You will learn much more about sketches and constraints in later chapters.

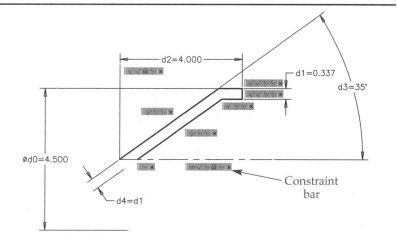

Figure 1-2.
This is a fully dimensioned and constrained 2D sketch.

Placed Features

Placed features are not based on a sketch. Instead, the designer uses tools to "place" features onto an existing part, which is a collection of features. Placed features are similar to machining processes and can be thought of as operations performed on an existing part. Examples of placed features include fillets, chamfers, and holes.

Work Features

Work features are used for construction purposes. Work features include work planes, work axes, and work points. In a 3D environment, sometimes there is nothing on which to base a 2D sketch except a work plane. The work plane serves as the "drafting table" for the sketch's "paper." A work axis can serve as a centerline of revolution. A work point can serve as an anchor point for a 3D path.

Parametric Modeling

As you have seen, an Inventor part is made up of features. This is why Inventor is considered a *feature-based* solid modeling program. However, Inventor is also a *parametric* solid modeling program. A parametric model contains parameters, or dimensions, that define the model. In Inventor, each feature that makes up a part has dimensions that control the size, shape, and location of the feature. By altering the dimensions (parameters), the feature is altered.

For example, suppose you need to design a rectangular plate and the preliminary design indicates the plate is 8″ × 4″ × .25″. To create the plate, you first sketch a rectangle that is roughly 8″ × 4″ on a sketch plane (the "paper"). Then, you apply a horizontal and a vertical dimension to the sketch and edit the values to 8″ and 4″. See **Figure 1-3.** The sketch completely describes the shape and size of the plate's top view. Next, you extrude the sketch .25″ to fully describe the part in three dimensions. See **Figure 1-4.** Since Inventor is a parametric modeler, you can now alter any or all of the three dimensions (parameters) to change the part.

A more powerful aspect of Inventor's parametric modeling is the ability to establish relationships between dimensions and features. For example, suppose the plate will always be half as wide as it is long, no matter what the length dimension is. Every dimension, or parameter, in Inventor has a unique name. By default, the first one is d0, second is d1, and so on. As you will learn, these can be renamed to descriptive names as needed. Now, since the vertical distance (d1) needs to be one-half of the horizontal distance (d0), instead of entering a number for the dimension, the simple equation d0/2 is entered for the value of d1. See **Figure 1-5.** As d0 (the horizontal

Figure 1-3.
A rectangle is sketched and dimensioned.

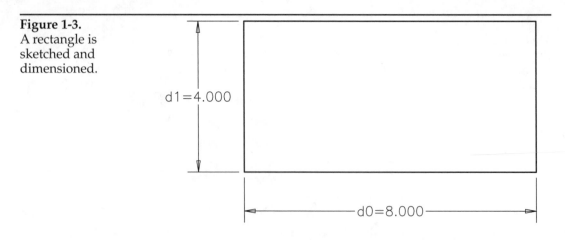

Figure 1-4.
The sketch from **Figure 1-3** is extruded .25" to create a solid part.

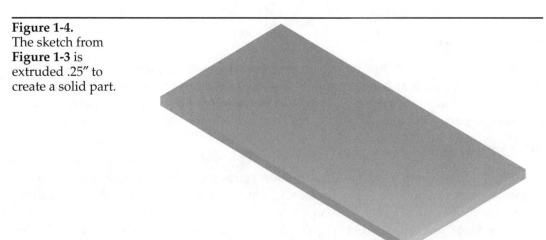

Figure 1-5.
An equation that is based on the length dimension is entered for the width dimension.

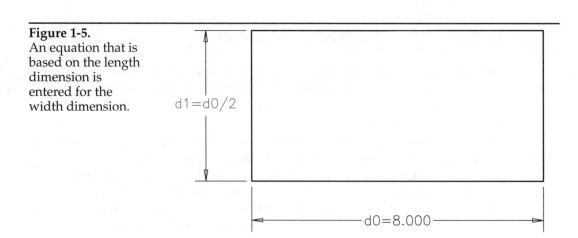

distance) is altered, d1 changes so the vertical distance is one half of the horizontal distance. This is known as *specifying design intent.* It is the intention of the designer that this plate always maintain its aspect ratio of 2:1 length to width. The sketch is thus dimensioned to reflect this design intent.

As mentioned, the default dimension names (d0, etc.) can be renamed. Descriptive names, such as Length and Width, are meaningful to the drafter and designer. A year after the part is created, anybody can open the part file and instantly know what the dimension controls. Alternately, designers and drafters are almost always part of a team. Using descriptive names allows other team members to interpret your design intent. This is further enhanced by adopting a standard naming convention in your department or company.

Figure 1-6.
By using descriptive names, you can build a spreadsheet that is used to "manufacture" different versions of the part.

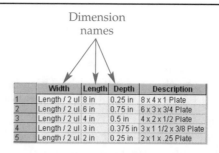

Dimension names

	Width	Length	Depth	Description
1	Length / 2 ul	8 in	0.25 in	8 x 4 x 1 Plate
2	Length / 2 ul	6 in	0.75 in	6 x 3 x 3/4 Plate
3	Length / 2 ul	4 in	0.5 in	4 x 2 x 1/2 Plate
4	Length / 2 ul	3 in	0.375 in	3 x 1 1/2 x 3/8 Plate
5	Length / 2 ul	2 in	0.25 in	2 x 1 x .25 Plate

Another use of named dimensions is to "manufacture" different versions of a part based on data in a spreadsheet. See **Figure 1-6.** As the part is "manufactured," the designer is prompted to enter values for various parameters. The prompts are based on the dimension names. Therefore, Pipe Length is meaningful where d0 is not.

Assembly Modeling

Inventor does a great job modeling parametric parts. However, one of the most powerful aspects of Inventor is the ability to create assemblies. As a part is a collection of features, an assembly is a collection of parts. See **Figure 1-7.** Each part is created and saved. Then, the parts are placed into an assembly file. Parts are only referenced in the assembly. The part files remain separate. Changes made to the part file are reflected in the assembly. Finally, the spatial relationships between parts are defined, or constrained, within the assembly file. An assembly file has an IAM file extension.

A partially-constrained part can be dynamically dragged to analyze its movement within the assembly. The extents of its movement is the part's work envelope in the assembly. You can also move parts in an assembly and then measure distances to determine design data for additional parts.

Figure 1-7.
This is a complex assembly. A number of individual parts were placed into the assembly and constrained to finish the assembly.

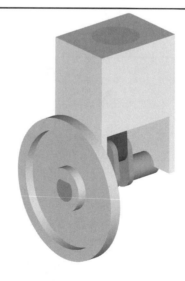

Modeling Motion

Assembly constraints can be "driven," or animated. The numeric values used in the assembly constraint can be dynamically changed over a specified range to model a part's movement within the assembly. This allows you to animate the motion of an assembly. Several types of motion can be animated:

- Rotational (gears).
- Rotational-translational (rack and pinion).
- Translational (cam and cam follower).

2D Drawings

Prints of 2D drawings are always required for the machinists, assemblers, and other workers in the shop. These drawings must follow accepted drafting conventions for lineweight, linetype, and symbol use. Inventor provides tools for creating 2D drawings of parts and assemblies. Inventor drawings have the IDW file extension. Parts and assemblies are referenced into the drawing and displayed using orthographic projection rules. Changes made to the part are automatically reflected in the drawing.

Presentations

If a picture is worth a thousand words, then how many words is a movie worth? A 2D drawing will always be required. However, you can capture to digital video the process for building an assembly. Inventor uses the IPN file extension for presentation files. The assembly file is referenced into the presentation and always reflects the current state of the assembly. The animated presentation can be distributed to others or saved as an AVI file to be viewed on workstations *not* equipped with Inventor. See **Figure 1-8.** Presentations provide a very effective form of communication.

Figure 1-8.
An animated assembly presentation, such as this one, can be saved as an AVI file and played in Windows Media Player.

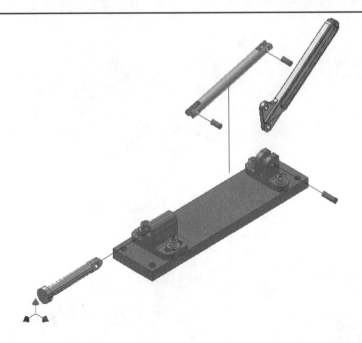

Figure 1-9.
The engineer's notebook can be used to share comments that are not actual drawing annotations.

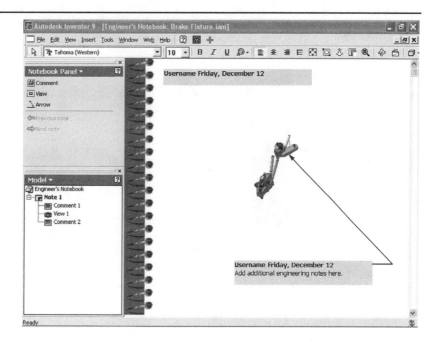

Engineer's Notebook

There is a handy little utility in Inventor that allows you to annotate a part in a notebook-style environment. This allows you to add notes to a part or feature that are not intended to be part of the final 2D drawing. See **Figure 1-9.** The engineer's notebook is meant to act as a communication tool between those collaborating on the design. Multiple notations may be made on a single part or assembly. The notes appear in the **Browser** in a separate branch in the part or assembly tree. An icon also appears in the graphics window near the annotated part. The engineer's notebook is discussed in detail in Chapter 2.

Design Assistant

In the process of creating parts, subassemblies, and an assembly, you may end up with hundreds of files. **Design Assistant** is a utility that acts as a document management system, **Figure 1-10.** It can track items such as revision number, status of design, and iProperties. It also allows printing from outside of Inventor and a very useful utility called **Pack and Go**. This utility allows you to select an assembly and **Pack and Go** will find all of the parts, subassemblies, design views, etc., in the assembly and copy them all to a destination folder. Then, you can zip all of the files and e-mail the zipped file. Accessed through Windows Explorer, **Design Assistant** a great utility for managing Inventor files.

Figure 1-10.
Design Assistant
serves as a
document
management
system.

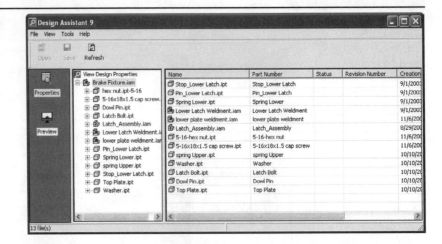

Chapter Test

Answer the following questions on a separate sheet of paper.

1. What is a *feature* in Inventor?
2. How is Inventor a feature-based modeling program?
3. What are the three types of features in Inventor?
4. Which type of feature is used for construction purposes?
5. What is the basic process for editing a part?
6. How is Inventor a parametric modeling program?
7. Give one example of why you would want to establish a relationship between a part's parameters.
8. What is an *assembly?*
9. How can motion be modeled in Inventor?
10. What is the purpose of the engineer's notebook?

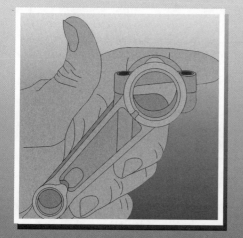

CHAPTER 2

The User Interface

Objectives

After completing this chapter, you will be able to:

* Understand the various components of Inventor's user interface.
* Locate the various components of the user interface.
* Open an existing file.
* Create a new file.
* Effectively use the user's interface to create and edit solid model geometry.

User's Files

*The following is a list of files that you will need to work through this chapter. These files can be found on the **User's Files** CD included with this text.*

Examples	Practices	Exercises
Example-02-01.ipt	Practice-02-01.ipt	Exercise-02-01.ipt
Example-02-02.ipt		Exercise-02-02.ipt
Example-02-03.ipt		Exercise-02-03.ipt
		Exercise-02-04.ipt

Opening an Existing Part File and the User Interface

The **Open** dialog box is accessed when Inventor is started. See **Figure 2-1.** When using Inventor, this dialog box can be accessed by selecting **Open** in the **File** pull-down menu or clicking the **Open** button on the **Inventor Standard** toolbar. Navigate to the folder containing the examples for Chapter 2. Click once on the Example-02-01.ipt part file. A preview of the part will be shown in the lower-left corner of the dialog box. To open the file, click the **Open** button at the bottom of the dialog box or double-click on Example-02-01.ipt.

The first thing that you will notice about the Inventor user's interface is the small number of menus and toolbars. See **Figure 2-2A.** By default, there are ten pull-down menus on the **Menu Bar** and the **Inventor Standard** toolbar showing. For the work in this textbook though, the **Panel Bar** and the **Browser** will be opened by picking them in the **Toolbar** cascading menu in the **View** pull-down menu. Resize, relocate, and dock the **Panel Bar** and the **Browser** so that the user's interface looks like **Figure 2-2B.**

Figure 2-1.
The **Open** dialog box showing highlighted file and file preview.

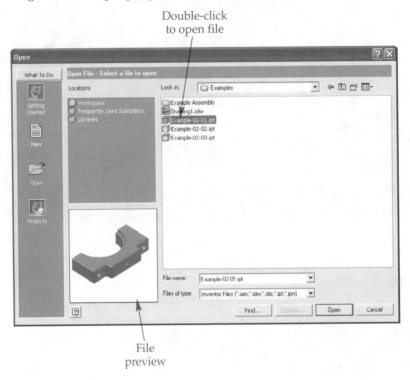

Double-click
to open file

File
preview

Figure 2-2.
A—The default Inventor user's interface. B— Inventor user's interface with the **Browser** and **Panel Bar** docked.

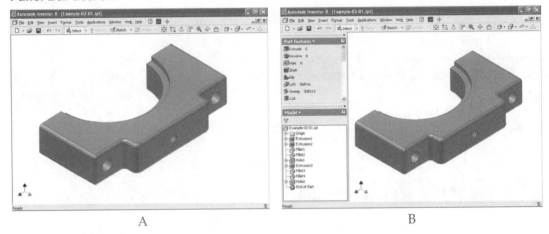

A   B

These are all context sensitive; that is they change as you work in different modes. The **Panel Bar** will change drastically displaying completely different tools and the others will change slightly. The modes for part modeling are Sketch, 3D Sketch, Features, Sheet Metal, and Solids. In the first ten chapters the Sketch, Features, and Sheet Metal modes will be used. The Solids mode is used for parts imported from other CAD systems and the 3D Sketch is for creating paths based on existing geometry for extruding sweep features such as pipes or tubes.

The system is now in the Feature mode. There are pull-down menu choices that are not standard Windows commands, are unique in that they do not have icons on the toolbars, and apply to part modeling.

Figure 2-3.
The **Inventor Standard** toolbar and its buttons.

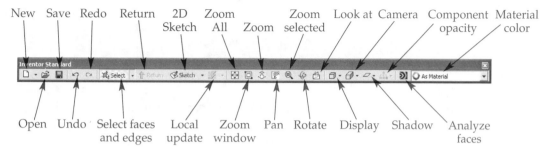

Inventor Standard Toolbar

The **Inventor Standard** toolbar is the main interface toolbar used in working with Inventor. See **Figure 2-3.** All of the common utility functions (**Save**, **Undo**, **Zoom Extents**, etc.) are located on this toolbar.

The New flyout menu

There are four options in the **New** flyout menu. See **Figure 2-4.** Selecting **Assembly** in the **New** flyout menu creates a new assembly file with the default name of Assembly1 with an .IAM file extension. Assembly files are collections of part files that are related to one another spatially and dimensionally so they are a cohesive whole. It will be based on the template that corresponds to the unit of measurement that was specified when Inventor was installed. The first time you save the file you will then have the opportunity to give it a more meaningful name.

Selecting **Drawing** in the **New** flyout menu creates a new drawing file with the default name of Drawing1 with an IDW file extension. Drawing files are used to create 2D orthographic views of your parts and assemblies.

Selecting **Part** in the **New** flyout menu creates a new part file with the default name of Part1 with an IPT file extension. It will be based on the default template for the default unit of measure. Part files are the basis of everything in Inventor. They are used to create assemblies as well as drawings.

Selecting **Presentation** in the **New** flyout menu creates a new presentation file with the default name of Presentation1 with an IPN file extension. Presentation files are used to document the steps in the assembly, or disassembly, of an assembly file.

Inventor is not unique in having different file types with different extensions. However, it can be new and perhaps disconcerting to those who have grown accustomed to AutoCAD and its DWG file extension.

The Open button

As previously mentioned, the **Open** button is used to find and open an existing Inventor file for editing. Pressing the button accesses **Open** dialog box. This dialog box uses the standard Windows folder navigation scheme.

Figure 2-4.
The **New** flyout menu.

Figure 2-5.
The **Files of type:**
drop-down list.

Near the bottom of the dialog box is the **Files of type:** drop-down list. There are a number of file types that can be opened from this list. See **Figure 2-5.** The first five listed are all Inventor file types. Some of the others, such as DWG or DXF, may be familiar but others may be more of a mystery. The IGES, SAT, and Step files are all for translation of 3D geometry from other CAD systems.

At the bottom of the dialog box is the **Find** button. Pick this button to access the **Find** dialog box. See **Figure 2-6.** Search the target location for files that meet the specified criteria. The list of files meeting the criteria are displayed in the **Files Found** dialog box.

The **Save button**

When working on a new file and the **Save** button is picked, the **Save As** dialog box is accessed. See **Figure 2-7A.** This is a standard Windows-type dialog box for saving Inventor files. Use the **Save in:** drop-down list to navigate to the folder where the file is to be saved. When specifying a filename, do not type in the extension. If this part is to be used in an assembly, the name of the component in the assembly will have the name specified here. Inventor knows what type of file it is and you will notice that is reflected in the **Save as type:** drop-down list. Once the file has a name, then subsequent saves will not access this dialog box.

If the **Options...** button is selected, the **File Save Options** dialog box is accessed. See **Figure 2-7B.** Check the **Save Preview Picture** check box and you have the choice of saving a thumbnail image with the file. The thumbnail can be viewed in Windows Explorer when thumbnail views are requested. When saving thumbnails, you can select one of the following for the thumbnail image source.

- Select **Active Window on Save** if you wish to capture whatever is in the active window.

Figure 2-6.
The **Find** dialog box is used to search a target location for files that meet specified criteria.

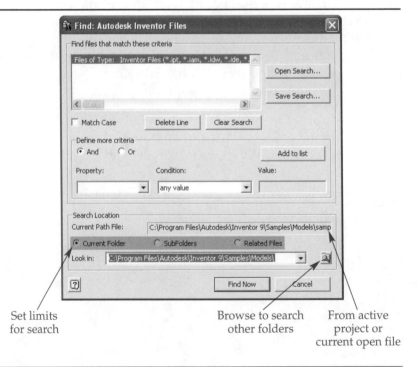

Set limits for search

Browse to search other folders

From active project or current open file

Figure 2-7.
A—The **Save As** dialog box. B—The **File Save Options** dialog box is used to save a thumbnail image with the saved file.

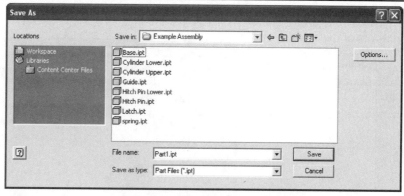

A

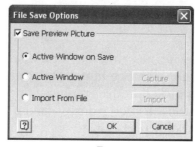

B

- Select **Active Window** if you wish to capture a predetermined image of the active window. If this option is chosen, then the **Capture** button is used to capture whatever image is in the active window at this moment. This will be used from then on whenever **Save** is pressed.
- Select **Import From File** if you wish to use an external 120×120 pixel bitmap image to be used as the thumbnail image. If this option is chosen, then the **Import** button is used to navigate to the location of the bitmap.

The Undo and Redo Buttons

Selecting the **Undo** button will undo all actions. View-related operations, such as **Zoom** or **Pan**, cannot be undone. Undo also acts on assemblies that have multiple part files open. Undo is extremely powerful. It will correctly undo changes made to the possibly hundreds of part files that may have been changed due to a change in the assembly. There is an *undo temporary file* maintained on the hard drive for these purposes. Increasing its size may help in cases of large assembly performance issues. To increase the file size, select **Application Options...** in the **Tools** pull-down menu. On the **General** tab, increase the file size shown for **Undo file size (MB)**. See **Figure 2-8**.

Selecting the **Redo** button will reverse the effects of the **Undo** button. A redo basically undoes the undo. As with undo, view-related operations are not affected by a redo.

NOTE	If the **Undo** button is selected enough times, the opened file will close. Selecting the **Redo** button will open that file again.

Figure 2-8.
The size of the undo temporary file can be increased using the **General** tab of the **Options** dialog box.

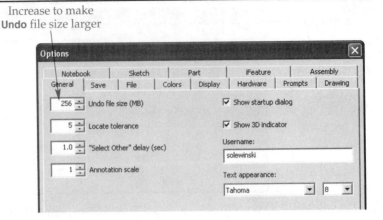

Increase to make **Undo** file size larger

The **Select** Flyout

The **Select** flyout has different options, depending on whether a part or an assembly is open for edit. See **Figure 2-9.** This flyout helps control what is selected when geometry is picked in the graphics window. In a large assembly with complex geometry, setting the selection priority can be very helpful. For example, if you need to select the edge of a part but are getting the part itself, try changing the select priority to **Select Faces and edges**.

The **Return** Button

When editing a sketch, the **Return** button is enabled. In general, the **Return** button allows you to return to a higher place in the database hierarchy. For example, if the button is picked, Sketch mode is exited and you are returned to Part mode. Also, if you are editing a subassembly and you pick the **Return** button, you will exit the subassembly and return to the next assembly up in the assembly hierarchy. It can be thought as being a "backing out" operation.

The **Sketch** Button

The **Sketch** button is used to create a sketch on the face of a part or on a work plane. While in Part mode, pick the **Sketch** button and then select a planar face on a part. This will initiate a new sketch in the **Browser** and activate the Sketch mode. The **Panel Bar** will also change to show the 2D sketch tools. Try it now by picking the **Sketch** button and then picking the top face on Example-02-01.ipt. You should be in Sketch mode and there is a new sketch listed at the bottom of the **Browser** and immediately above the **End of Part** feature. In this mode you could draw a new 2D shape. This shape could be constrained and extruded into a new 3D feature that will be added onto the part. This will be covered in greater detail in *Chapter 3*. To exit, pick the **Return** button.

Figure 2-9.
The **Select** flyout options help control what is selected when geometry is picked in the graphics window.
A—The flyout when working on a part.
B— The flyout when working on an assembly.

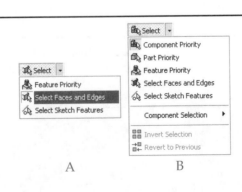

A   B

Figure 2-10.
The **Update** flyout is used to change the part to reflect the changes just made.

The **Update** Button

Inventor is a parametric modeling engine. Its parts and assemblies are designed so they can be changed—time and time again. Pick the **Update** button after a part has been modified, so the part will reflect the changes just made. See **Figure 2-10.** With Example-02-01.ipt still open, double-click on Extrusion1 in the **Browser**. The dimensions should appear in the graphics screen. Double-click on the 33.000 (mm) radius, change its value to 35 in the **Edit Dimension** dialog box, and pick the check mark button. Pick the **Update** button to update the part. After examining the change, pick the **Undo** button to undo the changes.

One of the main uses for this button is in Assembly mode when you are working on a large assembly with subassemblies. The **Local Update** button is for the subassembly you are currently editing. The **Global Update** button is to force the main assembly and all subordinate assemblies to be updated.

The **Zoom All** Button

The **Zoom All** button is used to show all of the part or assembly. The display will zoom out or in to fit all of the part or assembly in the graphics window. It also allows for a comfortable margin all around the edges.

The **Zoom Window** Button

After picking the **Zoom Window** button two points are used to define a window. Inventor zooms to that window to fill the graphics screen. The two points are defined by a click-and-drag of the mouse or by defining the window with two distinct mouse clicks. This also provides for a measure of panning before the mouse button is released. Try it now by picking one point and, before releasing the mouse, move the cursor around the graphics window. The zoom window will follow it around, constrained to the first pick point. Pressing [F5] function key will get you back to the previous view. Remember, **Undo** *does not* undo view-related operations.

The **Zoom** Button

After pressing the **Zoom** button the *zooming* icon appears in the graphics window. To zoom in, press and hold the left mouse button and drag the mouse *toward* you. Conversely, dragging the mouse away from you performs a zoom out and reduces the size of the image in the graphics window. Zoom may be used transparently at any time. If the [F3] function key is held down, the zooming icon appears and the zoom operation can be performed.

PROFESSIONAL TIP  A great trick is to simply use the mouse's roller wheel—roll the wheel toward you to magnify, away from you to reduce the image.

The **Pan** Button

The **Pan** button is used to shift the display within the graphics window. After pressing the **Pan** button the *panning* icon appears in the graphics window. To pan around the graphics window, press and hold the left mouse button and drag to obtain the desired display. **Pan** may be used transparently at any time. If the [F2] function key is held down, the panning icon appears and the panning operation can be performed.

The simplest way to pan, however, is to press the roller wheel down and move the mouse to shift the display. With some practice the roller wheel can be used for all zooming and panning operations and the toolbar accessed only on rare occasions.

The **Zoom Selected** Button

The **Zoom Selected** button is used to zoom in on a selected feature or face of a part. Pick a feature or a face of a part and then pick this button. It will fit the selected item within the graphics window.

The **Rotate** Button

This may be the most useful viewing tool in Inventor. It is sometimes called *3D Rotate*. With this tool the viewpoint can be moved around the part model in 3D space. Once the **Rotate** button is picked, the rotate circle appears on screen. See **Figure 2-11.**

Select the **Rotate** button and position the 3D rotate cursor inside the circle. Press and hold the left mouse button and move the mouse around. Your viewpoint should be swinging around the part. This is known as *free-rotate mode*. Release the mouse button and position the cursor outside the circle. Press and hold the left mouse button. Move the mouse in a large circle around the screen. This rotates the view about the view axis, which is the imaginary line coming out of the monitor at you. This is similar to a wheel rotating on an axis. Release the mouse button and position the cursor on top of the vertical axis. Press and hold the left mouse button. Move the mouse straight down the face of the screen. It seems that the part is rotating about an imaginary horizontal axis passing through the monitor. This is similar to a barbecue turning on a spit. Release the mouse button and position the cursor over the horizontal axis near the left-hand edge of the screen. Press and hold the left mouse

Figure 2-11.
The **3D Rotate** tool and the rotate circle as it appears in the graphics window.

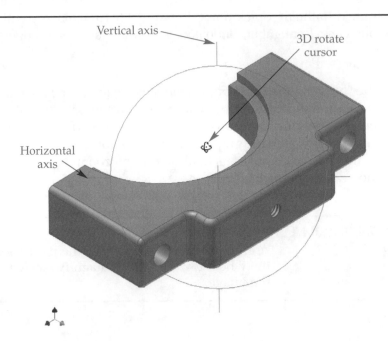

button. Move the mouse straight across the screen toward the right. It appears as if the part is rotating about the imaginary vertical axis passing through the monitor from top to bottom. Finally, with the cursor inside the circle, pick a point with a left-click of the mouse button. The view is panned to center the point in the view screen. Keep in mind as you use rotate, what you are actually doing is manipulating that point in space from which you are viewing the part—your viewpoint.

Rotate is an extremely effective visualization tool. It will be used every day you use Inventor so take the time to master its use. Just as the **Zoom** and **Pan** tools, this may be used transparently whenever the need arises. If the [F4] function key is held down, the rotate circle and the 3D rotate cursor icon appears and the rotating operation can be performed. With no command/tool active, right-click and choose **Isometric view** from the pop-up menu to revert back to the default view.

PROFESSIONAL TIP

One little-known trick is to create a continuous orbit. Pick the **Rotate** button from the toolbar and position the cursor inside the circle. Hold down the [Shift] key and the left mouse button, then "stroke" across the graphics screen. It takes some getting used to, but if you are successful you should have a spinning part. The length of the stroke determines the spin speed. The direction of the stroke determines the arbitrary axis about which the part will seem to spin. To cancel one spin and try another, left-click inside the circle. At any time, pressing the **Rotate** button or the [Esc] key will exit the entire operation. In a large model you will need a high-performance graphics card to perform a continuous orbit.

Common View option

Another option with the 3D rotate tool is the **Common View** option. Pick the **Rotate** button again and then press the *spacebar* on the keyboard. A translucent blue cube with green arrows appears in the graphics window. See **Figure 2-12.** At any time you can press the spacebar again to change the operation back to the other mode.

The green arrows represent view directions. If you click on a green arrow, then that view of the part is shown in the graphics window. Picking an arrow at a corner of the cube will result in the isometric view from that direction. For example, if you select the arrow at the top, back corner of the cube, the part will rotate to give you that isometric view. The green arrows on the flat faces of the cube correspond to the six standard flat orthographic views of the sides of a cube. Notice the coordinate system indicator in the lower-left corner of the graphics window as you pick different arrows. The **Common View** cube and arrows are oriented along the XYZ axes of the coordinate system. Pick the arrow for the top of the part. This will give you the top view or plan view in architectural terms. If you pick the same arrow again, the part rotates to give you the bottom view. You can do this for any of the arrows. Clicking twice on any view direction will give you the first view then its opposite view.

Click on the top arrow to give a plan view of the part. Now, hover over one of the edges of the cube. Notice that half of the line turns red. Click on this red half-edge and the view is rotated 90° to the right or left depending on the half selected.

Figure 2-12.
The **Common View** tool and the **Common View** cube.

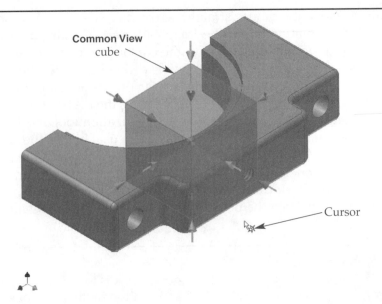

Common View cube

Cursor

Redefining the isometric view

Right-click to access the pop-up menu shown in **Figure 2-13.** Pick **Isometric View** to return to the default isometric view. Now pick one of the green arrows at a *corner* of the cube. Right-click and pick **Redefine Isometric** from the pop-up menu. Remember your current viewpoint. Press the [Esc] key to exit the entire operation. Using the [F4] key, simply rotate the part around to a new viewpoint. Now, right-click and choose **Isometric View** again. The view defined as the new isometric view should be shown in the graphics window.

Try this again on another part. Open the Example-02-02.ipt file. Right-click and choose **Isometric View**. Examine the orientation of the part in this view. Pick the **Rotate** button and then press the spacebar to activate **Common View**. Refer to **Figure 2-14** and pick the topmost green arrow pointing to the cube's corner. After the view rotates to the new viewpoint, right-click and choose **Redefine Isometric**. Hit the [Esc] key to exit the rotate operation. Use the [F4] key to rotate the viewpoint. Now, right-click and choose **Isometric View**. The viewpoint should rotate to the new isometric orientation. Close the part file and do not save changes.

Rotating around a large part

Sometimes when rotating a large part or assembly and you have zoomed into the area that you would like to view, the area rotates way out of the view as you rotate. To alleviate this problem, left-click inside the rotate circle to pan until the area that you would like to view is in the center of the circle. This will redefine the center of the view for the rotate operation to move around. Open Dogging Assembly.ipn found in the Example Assembly folder. Make sure that it is the IPN file, not the IAM file. You will learn more about this special type of assembly later in the textbook. To illustrate the concept, zoom in about halfway on the assembly. Then, pick the **Rotate** button

Figure 2-13.
The **Isometric View** option of the right-click pop-up menu.

Figure 2-14.
Pick the top arrow on the **Common View** cube.

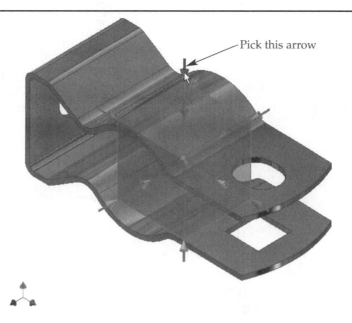

Pick this arrow

and rotate the viewpoint around whatever is in the center of the circle. To swing the viewpoint around the bronze-colored component, left-click to pan it into the center of the rotate circle. Click and drag the viewpoint and it will now rotate about the bronze component. Close the file without saving it.

The Look At Button

If you still do not have it open, open the Example-02-01.ipt file again. Pick the **Look At** button and then a face on the part. A full frontal view of that particular face is displayed. Try it again on the other faces on the part. The part sometimes turns around leaving it "upside down." Each face has its own internal coordinate system and Inventor is orienting that one with the screen's coordinate system. The **Look At** tool is useful when working on sketches where the sketch is usually plan to the graphics window, much like a piece of paper you are sketching on.

PROFESSIONAL TIP

Right-click and choose **Isometric View** from the pop-up menu. In the **Browser**, pick the plus sign next to Extrusion1 and click once on the Sketch1 feature. With it highlighted, pick the **Look At** button. You should have a plan view of the sketch in the graphics window.

The Display Flyout

The buttons on the **Display** flyout are used to change the display of the part. See **Figure 2-15A.** The **Shaded Display** shows the part as a shaded object. **Shaded with Hidden Edge Display** shows the part as a shaded object with hidden lines shown. **Wireframe Display** shows the part as a wireframe model only.

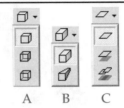

Figure 2-15.
A—The **Display** flyout is used to change the display of the part. B—The **Camera** flyout is used to determine the view method used for models. C—The **Shadow** flyout is used to set the type shadow used.

A   B   C

The Camera Flyout

"Camera mode" indicates the particular view method used for models shown in the graphics window. It does not mean you can record actions that take place in the graphics window. **Figure 2-15B** shows the buttons for the two camera modes. **Orthographic Camera** mode shows the part so all its points project along parallel lines to their positions on the screen. It is used most often. See **Figure 2-16. Perspective Camera** mode is very useful for presentations or whenever a more realistic picture of the part is desired. It provides a sense of how parts and assemblies would appear to the human eye when actually manufactured.

The Shadow Flyout

There are three shadow buttons. See **Figure 2-15C.** No shadow is shown when the **No Ground Shadow** button is selected. A simple shadow under the part is shown when the **Ground Shadow** button is selected. When the **X-Ray Ground Shadow** is selected, a ground shadow is shown with hidden lines within the shadow. It is a nice effect, but not very realistic. Pick the **Ground Shadow** button and use the [F4] rotate tool to see the effect. Shadows are useful in combination with perspective view for realistic images.

The Component Opacity Flyout

Component opacity controls the opacity of assembly components above the currently edited subassembly in the assembly hierarchy. It is used when editing assemblies and will be covered in greater detail in later chapters of this textbook.

The Analyze Faces Button

The **Analyze Faces** button is used to analyze surfaces for continuity or valid draft angles. It is to be used in conjunction with the **Analyze Faces** dialog box, which is accessed by picking **Analyze Faces...** in the **Tools** pull-down menu. The **Analyze Faces** button acts as a toggle, turning analysis on or off.

Open the Example-02-03.ipt part file and pick the **Analyze Faces** button. Then rotate the part to examine the surface continuity. See **Figure 2-17.** There will be more on analyzing surfaces for continuity in *Chapter 16*.

Figure 2-16.
A—The
**Orthographic
Camera** view.
B—The
**Perspective
Camera** view.

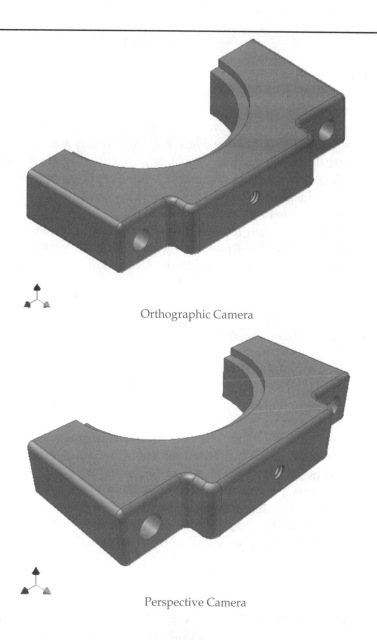

Orthographic Camera

Perspective Camera

Figure 2-17.
The part showing
the results of
analyzed surfaces.

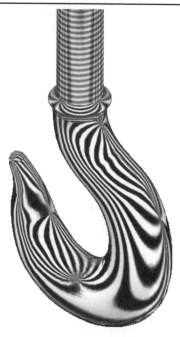

The **Panel Bar** is the main Inventor interface. You will use the **Panel Bar** as you create sketches or geometry on parts. There are a number of these panel bars. They automatically change as you move from mode to mode. With Example-02-03.ipt still open, notice the **Panel Bar** indicates Inventor is in Part Feature mode and the **Part Features** panel is active. See **Figure 2-18A.** This panel is complete with all of the feature creation tools needed to create a part. If you select the top face of the hook and pick the **Sketch** button, then the **2D Sketch Panel** will appear. See **Figure 2-18B.** This panel has all the tools necessary to draw and constrain a sketch. If you pick the **Return** button, the **Part Features** panel is accessed.

Inventor has a built-in sheet metal modeling module. Start a new file and select Sheet Metal.ipt from the **Open** dialog box. Pick the **Return** button and **Sheet Metal Features** panel is accessed. See **Figure 2-18C.** This panel contains all the tools necessary to complete a sheet metal part.

Inventor also has the ability to create 2D drawing views. Start a new file and select Standard.idw from the **Open** dialog box. The **Drawing Views Panel** is accessed. See **Figure 2-18D.** This panel contains all the tools necessary to create a number of types of 2D views.

The **Assembly Panel** is used to import, manipulate, and constrain part components into a cohesive assembly. See **Figure 2-18E.**

The **Presentation Panel** is used to import and manipulate components into a presentation. See **Figure 2-18F.**

Some of the buttons in the panel bars have submenus that are denoted by a small black arrow next to the button name. If you left-click on one of these arrows, then a submenu is accessed.

At any time you right-click on the background of a panel bar, a pop-up menu appears. If you pick **Expert,** the explanatory text for each button is toggled on or off. **Figure 2-19** shows the **2D Sketch Panel** with the **Expert** toggled on. Notice the text is gone and only the icon is shown on the button.

Resizing the Panel Bar

The **Panel Bar** can be resized by hovering over its border until the double arrow icon appears. Then, press-and-drag it to the new size. It can also be undocked by hovering over the two bar lines and pressing and dragging it away from the docked location. The same method is used for docking the **Panel Bar.** It can be closed by clicking on the **X** in the upper-right of each panel. To restore the **Panel Bar**, pick **Panel Bar** in the **Toolbar** cascading menu of the **View** pull-down menu.

Figure 2-18.
A—The **Part Feature** panel. B—The **2D Sketch Panel**. C—The **Sheet Metal Features** panel. D—The **Drawing Views Panel**. E—The **Assembly Panel**.
F—The **Presentation Panel**.

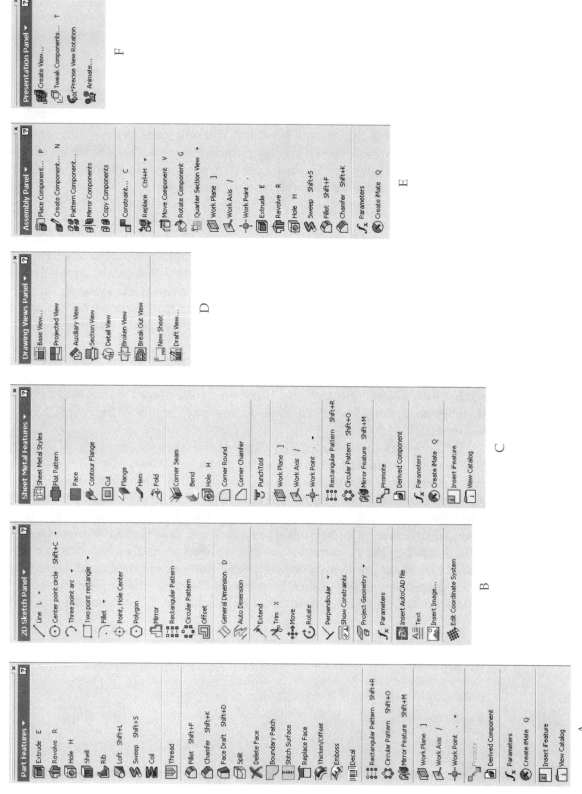

Figure 2-19.
The **2D Sketch Panel** with the **Expert** toggled on.

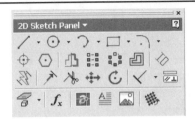

The Browser

The **Browser** is the heart of Inventor. You will spend a great deal of time using it to manage features. Inventor creates parametric parts that are based on parametric features. The features are, in turn, based on sketches. All of the items used to build your part are listed in the **Browser** as they are created.

Figure 2-20 shows the **Browser** listing for Example-02-01.ipt. If you move your cursor over the name of a feature, it highlights in red and the part in the graphics window is also highlighted. See **Figure 2-21.** Click on one of the features, such as Extrusion1, then it will stay highlighted in the graphics window. If you want to select more than one feature, hold down the [Shift] key to select consecutively listed features or the [Ctrl] key to select features randomly. In most CAD systems the geometry is selected in the graphics window only. In Inventor, the geometry can be selected in the graphics window or the **Browser**. This is extremely useful when you can *only* select items from the **Browser**.

Click on the plus signs next to Feature1 and the list is expanded to show the sketches that are the basis for that feature. Some of the features, such as the fillets, do not have plus signs next to them. These are called *placed features* and do not require sketches to define. The extrusions do use sketches and they are called *sketched features*.

Double-click on Sketch1 under Feature1. Your screen should look like **Figure 2-22.** Notice that the **Panel Bar** changed to the **2D Sketch Panel**. Also notice that all features in the **Browser** are grayed out to signify that they are not available. The graphics window changed to just show the 2D sketch that is the basis of Extrusion1. At this point the sketch can be edited—perhaps the dimensioning scheme needs to be changed or some geometry added or removed. You would not use this method to actually change the numeric values of the dimensions themselves, even though you could. Exit Sketch mode by clicking on the **Return** button or right-clicking anywhere in the graphics window and picking **Finish Sketch** in the pop-up menu. You should now be back in Part Feature mode. Notice the change in the **Panel Bar** and the **Browser**.

Figure 2-20.
The **Browser**
showing the tree
structure of a part.

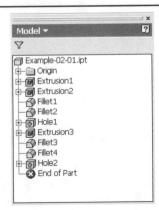

Figure 2-21.
Highlighting a feature in the **Browser**, also highlights the feature on the graphics window.

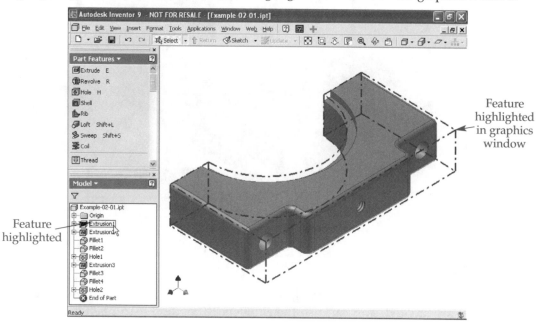

Feature highlighted

Feature highlighted in graphics window

Figure 2-22.
Double-click on the sketch in the **Browser** to change to Sketch mode.

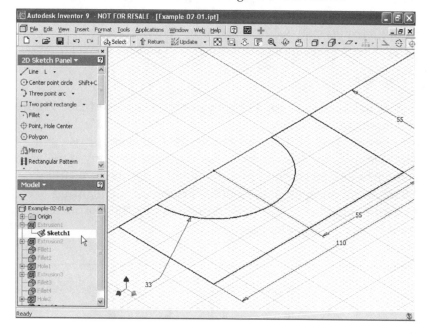

Changing Dimensions

To change the dimensions controlling a feature, right-click on the feature and pick **Show Dimensions** from the pop-up menu. For example, right-click on Extrusion1 and pick **Show Dimensions**. In the graphics window, find the 110 mm dimension and double-click on it. This accesses the **Edit Dimension** dialog box. Change the value to 130 mm and press [Enter] or click on the green check mark. See **Figure 2-23**. To update the part, pick the **Update** button on the **Inventor Standard** toolbar. The size of the part should change in the graphics window. This illustrates the parametric capabilities of Inventor.

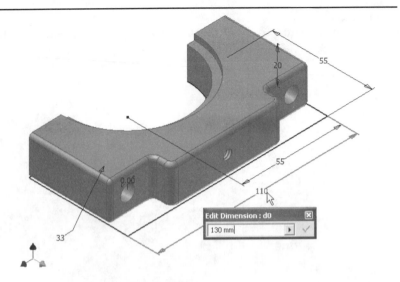

Figure 2-23.
Editing dimensions
on screen.

Renaming Features

The name Extrusion1 is not very descriptive. Inventor names features automatically as they are created. You can use the **Browser** to rename features. On Extrusion1 left-click twice—do not double-click. See **Figure 2-24**. The name will appear in a blue box. Now you can type in a new name. Type in Base Feature and press [Enter]. Renaming features is very important. For example, the hole features should be renamed so the name is more descriptive—perhaps revealing the size of the hole or if it is tapped. You do not want to have a finished part with 50 features with names Extrusion1 through Extrusion50. A little extra time renaming features can save hours of headaches later when you cannot figure out which feature is which.

The Pop-Up Menu

If you right-click on a feature, a pop-up menu is accessed. See **Figure 2-25**. The options in this menu are as follows:

- **Copy**. Places a copy of that feature into the Clipboard for use in another part.
- **Delete**. Removes the feature from the part and also any dependent features. Dependency will be covered in more detail later in this textbook.
- **Edit Sketch**. An alternative to double-clicking on the feature's sketch.
- **Edit Feature**. Enables you to edit more than just the feature's dimensions. Its use will be covered in later chapters.

Figure 2-24.
A—Renaming
features on the
Browser. B—The
renamed feature.

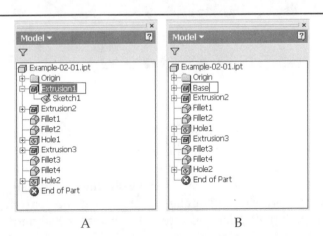

Figure 2-25.
The pop-up menu
and its options.

- **Create Note**. Activates the **Engineer's Notebook**. This will be covered later in this chapter.
- **Suppress Features**. Simplifies the part or temporarily removes a feature from the part. This will be covered in detail in later chapters.
- **Adaptive**. Turns on adaptivity for that feature. This is an extremely powerful capability of Inventor. When a feature is adaptive it does not use a fixed size, but instead is free to adapt its size to other parts in the assembly. This will be covered in more detail in the assembly chapters of this textbook.
- **Expand All Children**. Expands all plus signs revealing sketches and other subdata forms.
- **Collapse All Children**. Collapses all plus signs hiding sketches and other subdata forms.
- **Find in Window**. Zoom the graphics screen into a close-up of the feature.
- **Properties**. Accesses the **Features Properties** dialog box where the feature can be renamed and suppressed; the adaptivity changed to **Sketch**, **Parameters**, or **From/To Planes**; or the color of the feature changed so it has a different color than that of its part.

Working with Features

At the top of the **Browser** is the name of the currently open document. This is a part file because it has an IPT file extension. The part icon also identifies it as a part file. There is a different icon used for assemblies. Immediately below the part name is a folder called Origin. This is an extremely useful feature found in every part and assembly. It contains three work planes, three work axes, and one work point. See **Figure 2-26.**

Figure 2-26.
The **Browser**
showing common
features found in
every part and
assembly.

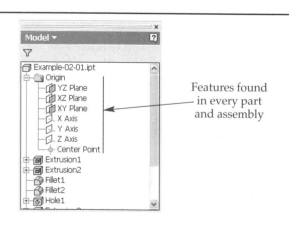

Features found
in every part
and assembly

The work point is at the coordinate system's origin 0,0,0. The work axes lie along the coordinate system's three axes X, Y, and Z. The work planes correspond to the coordinate system's planes that are formed from the X, Y, and Z axes. These work features are important in part modeling, as well as constraining assemblies together. Throughout this textbook there will be references to these work planes, work axes, and work point.

Beneath all of the features is a special feature labeled End of Part. This denotes the end of the feature listing, but it also has specific uses. For example, left-click and hold on End of Part and drag-and-drop it under the Origin folder. Notice that all features are grayed out in the **Browser**. The End of Part feature represents the end of the features list—the end of the database. After the move, it is above any of the features so they do not appear in the screen. Now, drag-and-drop the End of Part feature down below the first part feature, Extrusion1. The basic shape that was used to start this part should appear. Continue to drag-and-drop the End of Part feature down one feature at a time. This will reveal the method(s) used to construct this part. This is a great way to see how the part was built. A more powerful use of this important feature will be discussed in a later chapter of this textbook.

Resizing the Browser

The **Browser** can be resized by hovering over its border until the double arrow icon appears. Then, press-and-drag it to the new size. It can also be undocked by hovering over the two bar lines and pressing and dragging it away from the docked location. The same method is used for docking the **Browser**. It can be closed by clicking on the **X** in the upper-right of each of them. To restore the **Browser**, pick **Browser** in the **Toolbar** cascading menu of the **View** pull-down menu.

The Menu Bar

By default the **Menu Bar** is docked above the **Inventor Standard** toolbar. See **Figure 2-27**. It contains ten pull-down menus—**File**, **Edit**, **View**, **Insert**, **Format**, **Tools**, **Applications**, **Windows**, **Web**, and **Help**. These menus contain important tools that you need to become familiar with.

The File Pull-Down Menu

There are three saving options found in the **File** pull-down menu. If you pick **Save** while working on a new file that has not been saved, you will get the **Save As** dialog box. All saves after that will save the current file without opening the dialog box.

Save Copy As is different in that the current file does not change. A copy is saved with a new name using the **Save Copy As** dialog box. This is not the "save as" command, found in other software, that simply renames the current file.

PROFESSIONAL TIP

You can use the **Save Copy As** to write a BMP (raster bit map) file out to hard disk. Choose **Options** and specify the size of the image in pixels. For a very high-resolution image — one that will print well on a large-format plotter — use a size of 4000×4000 pixels. Adjust the pixel count to find a balance of resolution and file size that works for you.

Figure 2-27.
The **Menu Bar** and
its options.

File Edit View Insert Format Tools Applications Windows Web Help

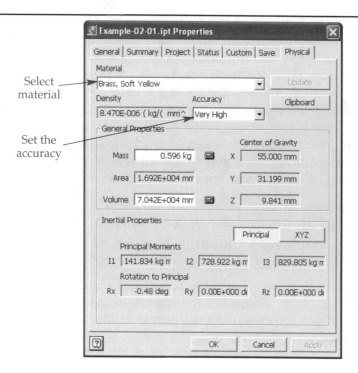

Figure 2-28.
The **Physical** tab in the **Properties** dialog box.

Select material

Set the accuracy

Save all will save the active file and all of its dependent files. For example, you are working on an assembly that is composed of dozens, perhaps hundreds of parts; picking **Save all** will save the assembly and all of its component parts.

The iProperties Option

Picking **iProperties** opens the **Properties** dialog box. See **Figure 2-28.** Most of the tabs are used to enter text information that will be saved along with your part. Some of this information will be used in the 2D drawing title block that is created of the part.

The **Physical** tab is used for the physical calculations of the part. If you select Very High for accuracy, the physical properties of this metric part will be displayed. The **Material** drop-down will let you change the part material and the values will change automatically. This topic will be covered in more detail in *Chapter 14*.

PRACTICE 2-1

❑ Open Practice-02-01.ipt and examine the part.
❑ Select **iProperties** from the **File** pull-down menu and then pick the **Physical** tab from the dialog box.
❑ Change the material to Brass, Soft Yellow and the accuracy to Very High. Immediately after you select the accuracy the calculation is done.
❑ You can save the output of this dialog box by picking Clipboard. This will place the report onto the Windows Clipboard, where you can paste it into a Word document.
❑ Save and exit the part file when you are done.

The View Pull-Down Menu

Selecting **Toolbar** in the **View** pull-down accesses a cascading menu showing the toolbars available while in a particular mode. For example, while in the Part Modeling mode the **Panel Bar**, **Browser**, and **Standard Bar** are checked. The **Part Features**, **Sketch**, **3D Sketch**, **Solids**, and **Sheet Metal Features** toolbars, or panels, are normally available in the **Panel Bar** as you need them. The **Inventor Collaboration** toolbar is for Net meetings.

If you are in Sketch mode you will find other toolbars that pertain to that mode. Open Example-02-01.ipt if you still do not have it open. In the **Browser**, click on the plus sign next to Extrusion and then double-click on Sketch1. You have now accessed the Sketch mode. Now look to see what toolbars are available in the **Toolbar** cascading menu. The **Inventor Precise Input** toolbar will be covered in the *Chapter 4*. Turn it on and dock it in the upper-right corner of your screen. The **2D Sketch Panel** toolbar is a duplicate of the panel shown in the **Panel Bar**. It is still very handy to have in the lower-right corner of the screen as you work on sketches. Place it in the lower-right corner, however, dragging it into two rows is best. To exit Sketch mode, click on the **Return** button on the **Inventor Standard** toolbar at the top of the screen.

The Status Bar

The **Status Bar** turns on and off the one-line bar at the bottom of the screen. This area serves as Inventor's equivalent to AutoCAD's prompts. As you work with Inventor, this area prompts you for what to do next.

NOTE   The rest of the options in the **View** pull-down will be covered using the button or the pop-up menus.

Inserting Files

There are three options in the **Insert** pull-down menu. **Object** is used to link or embed external files like Microsoft Word or Excel documents in an Inventor model or drawing.

Image is available only when in Sketch mode. It is used to bring in a raster BMP file for use in a sketch.

Import is available only when in Part mode. It is used to bring in a 3D geometry translation file such as IGES or SAT. These types of files are the only way to exchange 3D geometry with other diverse CAD systems.

The Format Pull-Down Menu

The **Format** pull-down menu is where you can create, edit, and activate styles. A style consists of selected lighting, material, and color settings. You can also copy the styles from one part or assembly to another. The following will explain each option in the **Format** pull-down menu.

Active Standard...

After the **Active Standard...** option is selected, the **Document Settings** dialog box appears with the **Standard** tab revealed. See **Figure 2-29**. This dialog box is also available under the **Tools** pull-down menu. The **Standard** tab presents two drop-down lists: Active Material Style and Active Lighting Style. When one of the drop-down arrows is picked, the styles available for use in the current drawing are displayed. Selecting a style from the list makes that style the default.

Styles Editor...

Selecting the **Styles Editor...** option displays the **Styles and Standards Editor** dialog box, where the designer can access and modify the Color Styles, the Lighting Styles, and the Material Styles. See **Figure 2-30**. In previous versions of Inventor these were three separate options on the Format pull-down menu. Now they have been incorporated into the new Style Editor. These are called *Style Definitions* and they are stored either locally in the part file or globally in style libraries. The top of the right side of the dialog box contains four buttons and a drop-down list. The buttons are labeled **Back**, **New**, **Save**, and **Reset**. Picking **Back** will revert the display below the buttons back to the color, lighting, or material that was previously displayed. **New**

Figure 2-29.
The **Standard** tab of the **Document Settings** dialog box is used to select the active styles.

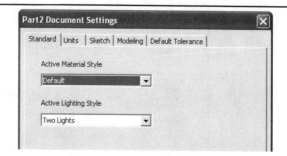

will create a new style based on the selected style. **Save** will save changes to the new or existing style. Pick **Reset** prior to **Save** if you want the settings to appear as they were prior to any changes. The **Filter Styles** drop-down list determines which styles will be available for selection. If **Local Styles** is selected, only the styles that have been applied to a part in the document or cached in the document are visible for selection. To *cache* (pronounced cash) a style into a document means the style is copied from the Style Library into the current document. The cached style will be in the list of local styles. In the beginning of a project, you may want to cache all of the colors, materials, and lighting that you expect to use. This way, when you need to select a style, you are able to select from the list of local styles — a much smaller list. This process will be shown, later in this chapter, in the section on the style pop-up menu.

On the left side of the dialog box is a compressed tree with **Color**, **Lighting**, and **Material** shown. Picking the plus symbol next to **Color** will show, in that same area, all of the colors available. Picking the word **Color** will show the same selections in the Color Style table on the right side of the dialog box. See **Figure 2-31**. Picking one of the color styles in the expanded tree or double-clicking on it in the table will open the color style for editing. This process applies to the **Lighting** and **Material** styles as well.

The Colors area displays four boxes that can be picked to modify the reflective qualities of a face. After picking one of the boxes, you can select a different color. If black is chosen, then that reflective quality has no effect on the model's display. The reflective qualities are:

- **Diffuse.** The color of the surface when seen in directed light. Unless you turn the lights off, all parts are seen in directed light by default.
- **Emissive.** The color the surface would appear if the part were lit from within. In the absence of directed lights or ambient lighting this color would appear.

Figure 2-30.
The **Styles and Standards Editor** dialog box is where styles are created and managed.

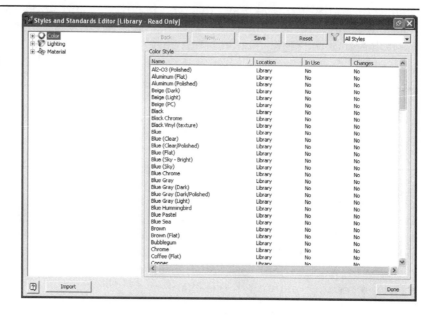

Figure 2-31.
Select a color style to edit or to create a new one.

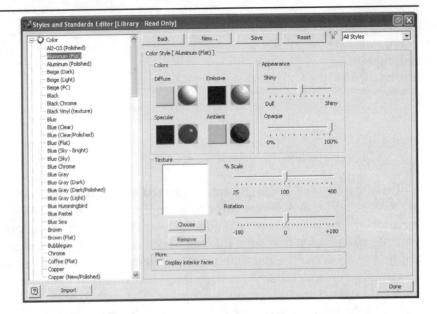

- **Specular.** The color that highlighted points, such as edges or corners, would use. The shininess setting determines the size and quality of the lit points. If the color style is shiny, the lit points would be small and sharp. If the style were duller, the points would be large and muddled.

- **Ambient.** This is the color that the part would appear if it were lit by ambient light only. Typically used to color the shaded areas of a part's surface.

The Appearance area contains two sliders that adjust the shininess and opaqueness of the color style. Take the time now to compare the values of Chrome, Lexan–Clear, and Metal–Steel. Notice the **Shiny** values for Chrome and notice the **Opaque** values for Lexan–Clear. Translucent (transparent) color styles have a 0% opaque setting. The part will disappear from the screen—effectively becoming invisible.

The Texture area allows you to specify a texture map to be included in the color style. For example, a color style meant to simulate wooden planking would use one of the wood grain textures defined in the texture database. Select Yellow (Flat) from the tree or table. Pick the **New** button, change the name to Wood, and press [Enter]. The new color should appear in the list. Change the color value for **Diffuse** to a defined custom color of Red 181, Green 158, and Blue 4. Change **Specular** to a defined custom color of Red 250, Green 247, and Blue 135. Also, change **Ambient** to a defined custom color of Red 183, Green 179, and Blue 0. Pick **Save** to save the changes and **Apply** to apply the color to your part. The value for **Emissive** was not changed. Now, change the **Shiny** setting to the dull end of the scale—wood is not very shiny unless it has been polished.

PROFESSIONAL TIP

To jump quickly to a style or texture within a long list, pick in the list to activate it and then type the first letter of style or texture you are seeking.

These color styles are also available from the drop-down list on the far right of the **Inventor Standard** toolbar. Picking one of them will change the color style that your part will use. It is important that you do not confuse these *color styles* with *materials*. This will be clarified later. Save your work under a different name if you want to keep the wooden part, but revert back to the previous color style (As Material) before proceeding on through this chapter.

Figure 2-32.
Each lighting style can have up to four direct lights and one ambient light.

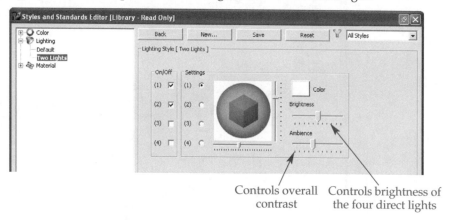

Controls overall contrast

Controls brightness of the four direct lights

Lighting is the next style listed. Pick the plus symbol and notice the lighting styles available are Default and Two Lights. See **Figure 2-32**. Selecting either one will open the **Lighting Style** area. This is where you can define the brightness, color, and position of one to four directed lights. This dialog box also controls the level of the ambient light, which is the pervasive light that comes from all directions. You can create styles and turn each of the directed lights on or off. By simply dragging a slider, dramatic changes in the lighting can be made to your part. The **Brightness** slider controls the level of intensity for all of the directed lights. Although you cannot control the brightness of each directed light by using the slider, you can do this by adjusting the color of each directed light. The **Ambience** slider controls the ambient lighting in your part by adjusting the overall contrast—the difference between the lighted and unlighted areas of the image. The slider to the right of the 3D cube controls the height of the current light source and the slider below moves the light source left or right.

To create a new lighting style, select **New** and change the name in the **Style Name** text box. Make your changes to the four directed lights and the slider bar settings, and pick **Save**. The new lighting style will be added to the list and will be available for this part only. Making styles available for other parts will be discussed in the Import/Export section of this chapter.

The third option is **Material**. See **Figure 2-33.** Select a material and notice the **Properties** area. All of this information is needed to properly define any material. These numeric values determine the results in the **Physical** tab of the **Properties** dialog box. For example, the density value is used to calculate the weight of the part. A thorough explanation of the values is beyond the scope of this text. Use the **Material** drop-down list shown in **Figure 2-28** to assign a new material to the part in Example-02-01.ipt. Try Lexan, Clear and other materials, remembering to pick **Apply**, to see the effect they have on the part's physical characteristics. Also, notice that associated with each material is the rendering style. This style determines the material's appearance on the screen. Now, set the part's material to Stainless Steel, 440C.

To create a new material, pick an existing material and then the **New** button. Now, pick in the **New Style Name** text box and enter a name, such as Test Substance. Notice when **OK** is picked, the new material is added to the list and the numeric properties remain the same as the original material. Make any changes to the material style and pick **Save**. Pick your new material from the **Material** drop-down list. This will apply the new material to your part. Change the rendering style to Metal–Red Hot and pick **Save**. The part should update to reflect the change in material. Change the rendering style back to Metal–Steel and pick **Save**. Keep the dialog box open.

Local Styles should be the current setting in the upper right-hand drop-down list. This filter allows only the materials that are stored in the part file to be displayed.

Figure 2-33.
After selecting a material, you may edit it or create a new one.

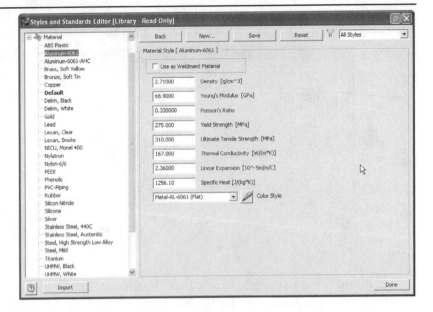

If you choose **All Styles** from the upper right-hand drop-down list, then you will see all of the styles defined in the global **Styles Library**. Try it now with Example-02-01 open. The list of **Local Styles** is shorter than that of **All Styles**. Choose **Local Styles** and right-click on **Stainless Steel** to access the context menu. Choose **Update Style** to over-write the local definition of **Stainless Steel** with that from the global **Styles Library**. With **Local Styles** chosen you will notice that there is no **Silver** defined in the part. Choose **All Styles**, right-click on **Silver**, and choose **Cache in Document**. This is the method used to load a material into the current part file. This style will now be avail-able from within the part file. The definition for a given material stored in the part can be different from that which is defined in the global library. Exit the dialog box by picking **Done**.

Update Styles...

The last three options are used to manage the local styles and those in the Styles Library. Picking **Update Styles...** opens the **Update Styles** dialog box listing all of the local styles that have settings that are different than those in the Styles Library. You have the choice of **Yes** or **No** to specify which local styles are updated. If any of the styles were edited locally you will be warned before overwriting them, as this will effectively eliminate any changes made to them.

Save Styles to Style Library...

This option will be available only if the **Use Styles Library** is set to **Yes** for the current project. This option is used to write the locally defined styles out to the global Styles Library. You would do this to have the local styles available for use in other parts or assemblies. Use this option with caution because you may be writing to the global library that others may use. This is why most companies and schools will have the **Use Styles Library** set to **Read Only** for most projects.

Purge Styles...

The final option allows you to remove any unused styles from the current part. You may have added a style to the part's cache or assigned a style to a feature that has since been deleted. Purging the unused style from the part will reduce the file size and is considered good housekeeping.

Import/Export

One final note, any of the styles discussed above may be exported to an external file stored on the hard drive by choosing **Export** from the right-click context menu. This file has an extension of ".styxml". Conversely, you may import any previously exported style by using the **Import** button in the lower left-hand corner of the **Styles and Standards Editor** dialog box. You would use this method to copy the .styxml file to another computer or to e-mail the style to another designer. This is not to be confused with the procedures in the following two sections.

The Tools Pull-Down Menu

The **Tools** pull-down menu has many uses. Many of the options will be covered in later chapters. The focus in this section is on the **Document Settings...** and **Application Options...** options.

Document Settings

Selecting **Document Settings...** in the **Tools** pull-down menu accesses the **Document Settings** dialog box. The settings in this dialog box affect the current document, whether it is a part, drawing, or assembly. The four tabs in this dialog box are as follows:

- **Standard tab.** Set the currently active Material and Lighting Styles.
- **Units tab**. Units of measure and the appearance of the dimension labels are set here. The Display as expression setting will be the most useful. See **Figure 2-34A.**
- **Sketch tab**. Settings pertinent to the sketch environment, such as the snap increment, and grid increment, are set here. See **Figure 2-34B.**
- **Modeling tab**. Contains settings related to the part-modeling environment. See **Figure 2-34C. Compact Model History** is used to condense the feature history of a part. This reduces part's file size, which is ideal for disk storage or e-mailing. However, this should not be used on all files because there is a significant delay when the part is opened for edit. The features have to be "rebuilt" and this takes time. If you do not need it, then do not use it.

   Advanced Feature Validation is only available for new parts created in Inventor 9. It uses a more complex and, therefore, slower method to compute part features. This can result in improved topology in rare circumstances. Not necessary to be used in regular work. **Allow Sectioning thru Part** is usually checked and need not be unchecked for most work. This will allow the part to be sectioned when creating a section view in a 2D drawing (IDW). **Tapped Hole Diameter** is set to **Minor** by default and need not be changed. This is support for legacy tapped holes from previous releases of Inventor.
- **Default Tolerance tab**. Inventor supports the use of tolerancing in part design. This tab is used to set up default tolerances. See **Figure 2-34D.** Use it only if your dimensioning scheme requires the modeling of tolerances.

Application Options

Selecting **Application Options...** in the **Tools** pull-down menu accesses the **Options** dialog box. See **Figure 2-35.** The settings in this dialog box affect Inventor itself and, subsequently, all files opened. The 12 tabs in this dialog box are as follows:

- **General tab**. Settings on this tab are for general utility. **Locate tolerance** decides whether Inventor finds an object underneath your pick point or not. **Undo file size** determines the size of the file that contains the number operations. With large assemblies, this size may need to be increased. **Username** shows up in the title block once a 2D drawing is created. If you do not like the **File Open** dialog box popping up when Inventor first starts, uncheck the **Show Startup Dialog** check box.

Figure 2-34.
A—The **Units** tab is used to change the units used for the part. B—The **Sketch** tab is used to change settings pertinent to the sketch environment. C—The **Modeling** tab is used to change settings related to the part-modeling environment. D—The **Default Tolerance** tab is used to set up default tolerances.

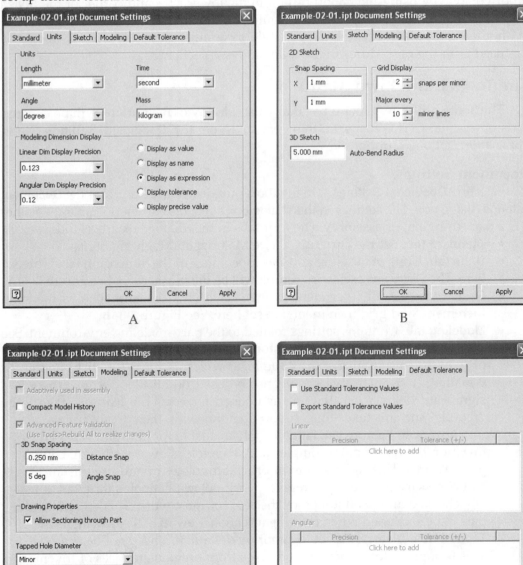

- **Save tab**. specifies various settings for saving files when minor changes have been made that do not affect the design intent of the part or assembly. In a multi-user environment you may not want to save the affected files because it will require that you check out the files in order to save them and this may interfere with a team member's more important design work.
- **File tab**. Inventor depends on many open files as it runs. This tab determines where they are located on the hard disk.
- **Colors tab**. Controls the appearance of the Inventor screen. The color is important for relief of eye fatigue. **Figure 2-36** shows one configuration.

Figure 2-35.
The **Options** dialog box changes Inventor itself and subsequently all files opened.

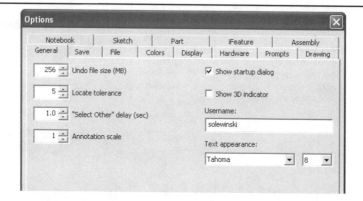

Figure 2-36.
The **Colors** tab is used to control the appearance of the Inventor screen.

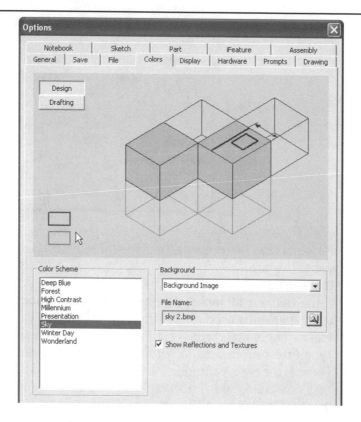

- **Display tab.** Contains many of the settings that affect how Inventor uses the screen. See **Figure 2-37.** The **Wire Frame Display Mode** area is used when displaying the part in wireframe. The **Shade Display Modes** area is used when displaying a shaded part. The **Display Quality** area is where the display quality is selected. Smooth quality will take longer to regenerate. The **View Transition Time** slider determines how quickly Inventor switches to a new view. The **Minimum Frame Rate (Hz)** slider controls the display of your model as you use certain display tools, such as **Rotate** using the [F4] key. If you move too fast, portions of the display drop out. This is the display trying to keep up with this setting. This will happen in a larger assembly or if your computer has a slower video card.
- **Hardware tab.** Because Inventor makes great demands on the video capabilities of your computer, this tab is used to tweak its performance or perhaps diagnose just what is causing any crashes that may occur. A major cause of problems in computers, besides poor programming, is the device drivers that the computer uses to drive the various components. The video card, mouse, monitor, and motherboard all have device drivers. The **Diagnostics** button is used to gather technical information about your video card.

Figure 2-37.
The **Display** tab is used to control the settings that affect how Inventor uses the screen.

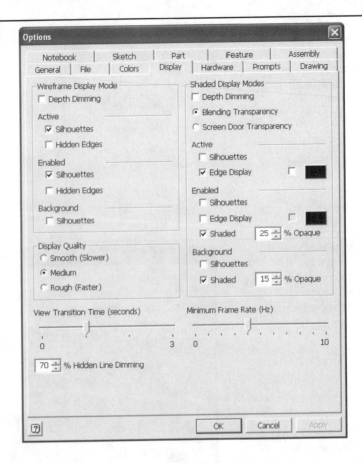

- **Prompts tab.** The prompts that appear in many boxes can be turned on and off using this tab.
- **Drawing tab.** This tab is used to control the settings when working with drawings. If the **Retrieve all model dimensions on view placement** check box is checked, the dimensions used to constrain the sketch(es) are placed on the 2D drawing views. There are various settings available under the icons in the **Dimension Type Preferences** area to control the linear, diametric and radial dimensions. There is a new area under the drawing tab titled **Line Weight Display Options**. You can choose how to display the different line weights or uncheck the **Display Line Weights** checkbox to display all lines at the same lineweight.
- **Notebook tab.** Used to determine the appearance and use of the **Engineer's Notebook.** This is a place for making notes with the intent of communicating information to coworkers. This is covered in more detail later in this chapter.
- **Sketch tab.** This should not be confused with the **Sketch** tab of the **Document Settings** dialog box. Remember, this **Options** dialog box controls Inventor on a global basis. The choices in the **Document Settings** dialog box control only the current document. This tab has many options, such as the display of grid lines and snap to grid. The **Edit dimension when created** check box is checked on by default. This means after picking the three points for a dimension you are prompted to change the default measured distance. Since Inventor is based on sketching, the measured distance almost always needs to be changed. The **Automatic reference edges for new sketch** check box allows you locate a new sketch in reference to existing part's edges or work features. This automatically creates reference geometry on the new sketch.
- **Part tab.** Offers control of a few of the universal options available when in Part Feature mode. The **Parallel view on sketch creation** check box switches the view to be plan to that of any new sketch that you create. Most people prefer to draw 2D sketches in plan view—much like drawing on a sheet of paper.

- **iFeature tab**. iFeatures are a 3D symbol library of features, such as slots, tabs, etc. They will be covered in *Chapter 14*. This dialog box determines file locations on the hard drive.
- **Assembly tab**. Contains many of the important options that govern how Inventor acts when creating assemblies. The **Defer Update** check box is unchecked by default. This means that updates to the assembly will happen automatically when individual parts are changed. This update of the assembly may take a long time, especially if the assembly is large. Deferring the assembly update lets you make changes to the parts without the assembly updating until you pick the **Update** button. Settings for the **Contact Solver** are at the bottom of the dialog box. In Inventor 9, there is a new option for the **Contact Solver — Surface Complexity**. This option is used to speed up the performance of the **Contact Solver** by ignoring complex surfaces. There are many additional settings available here that will be covered in later chapters.

The Applications Pull-Down Menu

This is used to switch between Modeling and Sheet Metal modes. If you try it, you will notice the **Panel Bar** changes accordingly.

The Windows Pull-Down Menu

As with all Windows-based software, Inventor takes advantage of the fact that several files can be open at the same time and managed via the **Windows** pull-down menu. See Figure **2-38**. Notice the tools available, as well as the three open files—two part files and one assembly.

The **New Window** option opens a copy of the current file in a new Window. Think of it as a clone. This file can be used to try out some radical procedure. The **Cascade** option arranges all open files like a deck of cards. See **Figure 2-39A**. The **Arrange All** option arranges all open files in a tiled configuration. See **Figure 2-39B**. Switching between the various open files is as simple as selecting one from this pull-down menu. One of the advantages of having multiple files open is that geometry can be copied from one to another. This is similar to the way any Windows program supports copying and pasting, to and from the Clipboard.

The Web Pull-Down Menu

This pull-down menu provides access to the Autodesk's Inventor-specific websites. See **Figure 2-40**. The **Autodesk Inventor** option accesses the Inventor home page. The **Streamline** option accesses an on-line collaboration tool for workgroups and vendors. The **Team Web** option opens a sample iDrop webpage that allows you to sample the iDrop technology of drag-and-drop geometry.

Figure 2-38.
The **Windows** pull-down menu and its options.

Figure 2-39.
A—The **Cascading** option showing four files open in the graphics window. B—The **Arrange All** option showing four files open in the graphics window.

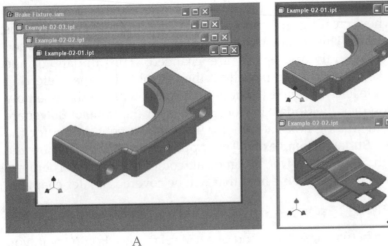

A   B

Figure 2-40.
The **Web** pull-down menu and its options.

The **Help** Pull-Down Menu

This pull-down menu provides access to a comprehensive help system. It gives a user a complete guide to the software. The options are as follows:

- **Help Topics**. Opens the Autodesk Inventor Help home page. This page contains an index, word search, links to tutorials, and other helpful features.
- **What's New**. Offers information on the new features in the latest release of the software.
- **Tutorials**. Provides samples procedures for working with Inventor tools and other options.
- **Help for AutoCAD Users**. Offers information designed to ease the transition of those veteran AutoCAD users to the new design software, Inventor.
- **Programming Help**. Offers information for users that need to work in Visual Basic for the purpose of extending the capabilities of Inventor.
- **Autodesk Online**. Offers access to the Autodesk home page.
- **Graphics Drivers**. Provides access to a website listing information on many of the video card drivers. Many of these drivers are downloadable here. This is the option used to try to rectify any video-related problems you might experience.
- **About Autodesk Inventor**. Provides information about the program's release number, build number, serial number, etc.

The Visual Syllabus

This tool is an excellent addendum to the help system. The Visual Syllabus icon is the last one *available* on the **Menu Bar** when the file is open. It provides an animated demonstration of certain topics listed graphically on the page. See **Figure 2-41.** The Part Modeling listing has 21 entries. Take a few moments and explore the various items. For example, look up how to apply hole features. There are animated demonstrations for Sheet Metal, Assembly Modeling, Presentations, and Drawings.

Figure 2-41.
A—The **Visual Syllabus** button is the last active button on the **Menu Bar.**
B—The **Visual Syllabus** is used to review techniques used to create standard parts.
C—The **Show Me** dialog box showing the animated demonstration of a part.

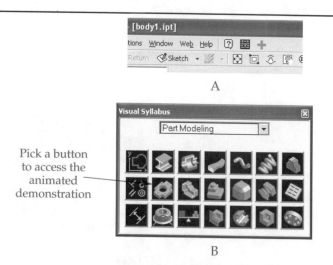

A

Pick a button to access the animated demonstration

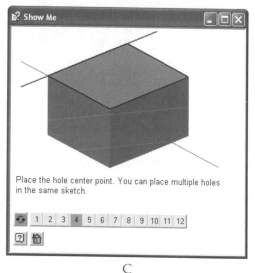

B

Place the hole center point. You can place multiple holes in the same sketch.

C

CAUTION

There is one more option on the **Menu Bar**. This is the cross, or large plus sign. This is the **Recover** tool and is not enabled. If it becomes enabled, there is a problem with your part design.

Engineer's Notebook

Have you ever wanted to make a note concerning a step in the design process? The **Engineer's Notebook** lets you do just that. Open Example-02-01.ipt if you still do not have it open. Left-click on Extrusion 1 (this was renamed Base Feature earlier in this chapter) in the **Browser**. This step is not entirely necessary, but it is nice to have it highlighted. Right-click on the Base Feature and choose **Show Dimensions**. Again, an unnecessary step; we just want them to show in the **Engineer's Notebook**. Right-click on Base Feature once again and choose **Create Note**. In the yellow comment box, enter the text shown in the **Figure 2-42.** This will document the change to the 110 mm dimension. Notice how the dimensions show up in the notebook graphic.

Figure 2-42.
Adding a note using the **Engineer's Notebook**.

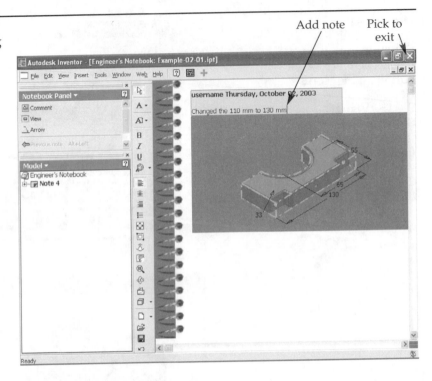

Expand the Note in the Browser and notice the View 1 and Comment 1 listing. A pop-up menu is accessed if you right-click on View 1. There are view controls that work the same in the **Engineer's Notebook** as they do in the regular Inventor graphics window. Pick the view and right-click to reveal these commands. **Delete** will remove the view. **Freeze** will lock the view at the current design phase. The **Display** flyout offers the three standard view modes. Try **Rotate** on the view. **Restore camera** acts as a zoom previous. **How To...** is the link to the Inventor help system.

Along the top of the screen is the **Menu Bar** that contains the pull-down menus. These menus contain many of the same items found in the standard Inventor menus, but some other entries pertain to the **Engineer's Notebook**. Below the **Menu Bar** is the **Inventor Standard** toolbar. This toolbar offers the common tools used for working with text entry, such as font choice; font point size; bold; italic; left, center, or right justification; bullets; etc. Also on this toolbar are buttons for pan, zoom, undo, and redo operations.

The **Comment** button in the **Notebook Panel** allows you to add an additional comment box. The **View** button in the **Notebook Panel** is used to add an additional view. The **Arrow** button in the **Notebook Panel** allows you to add an arrow in the view. When finished, exit out of the **Engineer's Notebook** by using the smaller **X** in the upper-right corner.

After you are done with the **Engineer's Notebook**, notice in the **Browser** that there is an entry for the note. It is, of course, listed under the feature it applies to.

If you look at the part in the graphics window, there is a yellow icon present. Hover the cursor over the icon and you should be able to read the note you just created. The visibility of this icon and the pop-up text can be controlled with the **Notebook** tab in the **Options** dialog box.

PROFESSIONAL TIP

It is possible to create an engineer's note by right-clicking on a face or edge of a part. Choose **Create Note** from the pop-up menu to access the **Engineer's Notebook**. A note added this way will be listed in the **Browser** under the *Origin* folder. Inventor does this to signify that an edge or face was selected and not a feature.

Figure 2-43.
The **Resolve Link** dialog box is accessed if a file that depends on other files is opened before a project is set up.

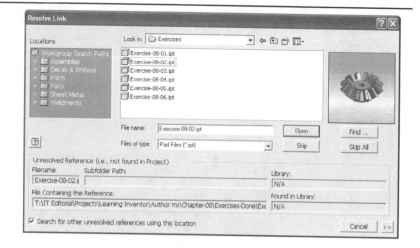

Projects

Many of the files used in Inventor are dependent on the existence of other files. For example, an assembly (IAM file) is composed of parts (IPT files). It requires that those parts be able to be located on the hard disk when the assembly is opened for editing. Another example is when a drawing (IDW file) is opened, the part or assembly that it depicts should be located easily. If the part or assembly cannot be located, then you have a serious problem. You may end up with nothing shown on the screen. Inventor uses a concept called *projects* to solve this problem. If you do not set up a project beforehand and you try to open a file that depends on other files, then you will encounter the **Resolve Link** dialog box. See **Figure 2-43**. It is asking that you specify the location of the needed file. You could navigate the folder structure and find the file for Inventor, but this is tedious and unnecessary.

Selecting a Project

You will see the benefits of using a project by selecting and examining one that has been created for this chapter. If you have not installed the Chapter 2 files from the Student CD that was included with this text, do so now. During the installation, folders were created to organize all of the files required for a project. Refer to **Figure 2-44** for an example of this layout taken from MS Windows Explorer.

Close all files, because you cannot activate a project with Inventor files open. This does not mean that you cannot work on files from another project, but you will have to browse through the folders for those files. Pick **Projects...** from the **File** pull-down menu to view the **Open** dialog box. Now pick **Browse...** and go to C:\Program Files\ G-W\Learning Inventor 9-User's Files\Chapter-02. If the Chapter 2 files were installed in another folder, go there. Open the project file Chapter 2.ipj. The IPJ file is a text file that stores the paths of folders for the project.

Figure 2-44.
MS Windows Explorer showing the folder/subfolder structure set up for all the files needed for a particular chapter's work.

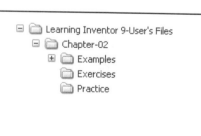

Figure 2-45.
The corresponding Inventor project set up for this chapter.

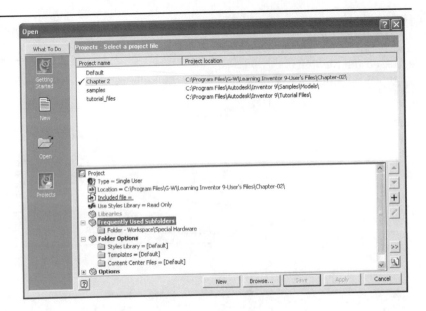

The **Open** dialog box should look similar to **Figure 2-45**. Pick the plus symbols in the lower area of the dialog box to reveal more information about the project. Notice the file path next to Location = . This is the project's main folder. Within this folder are subfolders that contain files for the project. One of the subfolders, Special Hardware, is listed under **Frequently Used Subfolders**. This shortcut was created by right-clicking **Frequently Used Subfolders** and adding the path.

Now that we have explored the project, let us see how this can help us locate files. If the check mark in the upper-half of the **Open** dialog box is not next to the Chapter 2 project, double-click Chapter 2. The project with the check mark next to it is the active project. Pick the **Open** icon in the left area of the **Open** dialog box. The dialog box should look similar to **Figure 2-46**, if not, pick the **Project** icon and activate the Chapter 2 project. Notice that the folder at the top of the dialog box is the project's main folder, Chapter-02, and the subfolders are listed below it. Pick on the Special Hardware folder listed under the **Frequently Used Subfolders** in the left of the dialog box. The right area quickly jumps to the contents of the Special Hardware subfolder.

Figure 2-46.
The **Open** dialog box with the project selected.

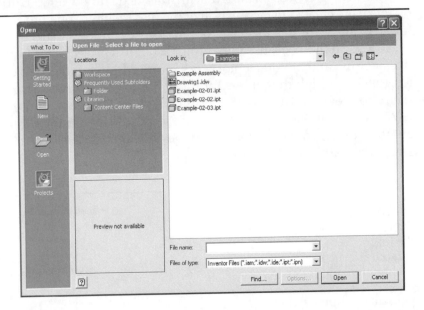

Figure 2-47.
A—Select the project type. B—Enter the name and folder location for the project. C—Include any library that may be needed in the course of working on the project.

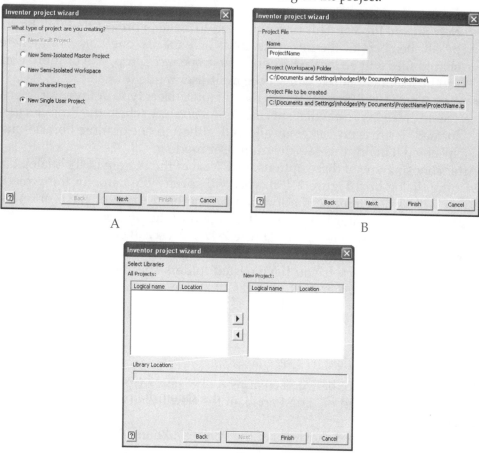

A

B

C

Creating a Project

You can create your own projects. This allows you the freedom to organize your work along project lines. Close all files, because you cannot create a new project with Inventor files open. To create a new project, pick the **Project** icon in the **Open** dialog box and then select the **New** button located near the bottom of the dialog box. This accesses the **Inventor projects wizard**. See **Figure 2-47.**

While reading the following, take note of the distinction between workspace and workgroup. Furthermore, each option allows control over the multiuser mode, which is a method by which access to one or more users is controlled. The options are as follows:

- **New Vault Project**. If Autodesk Vault is installed, this will be enabled. Allows the designer to create a new Vault project and one *workspace*. In addition, one or more libraries may be specified and sets the multiuser mode to Vault.

- **New Semi-Isolated Master Project**. Designed to manage and support the use of Inventor in a group environment of designers and engineers. It allows a master project to be set up that specifies one *workgroup* where the Inventor parts and assemblies used by the entire design team are stored. Allows the inclusion of one or more parts libraries and sets the multiuser mode to semi-isolated.

- **New Semi-Isolated Workspace**. This will create a personal project that will include a previously created master project. The *workgroup* path and library locations will be inherited from the master project. Allows you to specify one *workspace* and sets the multiuser mode to semi-isolated.
- **New Shared Project**. This option is used to create a project that will be shared by all members of the design team. It will specify one *workgroup* where the shared files will be located. There is no *workspace* setup, one or more libraries can be included, and it will set the multiuser mode to shared.
- **New Single User Project**. This is the default project type in Inventor. It enables you to create a project that specifies one *workspace* where the Project's files are located. There is no *workgroup* allowed, although one or more libraries may be specified. Finally, this sets the multiuser mode to Off.

After choosing one of these options, pick **Next** at the bottom of the dialog box. In the ensuing dialog box, **Figure 2-47B**, you will specify the name of the project file. Here you will also determine the location on the hard drive of the workgroup or workspace for the Inventor files that will be a part of the project.

An optional final step is seen in **Figure 2-47C**. This will enable you to include in the project search paths the location of any library that may be needed in the course of working on the project. If you have nested folders specified in the search paths, these folders will show up in red to indicate that they are subfolders.

Chapter Test

Answer the following questions on a separate sheet of paper.

1. What is the difference between creating a new part by picking **New** on the **File** pull-down menu, and picking **Part** from the **New** flyout on the **Inventor Standard** toolbar?
2. Why use the keyboard shortcuts of the **Pan** (F2), **Zoom** (F3), and **3D Rotate** (F4) tools?
3. Must a style be used in a part if that style is to be saved locally in that part?
4. Why would you move the End of Part feature higher in the model tree in the Browser?
5. When using the **Engineer's Notebook,** is there the capability of modifying a note by adding more comments or views? This does not mean creating an entirely new note.
6. After making a note in the **Engineer's Notebook,** a tiny yellow icon shows up on the part in the graphics screen. This can be distracting. Is there a way to turn its display off?
7. *True or False?* The **Visual Syllabus** will take over for you and finish your part if you run into a problem.
8. What does the **New Window** option in the **Windows** pull-down menu do?
9. *True or False?* On the **Standard Toolbar,** the **Return** icon is the same as pressing [Enter] on the keyboard. If false, then what is its use?
10. In the **Units** tab of the **Document Settings** dialog box, there is a setting Display as expressions. What does it do and why is it important?
11. Is it possible to activate a project when an Inventor part file is open?

Exercises

Exercise 2-1. *Ring Gear.* Go into **iProperties** and fill out each tab of the dialog box with information. This information is used in the title block of any drawing that you may make of this part and also is used in the parts list (BOM). Go to the **Physical** tab and calculate the mass properties of this part. Use the lighting control to change the lighting to brighten up the image. Also, change the color style to various ones, and finally leave it set to Rubber–Blue.

Exercise 2-2. *Swivel Yoke.* Use the **iProperties** dialog box to fill in the information for all of the tabs. On the **Physical** tab, change the Material to Steel, High Strength Low Alloy and calculate the mass properties of this part. Use the lighting control to swing the light around to the left, because the left-hand face is too dark. Rename the features in the **Browser** to those shown below.

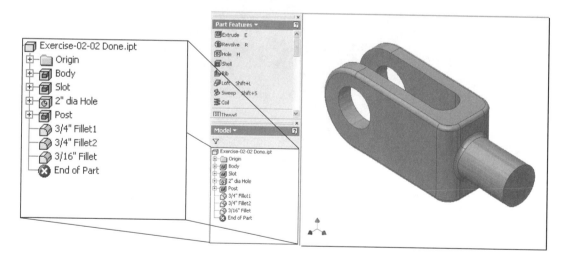

Exercise 2-3. *Piston.* Use the **iProperties** dialog box to fill in the information for all of the tabs. On the **Physical** tab, calculate the mass properties of this part—do not change the material. Use the lighting control to change the lighting. You will notice that no setting affects the screen image. Why? Could it be the chrome color style that is used? Rename the features in the **Browser** to those shown below. Add the **Engineer's Notebook** note as shown.

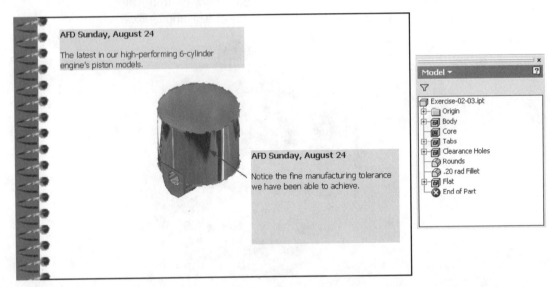

Exercise 2-4. *Gold Contact Bracket.* Use the **iProperties** dialog box to fill in the information for all of the tabs. On the **Physical** tab, calculate the mass properties of this part using Gold as the material. Finally, change the default isometric view to that shown below.

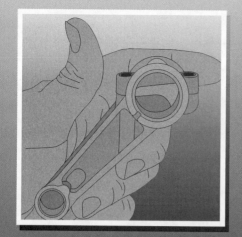

CHAPTER 3

Sketching, Constraints, and the Base Feature

Objectives

After completing this chapter, you will be able to:

* Describe the procedure for creating a base sketch.
* Sketch curves, including lines and arcs.
* Explain the geometric constraints that Inventor can apply.
* Apply and display geometric constraints.
* Add dimensions to constrain a sketch.
* Extrude solid parts from a sketch.

User's Files

*The following is a list of files that you will need to work through this chapter. These files can be found on the **User's Files** CD included with this text.*

Examples

Example-03-04.ipt   Example-03-05.ipt

The Process for Creating a Part

All parts in Inventor start with a base 2D profile that is extruded, revolved, or swept into a solid. The profile is constructed in Sketch mode, so it is called a *sketch.* A sketch is constructed from geometric shapes: lines, arcs, circles, and splines. It must be closed; an open sketch will extrude into a surface, not a solid. The closed sketch cannot cross over itself, but it can have interior islands or separated closed shapes. Valid and invalid shapes for solid profiles are shown in **Figure 3-1.** In this chapter, you will learn how to create single, closed shapes. These are called *unambiguous profiles.* You will also learn how to extrude these profiles into solid parts.

An Inventor sketch made up of six lines is shown in **Figure 3-2.** All lines, arcs, and circles in Inventor are considered curves. The curves (lines) in **Figure 3-2** are called ***intelligent objects*** because they each contain, or "know," unique information about themselves and each other. For example, Line AB was constructed horizontally and Inventor remembers this by automatically attaching a horizontal constraint to the

Figure 3-1.
The shapes on the left are valid 2D profiles for extruding. The shapes on the right are invalid.

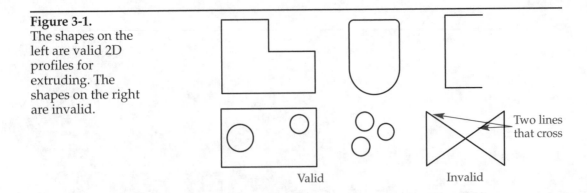

Valid   Invalid

Figure 3-2.
This is a sketch created in Inventor. All lines, arcs, and circles in Inventor are called curves.

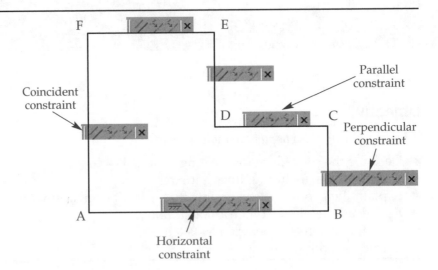

line. Line BC knows that it is perpendicular to Line AB and the perpendicular symbol is displayed on both lines. Line BC also knows that it is connected to Lines AB and CD and those symbols (coincident) are displayed on both pairs of lines. Line CD knows that it is parallel to Line AB.

Constraints define and maintain the relationships between the geometry in the sketch. By default, Inventor applies parallel and perpendicular constraints whenever possible in place of horizontal and vertical constraints. Horizontal, perpendicular, coincident, and parallel constraints are geometric constraints. There are a total of 11 geometric constraints. You can also apply dimension constraints.

The automatic constraint symbols for all six lines are displayed in the figure. The display of constraints can be turned on and off. Constraints can be manually controlled as well. How to display, add, and delete constraints is discussed later in this chapter.

The sketch in **Figure 3-2** was drawn in metric units. When line AB was constructed, it was not drawn to the exact length of 170 mm. Although it is possible to construct precisely, it is not necessary. After the sketch is drawn, dimensions can be added. Adjusting a dimension value changes the associated geometry. The fully dimensioned sketch is shown in **Figure 3-3.**

Even though you do not need to construct precisely, you do want to be reasonably close. Large changes in part size made by editing dimensions may distort the sketch beyond repair. You may be forced to undo the changes and start again. Based on these ideas, here are some sketching guidelines for this chapter:

- Sketch the size reasonably close.
- Sketch so the constraints are exact.
- Dimension the objects precisely.

Figure 3-3.
This is the sketch from **Figure 3-2** fully dimensioned and constrained.

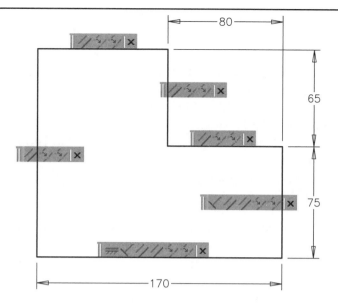

In addition, make the base sketch representative of the fundamental feature of the part, yet as simple as possible. This will give you control of all the added features for making quick changes. For example, a bracket and its base sketch are shown in **Figure 3-4.** The fillets, cutouts, and holes were all added as features.

In some cases, you may not add the dimensions. For example, this may be a preliminary design study and the dimensions may not be known. Of course, the dimensions can be added later. In another example, you may want to control the size of the part (its dimensions) by the position of other parts in the assembly. This is called an "adaptive" part and is discussed in *Chapter 12*.

Figure 3-4.
A—This is the completed part.
B—The base sketch used to create the part.

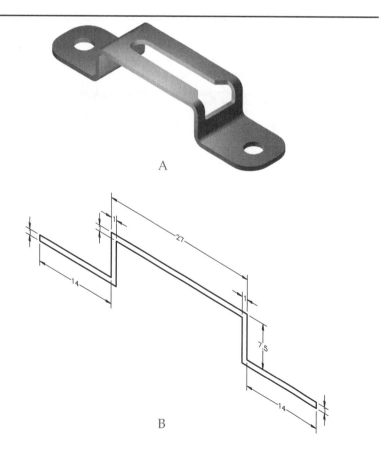

Now you will learn how to create a part. First, you will sketch the 2D profile shown in **Figures 3-2** and **3-3**. Then, you will extrude this sketch to create a part. Begin by starting a new metric part file:

1. Select **New...** from the **File** pull-down menu or pick the **New** button on the **Inventor Standard** toolbar.
2. Select the **Metric** tab in the **Open** dialog box that is displayed.
3. Select the Standard (mm).ipt template file and pick the **OK** button.

This brings up the graphics window with a grid and the coordinate system indicator (CSI) displayed in the lower-left corner of the grid. The CSI is not, however, at the origin (0,0). The screen shows the X and Y axes as black lines. The major horizontal and vertical grid lines are shown as light gray lines. The major grid lines are 20 mm apart. The minor gird lines are shown as dark gray lines and 2 mm apart; this is considered a 2 mm grid. See **Figure 3-5.**

The default name of the file is Part1 (or Part2, Part3, etc.). Later, when you save the file, it can be given a more logical name. You cannot save in Sketch mode, which is the current mode. Sketch mode is indicated by the **2D Sketch Panel** in the **Panel Bar**. Also, Sketch1 is highlighted in the **Browser**. If you try to save the file now, Inventor will ask if you want to exit this mode.

Since the part is 170 mm wide, you need to zoom out to display more of the grid. Pick the **Zoom** button on the **Inventor Standard** toolbar. Then, pick anywhere in the graphics window and drag the cursor up until ten major grid lines are displayed. Now, pick the **Pan** button on the **Inventor Standard** toolbar and pan the intersection of the X and Y axes to the lower-left corner of the screen so there are nine major grid lines to the right of the Y axis. If you have a roller wheel mouse, you can roll the wheel to zoom or hold the wheel down to pan.

Figure 3-5.
The Inventor screen after starting the new part.

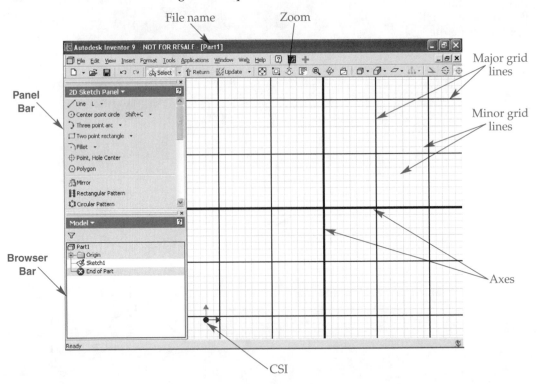

The intersection of the axes is called the *origin* and has the coordinates 0,0. In this example, you will start drawing the sketch at X=0, Y=0. Starting at the origin is not necessary, but you will find it useful when using work planes, which is introduced in Chapter 5.

Since Inventor is in Sketch mode, the 25 sketching tools with their names and buttons are displayed in the **2D Sketch Panel** in the **Panel Bar**. If you hover the cursor over a button, a tooltip is displayed next to the cursor and a help string is displayed at the bottom of the screen. Note that seven of the buttons are flyouts. For example, the **Center point circle** button is a flyout and contains the **Tangent circle** and **Ellipse** buttons as well. The button you pick from a flyout remains "on top" as the default button after you are done using the tool. Three of the tools have default hot keys; the **Line** tool is [L], the **General Dimension** tool is [D], and the **Circle** tool is Shift-C. Pressing the key on the keyboard activates the tool.

Now it is time to draw the first line, Line AB. Refer to **Figure 3-3.** First, activate the **Line** tool by picking the **Line** button in the **2D Sketch Panel** or pressing the [L] key on the keyboard. The **Line** button is depressed to indicate the tool is active. Also, the message Select start of line, drag off endpoint for tangent arc appears at the bottom-left of the screen.

Move your cursor to the origin and pick to set the first endpoint of the line. Now move the cursor to the right. A green line appears attached to the cursor and the first endpoint. A yellow dot appears under the cursor, which represents the second endpoint. Also, if you move the cursor horizontally, a small horizontal line appears under the cursor. This symbol indicates that the automatic horizontal constraint will be applied. If you move the cursor vertically, a small vertical line appears.

Notice that there are three windows in the lower-right corner of the screen. See **Figure 3-6.** The left-hand window displays the coordinates of the cursor position. The middle displays the length of the line from the first endpoint to the cursor location. The right-hand window displays the angle from the positive X axis to the line.

Move the cursor to the right so the line is horizontal and about 170 mm in length; *about* means between 160 and 180. Pick to set the second endpoint of the first line, Line AB. When the cursor is directly over the endpoint, the dot is gray. When you move the cursor off the endpoint, the dot is yellow. Now, draw the other five lines in the sketch. Remember, the values do not need to be exact at this point, rather *about* the correct value.

1. Move the cursor straight up. Notice that a perpendicular symbol appears next to the cursor and under Line AB. This indicates that the automatic perpendicular constraint will be applied. When the line is perpendicular and about 60 mm in length, pick to set the second endpoint of Line BC.
2. Draw Line CD by moving the cursor to the left. Notice that a parallel symbol appears next to the cursor and under Line AB. This indicates that the automatic

Figure 3-6.
As you draw, various information is displayed in the lower-right corner of the screen. Also, notice the constraint symbol that is displayed next to the cursor.

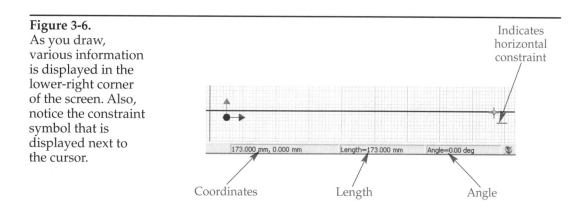

Indicates horizontal constraint

Coordinates   Length   Angle

parallel constraint will be applied. When the line is parallel and about 80 mm in length, pick to set the second endpoint of Line CD.

3. Move the cursor straight up. Notice that the automatic parallel constraint will be applied. Inventor prefers the parallel constraint over the perpendicular constraint. Therefore, Line DE will be constrained parallel to Line BC, not perpendicular to Line AB. When the line is parallel and about 65 mm in length, pick to set the second endpoint of Line DE.

4. Move the cursor to the left. Notice that the automatic parallel constraint (parallel to Line AB) will be applied. When you move the cursor above the first endpoint of Line AB, a dotted vertical line appears indicating what Inventor thinks is a logical next line. This helps you locate the second endpoint of Line EF. When the dotted line appears and Line EF is parallel, pick to set the second endpoint of Line EF.

5. Now, move the cursor to the first endpoint of Line AB, which is in this case the origin. The dotted line appears as you move the cursor straight down. Also, the automatic parallel constraint (parallel to Line BC) will be applied. When the cursor is directly on top of the first endpoint of Line AB, the yellow dot changes to green. Also, the coincident constraint symbol appears next to the cursor that indicates the two endpoints will be connected. Pick on the first endpoint of Line AB to draw Line FA and close the sketch.

6. Exit the **Line** tool by pressing the [Esc] key or right-clicking in the graphics window and selecting **Done** from the pop-up menu.

If you make a mistake while drawing the profile, you can pick the **Undo** button on the **Inventor Standard** toolbar, press the [Ctrl]+[Z] key combination, or select **Undo** from the **Edit** pull-down menu. This deletes the last line segment and cancels the **Line** tool. To continue drawing the profile, pick the **Line** button, move the cursor to the endpoint of the last line, and pick when the yellow dot becomes green. Then, finish sketching the profile.

Constraints

Constraints define and maintain the relationships between the geometry in the sketch. A constraint can be geometric, such as parallel or perpendicular. In addition, dimensions are used to constrain sketches. Dimensional constraints are discussed in the next section.

As you saw in the previous example, you can automatically place geometric constraints on the sketch as you draw. If for some reason you want to prevent the automatic constraint from being applied, hold down the [Ctrl] key as you draw.

To show how the geometry is constrained, pick the **Show Constraints** button in the **2D Sketch Panel**. Then, pick on the geometry. The constraints for that geometry appear in a small bar, called a *constraint bar,* next to the geometry. Review the constraints by placing the cursor over any of the constraint symbols. A red box appears around the symbol and the geometry to which the constraint applies is highlighted in red. To exit the **Show Constraints** tool, press the [Esc] key.

The order in which symbols appear in the bar, from left to right, is the order of precedence for the constraints. To delete a constraint, right-click on the symbol in the constraint bar and select **Delete** from the pop-up menu. To move the constraint bar, move the cursor to the double line at the left end of the bar. The "move" cursor appears. Pick and drag the bar to a new location. To turn off the display of (close) a constraint bar, pick the X at the right end of the bar.

You can show the constraints for all geometry by right-clicking in the graphics window when no tool is active and selecting **Show All Constraints** from the pop-up menu. To turn off the display of all constraint bars, right-click in the graphics window. Then, select **Hide All Constraints** from the pop-up menu.

Dimensions

Since there are currently no dimensions on the sketch, you can dynamically change the geometry in the sketch. Select a line or an intersection of two lines, hold down the mouse button, and then drag the geometry to the new location. Try this by selecting any corner in your sketch and moving it to a new location. However, notice that as you move the corner, the two lines attached to it remain constrained by the geometric constraints that were automatically placed as you drew the profile. This is a useful technique for preliminary design work.

You have two choices for applying the dimensions needed to accurately size the part. You can apply automatic dimensions, in which case Inventor decides where the dimensions should be placed. Alternately, you can manually apply and place each dimension.

To manually apply a dimension, pick the **General Dimension** button in the **2D Sketch Panel**. Then, pick the geometry to which you want the dimension applied. Finally, drag the dimension to the desired location. To cancel the **General Dimension** tool, press the [Esc] key.

For this example, first apply automatic dimensioning. Pick the **Auto Dimension** button in the **2D Sketch Panel**. The **Auto Dimension** dialog box shown in **Figure 3-7** is opened. All curves (lines) in the sketch are automatically selected. Notice the dialog box reports that six dimensions are required. This means six dimension constraints are required to fully constrain the sketch. The number of dimensions needed to fully constrain the sketch is based on the geometric constraints currently applied to the sketch. If you remove some of the geometric constraints, additional dimensions may be needed. Now, pick the **Apply** button and Inventor constructs *four* dimensions.

Why four? Notice the dialog box reports that two dimensions are still required. These two dimensions are required to constrain the sketch to a location in 2D space. These cannot be applied automatically. For this example, you will apply geometric dimensions rather than locational dimensions. You will do this in the next section.

Dimensions are what allow you to draw the sketch at its approximate size. Once placed, a dimension value can be edited to an exact value. Since the associated geometry is constrained by the dimension, it is updated to match the exact value. First, close the **Auto Dimension** dialog box by picking the **Done** button. Then, double-click on the dimension across the bottom that is supposed to be 170 mm. The **Edit Dimension** dialog box shown in **Figure 3-8** is displayed. Each dimension has a name that is, initially, assigned by Inventor. In this case, the name is d1, as reflected in the title bar. Depending on how you drew your sketch, this name may be different but it will start with the letter d. In the next chapter, you will learn how to rename dimensions and relate dimensions to each other. For now, enter the exact value of 170 and pick the check mark button. Edit the remaining dimensions to the exact values shown in **Figure 3-3**.

You do not need to enter the "mm" for millimeters as this is a metric part. If other units are entered, you need to include the unit abbreviation, such as 6.6929 in, 0.558 ft, or 17 cm. Any of these entries will change the dimension to 170 mm. Regardless of the units entered, the dimension on the sketch will always display the millimeter value.

Figure 3-7.
You can
automatically
dimension a sketch.

Number of
dimensions
needed to fully
constrain the
sketch

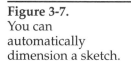 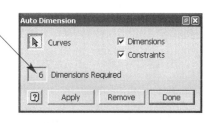

Figure 3-8.
By editing a
dimension, you can
alter the associated
geometry.

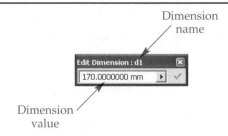

Dimension
name

Dimension
value

Once a dimension is placed, you can reposition it. Simply pick on the number when the "move" cursor is displayed. Then, drag it to a new location. The **Auto Dimension** and **General Dimension** tools must be inactive to do this.

> **PROFESSIONAL TIP**
>
> You can use the **Auto Dimension** tool at any time to check for the number of dimension constraints that are still required to fully constrain the sketch.

Applying Fix Constraints

The sketch is now dimensionally and geometrically constrained. However, it is not locked to the coordinate system. This means that the geometry can be moved around in the XY plane of the coordinate system. You can fix the sketch to the coordinate system in order to take advantage of the existing work planes. To fix this sketch, you will apply a fix geometric constraint to Line AB and Line AF.

The default constraint in the **2D Sketch Panel** is the perpendicular constraint. However, this is a flyout. Pick on the arrow next to the button name. Buttons for all eleven geometric constraints are displayed, as shown in **Figure 3-9.** Pick the **Fix** button to make it active. Then, pick Line AB to apply the fix constraint. Notice that the line changes from green to black. Also, Line CD and Line EF are displayed in black. This is because they are constrained parallel to Line AB, which is now constrained to the Y axis.

If you exit the **Fix** tool by pressing the [Esc] key, you can pick on the sketch and move it left and right. However, it cannot be moved up and down. This is because the Y values of Line AB are constrained, or locked, to the coordinate system.

Pick the **Fix** button again and pick Line AF. Now, all lines in the sketch are displayed black, indicating that the sketch is fully constrained. Press the [Esc] key to exit the **Fix** tool. Now, none of the lines can be moved.

Figure 3-9.
The constraint
flyout contains the
eleven geometric
constraints.

Pick

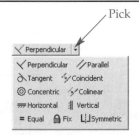

Extruding the Part

Now that the sketch is fully constrained, it can be extruded into a solid. First, display the isometric view by right-clicking in the graphics window and selecting **Isometric View** from the pop-up menu. It is easier to see the extrusion direction choices in this view. Right-click again and select **Finish Sketch** from the pop-up menu, or press the [S] key. You can now save the file as Example-03-01.ipt; remember, you cannot save in Sketch mode.

Once you have "finished" the sketch, the **2D Sketch Panel** in the **Panel Bar** is replaced by the **Part Features** panel. There are 32 tools in the **Part Features** panel. The tool you will use is **Extrude**, which has a hot key of [E]. This tool does several things:
- Since there is only one unconsumed and unambiguous sketch on the screen, it selects the sketch as the profile and shades it in cyan.
- It displays the **Extrude** dialog box shown in **Figure 3-10.**
- It displays a wireframe preview in the workspace of the profile extruded with the default settings.

The **Shape** tab contains three buttons in the middle of the tab. These are **Join**, **Cut**, and **Intersect**. Since this is the first profile, only **Join** is allowed.

The middle text box on the right side of the **Shape** tab has a value of 10 mm by default. This is the thickness of the extrusion. For this example, change the value to 60 mm; do not press [Enter]. The wireframe preview changes to reflect the value. Pick one of the top edges in the wireframe and drag it up and down (or back and forth). The distance value changes with the wireframe movement. The moving wireframe may be hard to see because of the color. Change the direction of the extrusion to the negative Z axis by picking the center button of the direction indicators. Change it to mid-plane with the third choice. In this case, one-half of the extrusion distance is applied to each side of the sketch.

The extrusion is by, default, a solid object. You can extrude to a surface object by picking the **Surface** button in the **Output** area of the **Shape** tab. However, most often the **Solid** button is used.

Pick the **More** tab and enter –20 in the **Taper** text box. This is a degree value for a taper angle. A negative angle makes the extrusion smaller as it is extruded, a positive angle makes it larger. The wireframe does not change, but a small line and arrow appear on the sketch showing the direction of the taper.

When you are finished making settings, pick the **OK** button. The part is extruded with a taper angle on all faces, as shown in **Figure 3-11.**

Figure 3-10.
Extruding a sketch into a solid.

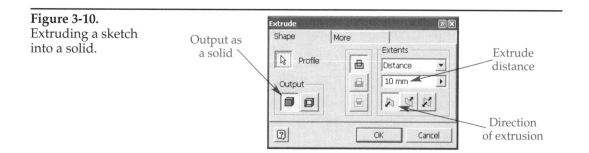

Figure 3-11.
The completed part. Notice the taper angle.

Editing the Feature and the Sketch

The tree for Part1 in the **Browser** now includes Extrusion1. If you expand this branch, Sketch1 appears below Extrusion1. Right-click on Extrusion1 to display a pop-up menu with many options, **Figure 3-12.** There are three options that allow you to change the size and shape of the part. Select **Show Dimensions** from the pop-up menu and all the dimensions are displayed on the part, including the extrusion height and angle. Now, double-click on the 170 mm dimension and change it to 150 mm. When you pick the check mark or press [Enter], the dimension changes but the part does not. You must "update" the part by picking the **Update** button on the **Inventor Standard** toolbar, **Figure 3-13.** You can change as many dimensions as you wish and update them all at once, if you like.

Figure 3-12.
This pop-up menu is displayed when you right-click on the extrusion name in the part tree in the **Browser Bar**.

Figure 3-13.
Updating the part.

Pick to update
the part

The second option that allows you to modify the part is **Edit Sketch**. This option puts you in Sketch mode. Now, change the 150 mm dimension back to 170 mm. To complete the edit, right-click and select **Finish Sketch** from the pop-up menu. This applies the changes and redisplays the extruded part.

The third option that allows you to modify the part is **Edit Feature**. This option displays the **Extrude** dialog box. Make changes, such as changing the taper. Pick the **OK** button and the changes are displayed. Using this method, change the taper to 0°.

Circles, Tangent and Horizontal Constraints, and Trimming

In this section, you will use the **Center point circle** and **Line** tools to construct the part shown in **Figure 3-14A.** You will also apply tangent and horizontal constraints to the sketch before extruding the profile. Start a new Standard (mm).ipt file. With the **Center point circle** tool, construct two circles as shown in **Figure 3-14B.** Pick once to place the center of the circle. Place the center of the small circle on the origin of the coordinate system. Then, move the cursor and pick to set the diameter. The radius of the circle is displayed at the bottom-right corner of the screen where the length and angle of a line is displayed.

Apply dimensions and edit them to the exact values. To put a diameter dimension on a circle, pick the **General Dimension** button in the **2D Sketch Panel**. Then, pick anywhere on the circumference of the circle and drag the dimension to the desired location. You can pick on either the circumferences or the center points of the circles to place the 55 mm linear dimension. Remember, you can draw the circles to an approximate size and edit the dimensions to exact values.

Now, place a horizontal constraint between the center points of the two circles to align them vertically. Pick the **Horizontal** button in the constraint flyout in the **2D Sketch Panel**. Then, pick the center point of the small circle. Finally, pick the center point of the large circle. If the circles were drawn so their centers are not horizontal (different Y values), the large circle moves to reflect the constraint.

To complete the sketch, you need to draw two lines and trim the circles. Pick the **Line** button in the **2D Sketch Panel**. Then, pick the first endpoint of the line anywhere on the top of the smaller circle. The cursor will show the coincident constraint symbol. Now, put the cursor on the larger circle and move it around until you see both the coincident and tangent constraint symbols. Pick this point to finish the line.

The line is tangent to the large circle, but not to the small circle. To constrain the line tangent to the small circle, pick the **Tangent** button in the constraint flyout. Then, pick the line and then the circle. The first endpoint of the line moves to a location so that the line is tangent to the small circle. The second endpoint of the line also moves slightly so the line remains tangent to the large circle.

Use a similar process to draw a line across the bottoms of the circles. Press [Esc] to end the **Line** tool. Then, display the constraints for the sketch. Right-click in the graphics window and select **Show All Constraints** from the pop-up menu. Both lines have two coincident and two tangent symbols displayed in their constraint bar. This is because each endpoint is on (coincidental to) a circle and each line is tangent to each circle.

Currently, the sketch is an ambiguous profile. Ambiguous profiles are discussed in the next chapter. To make the sketch unambiguous so it can be extruded, the inner portions of the circles need to be removed. The **Trim** tool is used to do this. Pick the **Trim** button in the **2D Sketch Panel**. Then, place the cursor over the inside edge of the small circle. The portion that will be removed by the tool is displayed as a dashed red line. When you pick that portion of the circle, it is trimmed (removed). Also, trim the inside edge of the large circle, **Figure 3-14C**.

Figure 3-14.
A—The completed part. B—Start the sketch used to create the part by drawing two circles.
C—The sketch is completed and fully constrained.

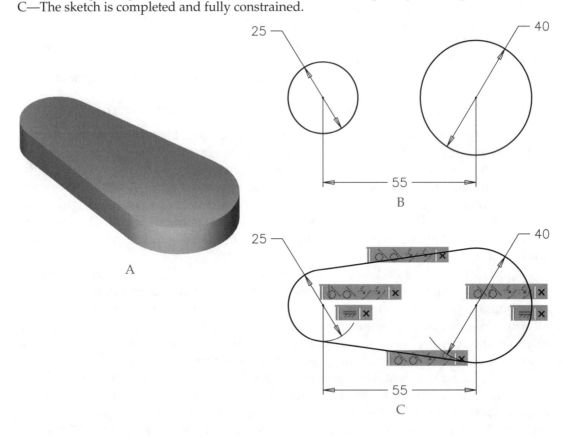

After using the **Trim** tool, you may still see a portion of an arc. This arc is actually the dimension line for the diameter dimension. Since the dimension is not part of the geometry, this does *not* create an ambiguous sketch.

The sketch is complete, fully constrained, and unambiguous. You can now extrude it into a solid part. Display the isometric view by right-clicking anywhere in the graphics window and selecting **Isometric View** from the pop-up menu. Then, finish the sketch by right-clicking again and selecting **Finish Sketch** from the pop-up menu. Also, save the file now as Example-03-02.ipt. Finally, extrude the profile 10 mm with a 0° taper using the **Extrude** tool in the **Part Features** panel.

Arcs and More Constraints

Five lines and an arc are required to construct the part shown in **Figure 3-15A.** Three constraints will be used to locate the arc: equal, colinear, and coincident. Start a new English Standard (in).ipt file. Using the **Line** tool, draw the five lines shown in **Figure 3-15B.** Make sure to leave the gap between Lines AB and CD. Apply a fix constraint to Lines AB and AF, which will keep them from moving as other lines are dynamically moved. Now, dimension the sketch as shown in **Figure 3-15B.**

Show all constraints. Notice that the only relationship between Lines AB and CD is that they are parallel. In fact, if you pick on Line CD, you can drag it independent of Line AB and Line DE. By adding a colinear constraint to Lines AB and CD, the lines are constrained to the same Y values. If the lines were vertical, they would be constrained to the same X values. If the lines are angled, they would be constrained to the same "slope" line. Pick the **Colinear** button in the constraint flyout. Then, pick Line AB and then Line CD. In this case, the order is not important. However, for angled lines, the second line picked is constrained colinear to the first line picked.

Figure 3-15.
A—The completed part. B—Start the sketch used to create the part by drawing the line segments shown here.

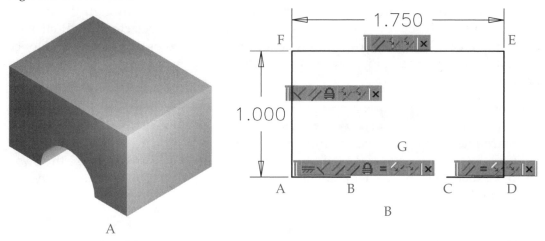

Now, the arc needs to be added to the sketch using the **Three point arc** tool. Pick the **Three point arc** button in the **2D Sketch Panel**. Three point arcs are drawn by selecting the start point, endpoint, and then middle point (not center point). Pick the right endpoint of Line AB; the yellow dot will change to green. Then, pick the left endpoint of Line CD. Finally, pick at approximately Point G. The Y coordinate of the center of the arc is approximately the same as the Y coordinate for Lines AB and CD.

An equal constraint applied to Lines AB and CD will center the arc horizontally between the lines. This is because the endpoints of the arc are constrained to the endpoints of the lines. Pick the **Equal** button in the constraint flyout, pick Line AB, and then pick Line CD.

Add another equal constraint to the two vertical lines, AF and DE. You should receive a message indicating that applying the constraint will over-constrain the sketch. See **Figure 3-16.** Inventor does not allow you to over-constrain a sketch. Line ED is already constrained by the 1.000 dimension, the coincident constraints, and the collinear constraint of AB and CD. Pick the **Cancel** button in the message box to close it. Press the [Esc] key to cancel the **Equal** constraint tool. Because the first equal constraint is valid, it remains to control lines AB and CD.

The arc now needs to be constrained vertically. This can be done by applying a coincident constraint to the center of the arc and Line CD (or Line AB). Pick the **Coincident** button in the constraint flyout. Then, pick the center of the arc (the dot) and then Line CD. Make sure you pick when Line CD is highlighted red.

The last constraint that needs to be added to the sketch is a dimension constraint for the size of the arc. This will constrain not only the size of the arc, but the length of Lines AB and CD. Pick the **General Dimension** button in the **2D Sketch Panel**. Then, pick the arc and drag the dimension to the desired location. Finally, edit the dimension and enter a radius value of 0.40.

Figure 3-16.
If you attempt to apply a dimension or geometric constraint that will over-constrain the sketch, a warning appears. Inventor will not allow you to over-constrain a sketch.

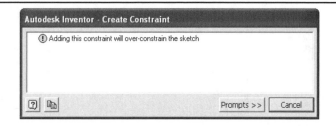

The sketch is complete, fully constrained, and unambiguous. Display the isometric view and finish the sketch. Also, save the file now as Example-03-03.ipt. Complete the part by extruding the sketch 1.25 inches with a 0° taper.

The **Line** tool and the 11 constraints can also be accessed through the pop-up menu. Press the [Esc] key to cancel any tools you may have selected. Right-click on an empty area of the graphics window to see the pop-up menu, **Figure 3-17.** Notice the **Line** tool is available. Pick on **Create Constraint** to view all of the constraints that are available.

PRACTICE 3-1

❑ Start a new English Standard (in).ipt file.
❑ Create the same solid described in the previous section. However, use the **Two point rectangle** and **Center point circle** tools. Use the **Trim** tool to remove lines as needed.
❑ Display the constraints and determine which ones you need to add.
❑ Finish the sketch and extrude the part.

Drawing an Arc from within the Line Tool

Tangent and perpendicular arcs can also be constructed from *within* the **Line** tool. This is a little tricky but once mastered, you will find it very useful. For example, you can construct the sketch shown in **Figure 3-18,** not including dimensions, using only one session of the **Line** tool. For AutoCAD users, the procedure is similar to constructing polyarcs within the **POLYLINE** command, but without the typing.

Start a new metric Standard (mm).ipt file. Draw a line from Point A to Point B about 170 mm long. Now, move the cursor over Point B and let it "hover." The yellow dot will change to gray. Pick and drag the cursor to the right and up, as if you were sketching an arc. Release the mouse button when the radius is about 40, as shown in the lower-right corner of the screen. The arc segment is constructed. Then, draw Line CD normally. Finally, use the same "drag off" method to draw the second arc from Point D to Point A.

Show all constraints. Make sure both circles are tangent to both lines. If not, add the necessary tangent constraint. Now, if you were going to finish the part, you can add dimensions and adjust their values. Then, finish the sketch and extrude the part.

PRACTICE 3-2

❑ Start a new English Standard (in).ipt part.
❑ Create the sketches shown below. Use only the **Line** tool to create the sketches.
❑ Approximate the dimensions. You do not need to dimension the sketches and enter exact values.
❑ Create other sketches of your own design until you are comfortable with the "drag off" method of creating arcs from within the **Line** tool.

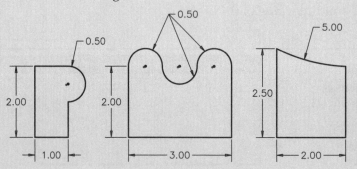

Figure 3-17.
Lines and constraints can be added by right-clicking on the graphics window

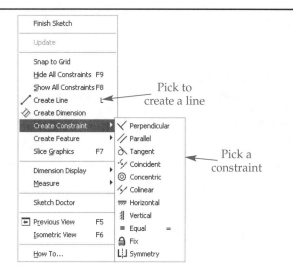

Pick to create a line

Pick a constraint

Figure 3-18.
All curves in this sketch can be drawn in a single session of the **Line** tool.

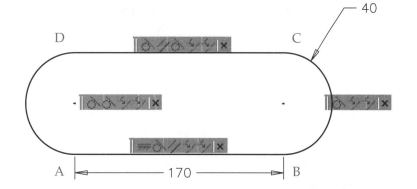

Things That Can Go Wrong with Sketches

There are several mistakes you can make in the construction of sketches that will generate errors when you attempt to use the **Extrude** tool. The two most common errors are having gaps at corners and having overlapping lines. For example, open the Example-03-04.ipt file. There is a very small gap between the two lines in the upper-right corner, as shown in **Figure 3-19.**

Pick the **Extrude** button in the **Part Features** panel. In the **Extrude** dialog box, notice that the **Surface** button is on in the **Output** area. Because of the gap, Inventor assumes this sketch is to be extruded into a surface. However, you want this to be a solid, not a surface, so you pick the **Solid** button. A new button—with a red cross—appears below the **Output** area. See **Figure 3-20.** This button indicates a solid cannot be created because a problem exists. If you hover the cursor over the button, the tooltip is **Examine Profile Problems**.

Figure 3-19.
At a "normal" zoom level, the sketch on the left appears to be closed. However, if you zoom in close on the upper-right corner of the sketch, there is a small gap, as shown in the detail.

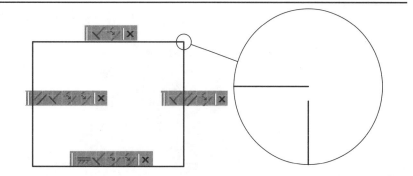

Figure 3-20.
If you attempt to extrude an open sketch into a solid, the **Examine Profile Problems** button appears in the **Extrude** dialog box. Pick this button to open the **Sketch Doctor** wizard.

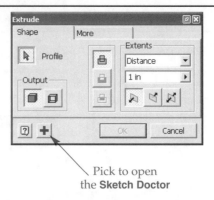

Pick to open
the **Sketch Doctor**

Now, pick the **Examine Profile Problems** button and the **Sketch Doctor** wizard appears, **Figure 3-21.** A description of the problem and possible solutions is provided on the first page of the wizard. In this case, the problem is indicated as an open loop. Also, the problem is indicated on the sketch by a green dot.

Pick the **Next>** button to display the next page of the wizard. On this page, you are given options for fixing the problem, **Figure 3-22.** Inventor refers to these as "treatments." Highlight the **Close Loop** treatment and pick the **Finish** button. In this case, a message box appears indicating there is a gap and asking if you would like Inventor to close the gap between the highlighted lines. In other instances, this message may be different, depending on the problem. Pick the **Yes** button and Inventor will automatically close the sketch. Another message box appears indicating that the problem was successfully fixed. Now, you can "finish" the sketch and extrude it into a solid.

Figure 3-21.
The first page of the **Sketch Doctor** wizard.

Description of the problem and suggested solutions

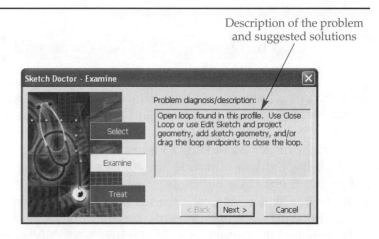

Figure 3-22.
Selecting a treatment option in the **Sketch Doctor** wizard.

Select a treatment

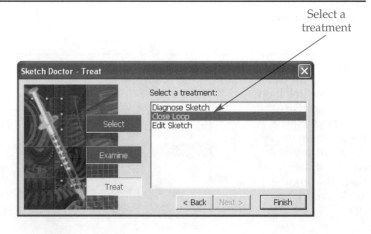

Figure 3-23.
When diagnosing a
sketch using the
Sketch Doctor
wizard, you can
choose which tests
to perform.

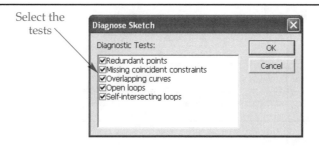

For another example of a sketch with a problem, open the Example-03-05.ipt file. The bottom line in this sketch is actually two lines that overlap. Pick the **Extrude** button in the **Part Features** panel. Once again, Inventor defaults to creating a surface instead of a solid. Pick the **Solid** button in the **Output** area. Then, pick the **Examine Profile Problems** button to open the **Sketch Doctor** wizard. The first page of the wizard indicates the problem. Pick the **Next>** button to continue.

On the second page of the wizard, highlight the **Diagnose the Sketch** treatment. Then, pick the **Finish** button. The **Diagnose Sketch** dialog box appears, **Figure 3-23.** This dialog box allows you to choose which tests to perform. Check all of the tests and pick the **OK** button to continue.

The sketch is analyzed and the **Sketch Doctor** wizard is redisplayed. There are three problems listed in the wizard—a missing constraint, overlapping lines, and a gap (open loop). See **Figure 3-24.** In addition, as you highlight a problem in the wizard, that problem is highlighted on the sketch. In this case, the multiple problems make it hard for Inventor to automatically fix the sketch. The easiest fix is to cancel the wizard, delete the bottom lines, and properly draw a new line.

A Review of All Constraints

There are eleven geometric constraints available in the constraints flyout in the **2D Sketch Panel**. The types of objects or "picks" required when applying each constraint varies. In addition, the symbol that appears in the constraint flyout is the same symbol that appears in the constraint bar for geometry. The chart on the next page reviews the choices for all constraints. In order to become more familiar with the available constraints, copy this chart by hand to a blank sheet of paper. Be neat so you can use the sheet as reference as you work in Inventor. In the fourth column, which is currently blank, sketch the symbol for the constraint.

Figure 3-24.
After diagnosing
the sketch, the
Sketch Doctor
wizard displays the
problems it found.

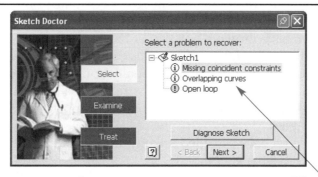

Diagnosed
problems

Name	1st Choice	2nd Choice	Symbol
Perpendicular	Line	Line	
Parallel	Line	Line	
Tangent	Line, arc, or circle	Line, arc, or circle	
Coincident	Centers of arcs or circles Endpoints of lines or arcs	Centers of arcs or circles; endpoints of lines Endpoints of lines or arcs	
Concentric	Circle or arc	Circle or arc	
Colinear	Line	Line	
Horizontal	Line Center or endpoint	*(none)* Center or endpoint	
Vertical	Line Center or endpoint	*(none)* Center or endpoint	
Equal	Line, circle, or arc	Line, circle, or arc	
Fix	Line, arc, circle, or spline	*(none)*	
Symmetric	Line, arc, circle, or spline	Line *(normal, construction, or center)*	

Chapter Test

Answer the following questions on a separate sheet of paper.

1. Define *sketch* in Inventor.
2. All lines, arcs, and circles in Inventor are considered _____ objects.
3. What is a *constraint* in Inventor?
4. By default, how are the X and Y axes displayed in Inventor?
5. What is the hot key for the **Line** tool?
6. When using the **Line** tool, what information is provided in the three displays at the lower-right corner of the Inventor screen?
7. What are the two basic types of constraints?
8. How do you display the constraints for all geometry in the sketch?
9. When constraints are displayed for geometry, how are they represented on-screen?
10. If you have drawn a line that is 150 mm in length, but it should be 148 mm in length, how can you correct this problem in Inventor?
11. Which tool allows you to manually apply a dimension?
12. What function does the fix constraint serve?
13. Briefly describe how to extrude a fully constrained, unambiguous sketch into a solid part.
14. Once you have extruded a sketch into a solid part, how can you edit the original sketch to change the part?
15. Briefly describe how to draw an arc from within the **Line** tool. Assume you have drawn one straight line segment.

Exercise 3-1. Start a new English Standard (in).ipt file. Construct the test sample shown below. Apply dimensions and geometric constraints to the sketch. Extrude the profile into a solid part. Do not create a title block drawing.

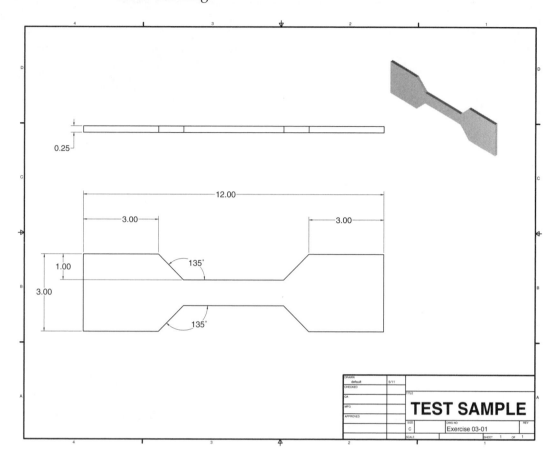

Exercise 3-2. Start a new English Standard (in).ipt file. Construct the clamp block shown below. Apply dimensions and geometric constraints to the sketch. Extrude the profile into a solid part. Do not create a title block drawing.

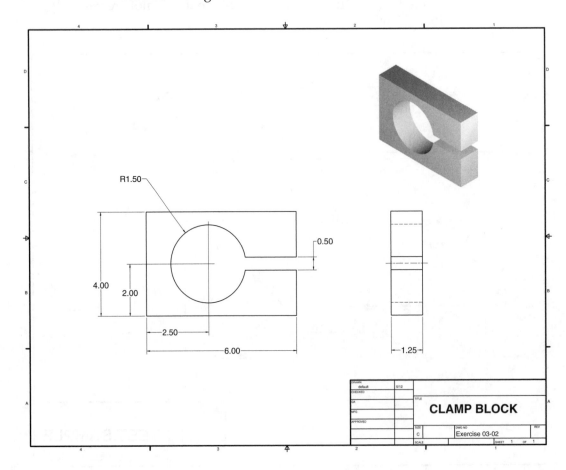

Exercise 3-3. Start a new metric Standard (mm).ipt file. Construct the plate shown below. Apply dimensions and geometric constraints to the sketch. Extrude the profile into a solid part. Do not create a title block drawing.

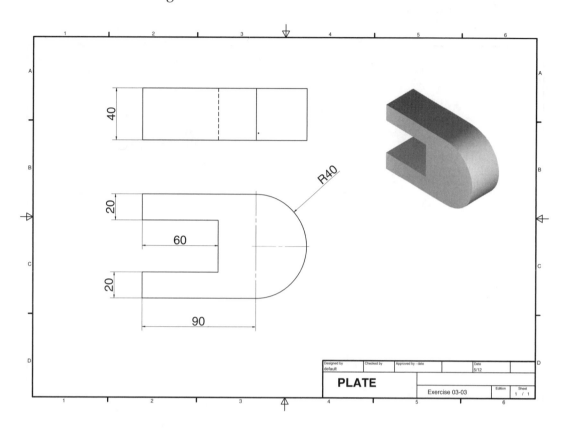

Exercise 3-4. Start a new metric Standard (mm).ipt file. Construct the pipe bracket shown below. Try creating the sketch using only one session of the **Line** tool. Apply dimensions and geometric constraints to the sketch. Extrude the profile into a solid part. Do not create a title block drawing.

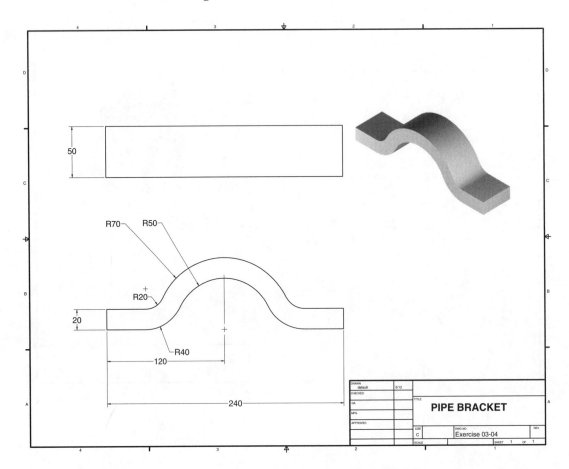

Exercise 3-5. Start a new metric Standard (mm).ipt file. Construct the shear blade shown below. You can draw lines, arcs, and circles to help you locate geometry and then delete them. Apply dimensions and geometric constraints to the sketch. Extrude the profile into a solid part. Do not create a title block drawing.

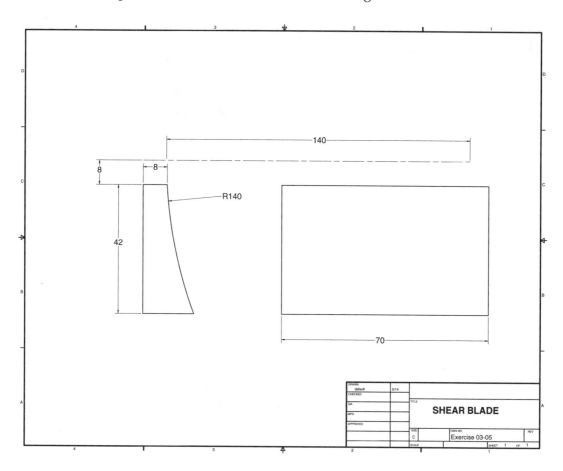

Exercise 3-6. Start a new metric Standard (mm).ipt file. Construct the taper wedge shown below. Apply dimensions and geometric constraints to the sketch. Extrude the profile into a solid part. Make sure to use the proper taper angle. Do not create a title block drawing.

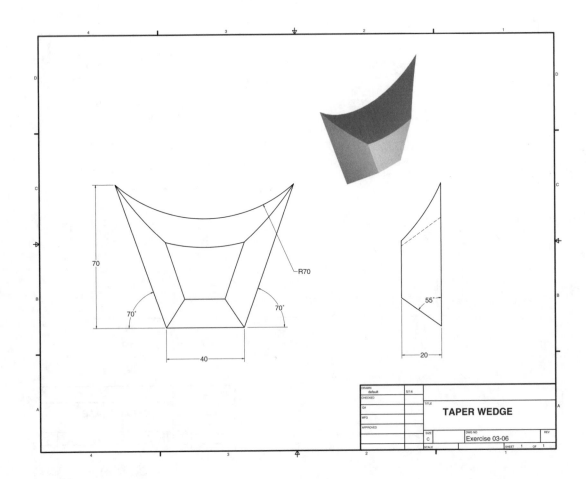

Exercise 3-7. Start a new English Standard (in).ipt file. Construct the U-bracket shown below. Try creating the sketch using only one session of the **Line** tool. Place dimensions on the sketch. Also, apply colinear and equal constraints on the lines. Extrude the profile into a solid part. Do not create a title block drawing.

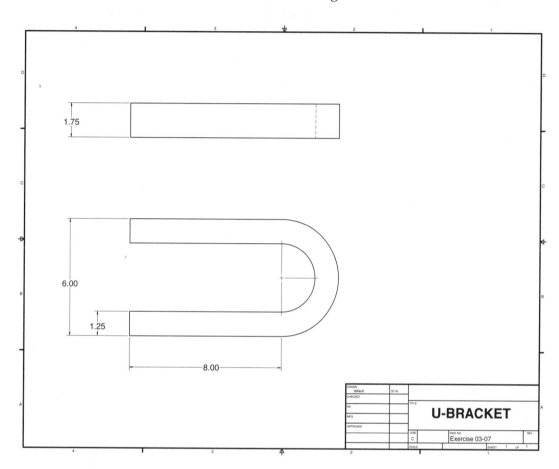

Exercise 3-8. Start a new metric Standard (mm).ipt file. Construct the cover plate shown below. Apply dimensions and geometric constraints to the sketch. Extrude the profile into a solid part. Do not create a title block drawing.

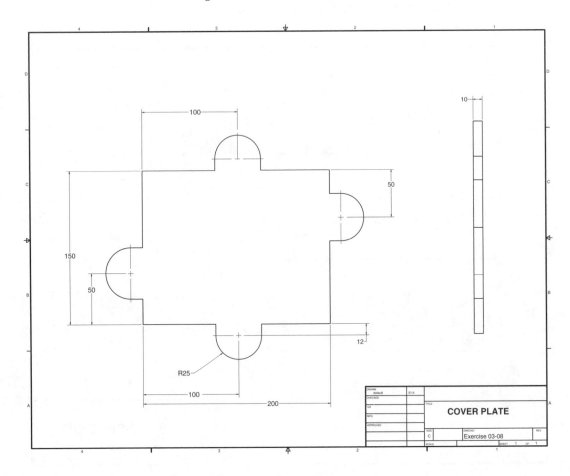

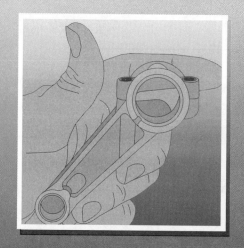

CHAPTER 4

More Complex Sketching, Constraints, Formulas, and the Construction Geometry

Objectives

After completing this chapter, you will be able to:

* Explain how to create ambiguous profiles and select the profile.
* Use the d0 variables to create dimensions with equations.
* Create and use construction geometry and centerlines to locate sketch objects.
* Use the **Sketch Mirror** tool.
* Use the **Revolve** tool and understand its features.
* Explain and use the **Inventor Precise Input** toolbar.

User's Files

*The following is a list of files that you will need to work through this chapter. These files can be found on the **User's Files** CD included with this text.*

Examples

Example-04-01.ipt   Example-04-03.ipt
Example-04-02.ipt

Creating More Complex (Ambiguous) Profiles

The sketches you created in *Chapter 3* had no interior geometry or "islands." There was only one choice for the profile, therefore, they were unambiguous. When you initiated the **Extrude** tool, you did not have to select the profile. Inventor selected the only possible profile for you. In this chapter, you will work with sketches that have several possible profiles. These are called *ambiguous profiles*. You must select the specific profile you want to extrude. The islands in the sketch will extrude as *cutouts*. They are not called holes because there is a **Hole** tool, which is discussed in the next chapter.

An Inventor sketch made up of four lines and two circles is shown in **Figure 4-1.** This geometry presents seven possible profiles that can be extruded, which are shown in the figure. You can select the profile to extrude by picking the **Profile** button in the **Extrude** dialog box, then pick within the areas you want to extrude. As you move the

Figure 4-1.
There are seven possible profiles from the sketch shown on the left. The seven are shown on the right. The area that will be extruded is represented by the hatch lines.

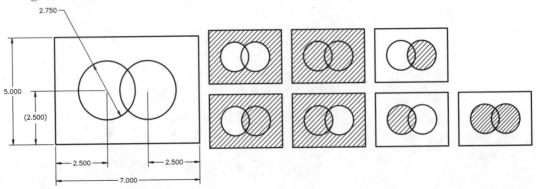

cursor over an area, the area that will be added to the selection is highlighted in red. The total area in the selection is highlighted in cyan. This represents the profile that will be extruded. You can remove an area from the selection by holding down the [Shift] key and picking in that area again. The area that will be removed is highlighted in red before you pick. To see that the area has been removed, move the cursor off of the sketch. Only the profile that will be extruded is highlighted.

Open the file Example-04-01.ipt. Pick the **Extrude** button in the **Part Features** panel. In the **Extrude** dialog box, notice that the **Profile** button is selected (depressed). However, no part of the sketch is highlighted. This is because the sketch is ambiguous. With the button selected, move the cursor over the sketch and pick the appropriate areas to create the 1" thick part shown in **Figure 4-2A.** Create the part. Then, right-click on Extrusion1 in the **Browser** and select **Edit Feature** from the pop-up menu. Pick the **Profile** button in the **Extrude** dialog box and select or deselect the areas to create the extruded part shown in **Figure 4-2B.** Pick the **OK** button to update the part.

Figure 4-2.
A—One possible part produced by extruding the sketch shown in **Figure 4-1.** B—Another possible part.

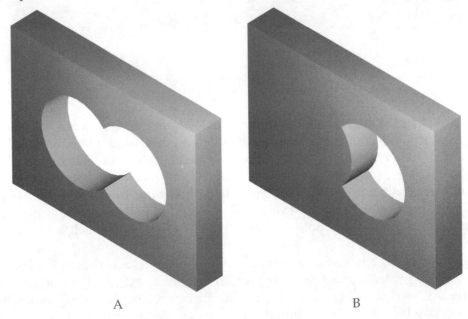

A   B

The **Two point rectangle** Tool and Precise Input

Now, you will create a part that is similar to the one in the last section. You will use the **Rectangle** tool, precise size input, and equations in the dimensions. You can input precise sizes with the **Inventor Precise Input** toolbar. Display this toolbar by selecting **Toolbar** from the **View** pull-down menu. Then, select **Inventor Precise Input** from the cascading menu. You can also right-click on a blank area of any toolbar and select **Inventor Precise Input** from the pop-up menu. Note: You must be in Sketch mode for this option to appear.

Start a new English Standard (in).ipt file. Now, pick the **Two point rectangle** button in the **2D Sketch Panel**. Pick the first point of the rectangle at the origin. The **Inventor Precise Input** toolbar will look like **Figure 4-3** with the vertical blinking bar in the **X:** text box. You can now type a value of 5 for the X coordinate of the second corner of the rectangle. Then, either press the [Tab] key to move to the **Y:** text box or pick in that text box with the cursor. Now, enter a value of 4 for the Y coordinate of the second corner of the rectangle. As you enter values, the second corner of the rectangle moves in the graphics window. When the coordinates are correct, press the [Enter] key to finish the rectangle.

Pick the **Zoom All** button to see the entire rectangle. Then, pick the **Center point circle** button in the **2D Sketch Panel**. Using the **Inventor Precise Input** toolbar, enter X=2.5 and Y=2.0 for the center point. Enter X=3.5 and Y=2.0 for a point on the circumference of the circle to set the size. Close the **Inventor Precise Input** toolbar.

Now, you can finish the sketch. Using the **Extrude** tool, select the proper profile and create the 1″ solid part.

Precise input is a nice technique that you might find useful for establishing the start of a sketch. For this example, using precise input to draw the rectangle may have been useful. Using precise input to construct the circle was probably counterproductive.

Figure 4-3.
You can use the **Inventor Precise Input** toolbar to enter exact coordinates.

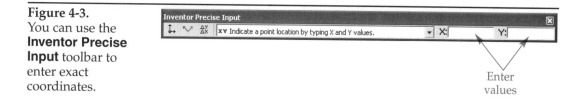

Enter values

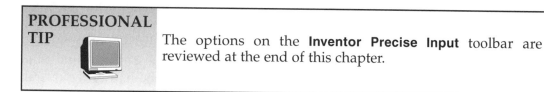

PROFESSIONAL TIP

The options on the **Inventor Precise Input** toolbar are reviewed at the end of this chapter.

Using the d0 Model Parameters in Equations in Dimensions

In *Chapter 3*, you learned how to add dimensions to constrain a sketch. Now, you will learn how to relate dimensions to each other. For example, in the previous section you drew a circle in the center of a rectangle. Using dimensions, you can constrain the circle so it is always in the horizontal center of the rectangle.

With the extruded part from the previous section still open, expand Extrusion 1 in the **Browser** and right-click on Sketch 1. Select **Edit Sketch** in the pop-up menu. Now, using the **General Dimension** tool, place a dimension across the bottom of the part. If the X and Y values were correctly input in the previous section, the value will be 5.000

inches. Then, place a dimension from the center of the circle to the left-hand edge of the part. Press the [Esc] key to cancel the **General Dimension** tool.

Every dimension in a part has a unique name starting with a lowercase "d" and ending with a number; the first dimension in the first sketch will be d0 and the second d1 and so on. We will use these names to locate the center of the circle at the horizontal midpoint of the rectangle, regardless of the rectangle's width.

Double-click on the 2.5 dimension to bring up the **Edit Dimension** dialog box. While the 2.5in dimension is highlighted, click on the 5.0 dimension. The name of the 5.0 dimension, in this case d2, will appear in the dialog box. Type /2 after the d2 so the equation reads d2/2, **Figure 4-4.** This is an equation that means the value of d3 is the value of d2 divided by 2. In other words, the value of dimension d3 will always be one-half of the value of dimension d2. Try this by editing dimension d2 to a value of 10. The circle remains horizontally centered because dimension d3, which horizontally constrains the center of the circle, is defined as one-half of d2. Edit dimension d2 to a value of 5 and the circle returns to its original position.

Figure 4-4.
By entering the equation in place of the 2.500 value, the center of the circle will always be at the horizontal center of the part, even if the 5.000 dimension is changed.

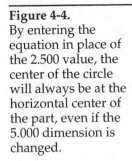
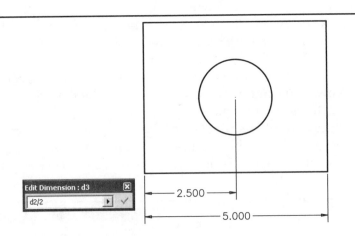

Now, using a similar method, vertically constrain the center of the circle. Place a dimension on the left side of the part. Then, place a dimension from the center of the circle to the bottom edge of the part. Now, pick the vertical dimension to the center of the circle to edit it. The value should be 2.000 inches. Notice when the **Edit Dimension** dialog box is first opened that the value is highlighted. With the value highlighted, pick anywhere on the overall vertical dimension on the sketch. The highlighted dimension value in the **Edit Dimension** dialog box is replaced with the name of the overall vertical dimension, which should be d4. Now, type /2 after d4 and press [Enter]. The circle's vertical dimension is now constrained to one-half of the overall vertical dimension.

Now, you will constrain the size of the circle based on the overall size of the part. First, place a diameter dimension on the circle. Then, edit the dimension value and enter the equation d4*0.50, where d4 is the name of the overall vertical dimension. See **Figure 4-5.** This equation is the same as d4/2. Now, any time the overall vertical dimension is changed, the circle will remain centered vertically *and* the diameter of the circle will be one-half of the vertical distance.

Press [Esc] and finish the sketch. With the dimensions related to each other through the use of equations, the part can be easily updated in the future. You could even take it one step further and constrain the overall vertical dimension to 80% of the overall horizontal dimension, if needed. The value entered in the **Edit Dimension** dialog box can vary from a plain number to an arithmetic, trigonometric, or algebraic equation with the dimension names used as variables.

Figure 4-5.
Entering an equation to constrain the diameter of the circle to one-half of the overall vertical dimension.

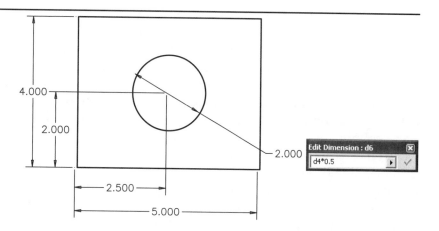

More on Equations

Inventor expects the result of an equation to have the correct units. For example, the equation d0/d2 would not be accepted for the diameter of the circle. This equation results in a *unitless* solution and Inventor expects a solution in inches. The solution is unitless because d0 is inches and d2 is inches. When inches are divided by inches, the units cancel each other out and the result has no unit. However, the equation (d0/d2)*1 in is acceptable because the solution has units. An unacceptable equation is highlighted in red in the **Edit Dimension** dialog box.

In the previous section, you entered the equation d4*0.5 for the dimension value of the circle diameter. Verify this equation by editing the dimension. If you are not in Sketch mode, right-click on the extrusion name in the **Browser** and select **Show Dimensions**. Then, double-click on the diameter dimension to open the **Edit Dimension** dialog box. Notice that Inventor has rewritten the equation you entered to d4/2 ul. Inventor has guessed that the constant 0.5 is *unitless* and automatically indicates this by adding the "unit" ul. Since the dimension d4 is in inches, the result of the equation is in inches and the equation is acceptable.

For these simple equations, assume that dimension d3 has a value of 175 mm and dimension d4 has a value of 50 mm. The five arithmetic operators available in Inventor, and some sample equations, are:

Operation	Operator	Equation	Result
Addition	+	d3+d4	225 mm
	+	d3+100ul	275 mm
Subtraction	–	d3–d4	125 mm
Multiplication	*	0.5ul*d4	25 mm
	*	d3*d4	Error
Division	/	d4/10ul	5.0 mm
		(d3/d4)*1mm)	3.5 mm
Exponent	^	d3^2	Error
		(d4^2)/100mm	25 mm

To see all of the dimensions and their values in a chart similar to a spreadsheet, select **Parameters** from the **Tools** pull-down menu. The **Parameters** dialog box is shown, **Figure 4-6.** This dialog box shows the dimension names on the left side in the **Parameter Name** column. The unit that Inventor expects is listed in the **Unit** column. The value of the dimension, whether a numeric value or an equation, is listed in the **Equation** column. Notice that all constants have been assigned the "unit" ul to indicate they are unitless. You can also add comments about a dimension using the right-hand column.

You can edit the dimension name, value, or comment for a dimension by picking in that cell and retyping. When done typing, press [Enter]. For example, change the name of d2 to Long. Notice that after you press [Enter] all equations that referenced d2 are updated to reference Long. The dimension name is case sensitive; LONG is different than Long and long. You can use letters and numbers, and some symbols such as the underscore. You cannot use arithmetic operators, such as the plus sign, or symbols that may be used in an equation, such as a period. If you make a mistake, Inventor will give an error message: Variable name contains invalid characters.

Now, add Length of part to the Comment cell of the Long dimension. Notice in **Figure 4-6** that there are six dimensions and the sketch only had four. The dimensions d0 and d1 are related to the extrusion and were automatically added after using the **Extrude** tool. Change the name of dimension d0 to Extrusion Height and dimension d1 to Taper Angle. If you like, you can add comments to these dimensions as well.

The width of a column can be widened or narrowed. To do so, move the cursor to the line separating the column headings; a double-arrow cursor appears. Drag the column border left or right. This is the same as in most Windows-based software.

Figure 4-6.
The **Parameters** dialog box can be used to rename dimensions and enter equations.

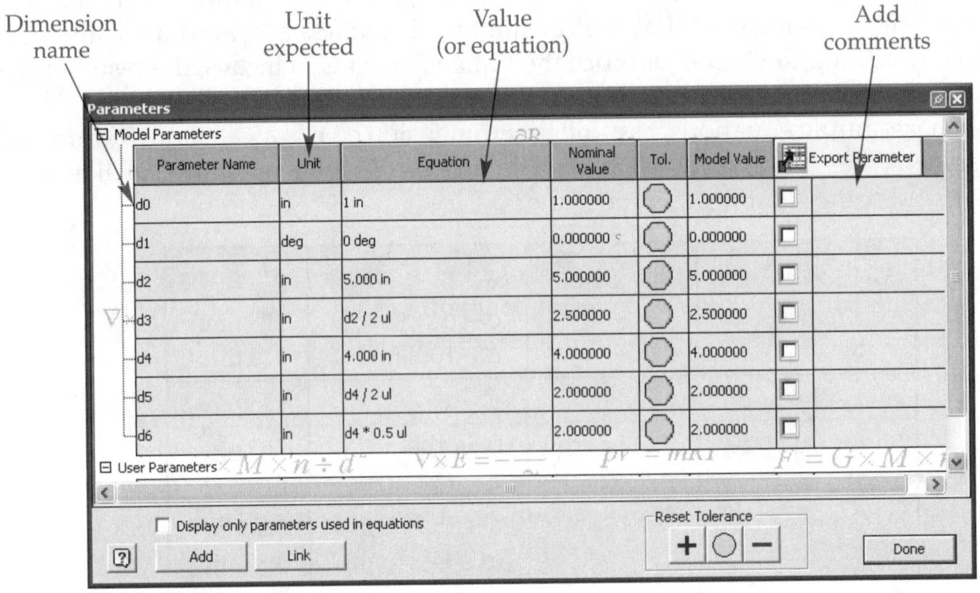

Construction Geometry

You can sketch lines in different styles. So far, you have only sketched in the Normal style. This style creates solid (continuous) curves that are evaluated as part of the profile when the sketch is "finished." In this chapter, you will learn about two other styles: Construction and Centerline. Geometry created in the Construction style can be used with constraints to locate geometry created in the Normal style without the use of dimensions. Construction geometry is not part of the profile and disappears from the model view when the profile is extruded. In this section, you will locate a circle at the center of a rectangle using a line drawn in the Construction style. The next section discusses the Centerline style.

Start a new metric Standard(mm).ipt file. Construct a 100 mm × 70 mm rectangle with one corner at the origin. Dimension the bottom and the left side. Right-click in the graphics window and pick **Done [Esc]** in the pop-up menu. There are four buttons near the right end of the **Inventor Standard** toolbar. The first two control the Construction and Centerline styles. See **Figure 4-7.** The first is the **Construction** button. This is a toggle button; therefore, when it is pressed it stays depressed until it is pressed again. Press this button to toggle it on.

Figure 4-7.
The Construction and Centerline styles can be changed using the buttons on the Inventor Standard toolbar.

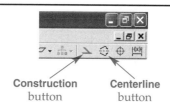

Construction button

Centerline button

Draw a line from the lower-left corner of the rectangle to the upper-right corner. Notice that the line is dashed, thinner than the other lines and displayed in yellow, not green. These changes indicate that the line is in the Construction style. Press [Esc] to exit the **Line** tool. In this particular application, the construction line can also be drawn as a horizontal line between the midpoints of the vertical sides or a vertical line between the midpoints of the horizontal sides.

Change the current style to Normal by pressing the construction style button. Make sure nothing is selected when you change the style. Then, pick the **Center point circle** button in the **2D Sketch Panel**. To locate the first point, which is the center, right-click in the graphics window to display the pop-up menu. Select **Midpoint** to activate the midpoint object snap. Then, pick the construction line. The center of the circle is placed at the midpoint of the construction line. Since the construction line bisects the rectangle, this point is also the geometric center of the rectangle. Now, pick a point for the diameter. Then, dimension the circle's diameter and change the value to 40 mm.

Change the length of the part from 100 mm to 150 mm. Notice that the circle stays in the center of the rectangle. Change the dimension back to 100 mm; the circle moves to stay in the center of the rectangle. Pick the **Show Constraints** button in the **2D Sketch Panel**. Then, pick the construction line, circle, and center of the circle. See **Figure 4-8.** Press [Esc] to exit the **Show Constraints** tool. The reason the circle stays at the center of the rectangle is that there is a coincident constraint automatically applied between the center of the circle and the midpoint of the construction line. The endpoints of the construction line are similarly constrained to the corners of the rectangle. Therefore, as the corner of the rectangle is moved, the endpoint of the construction line moves. In turn, the center of the circle follows the midpoint of the construction line.

Figure 4-8.
The constraints are shown for the construction line, circle, and circle center.

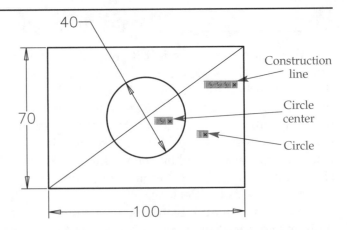

To complete this example, you will use the circle to construct a five-sided (pentagon) shape. Since the circle will not be part of the profile, it can be changed to the Construction style. First, select the circle. Selected curves are displayed in cyan. With the circle selected, press the construction style button. The style of the circle is now Construction. Pick anywhere in the graphics window to unselect the circle. It is now displayed as a thin, dashed, yellow line like the line you drew from corner to corner.

Now, you can construct the pentagon. Each segment of the pentagon will be constrained tangent to the circle and equal to the other segments. See **Figure 4-9A**. However, before you draw anything, make sure the style is Normal.

With the **Line** tool, draw the left side of the pentagon vertical. Remember, the starting point and length can be "about" correct. Inventor will apply a parallel constraint between the line and the left side of the rectangle. Then, holding down the [Ctrl] key to prevent automatic constraints, sketch the other four sides at approximate sizes and locations. Make sure the last endpoint of the last segment is connected to the first endpoint of the first segment. Now, apply a tangent constraint to each of the five line segments in the pentagon so they are tangent to the circle. Then, apply an equal constraint to the second through fifth line segments in the pentagon so they are equal to the first line segment. Finally, "finish" the sketch and extrude the profile 10 mm. Since this is an ambiguous profile, you will need to select the proper profile to create a part with a cutout (hole). See **Figure 4-9B**.

PROFESSIONAL TIP

You can quickly change the style of an existing curve by selecting it and then pressing the style button. Multiple curves can be changed in one step by selecting all curves to change and then selecting the style in the drop-down list.

Figure 4-9.
A—Using construction lines and constraints to construct a circumscribed pentagon. B—The sketch is extruded into a part.

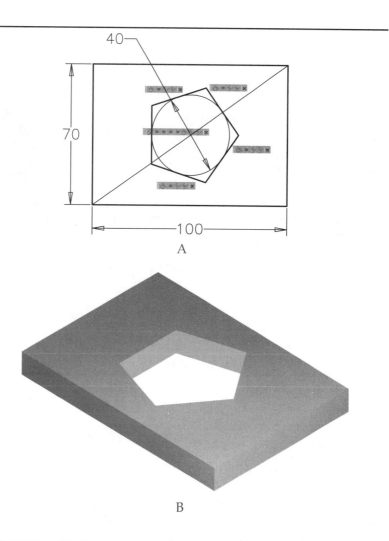

A

B

PRACTICE 4-1

❏ In this practice, you will construct a rectangular part with a slot, as shown at the top of the next page. Start a new English Standard (in).ipt file.

❏ Draw the rectangle first.

❏ Construct the arcs with the **Center point arc** tool. Set the arcs equal by applying an equal constraint. Then, construct lines for the sides of the slot.

❏ Draw two construction lines from the midpoints of Lines AB and CD to the centers of the arcs. Use the midpoint and center object snaps.

(Continued)

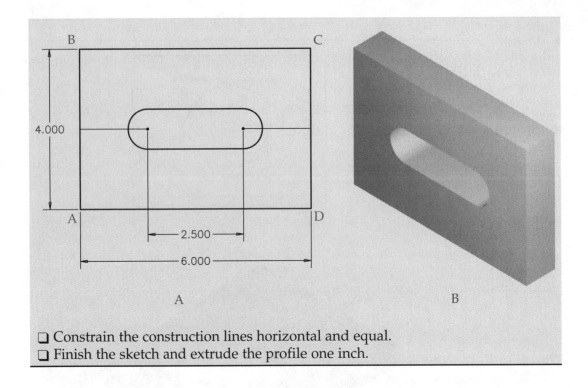

❏ Constrain the construction lines horizontal and equal.
❏ Finish the sketch and extrude the profile one inch.

The Sketch **Mirror** Tool

The **Mirror** tool is used to create a mirror of geometry or objects by reflecting about a centerline. There are actually two mirror tools in Inventor. One is found in the **Part Feature** panel and called the **Mirror Feature** tool. The other is found in the **2D Sketch Panel** and called the **Mirror** tool. In this section, you will look at the sketch **Mirror** tool. It is used to mirror sketched curves about any line in the sketch. The style of the line used as the mirror line can be either Construction or Normal. Open the Example-04-02.ipt file. Right-click on Sketch1 in the **Browser** and select **Edit Sketch** from the pop-up menu. Then, select the **Look At** button on the **Inventor Standard** toolbar and pick any line. Zoom extents if needed. See **Figure 4-10.**

In this example, you will mirror the arc and two short lines about a line from the midpoint of Line AD to the midpoint of Line BC. First, using the Line tool, draw a line from the midpoint of Line AD to the midpoint of Line BC. The style can be Normal, Centerline, or Construction. Use the midpoint object snap. Press the [Esc] key to end the Line tool.

Figure 4-10.
The "tab" feature on this sketch will be mirrored to create a second tab.

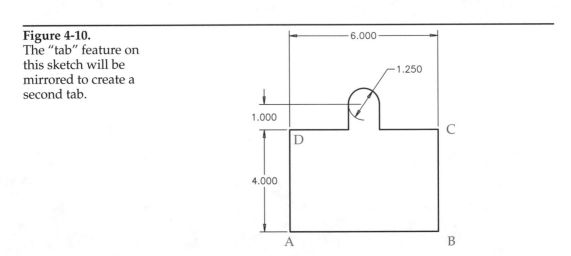

Now, you will use the Mirror tool. Pick the Mirror button in the 2D Sketch Panel. The Mirror dialog box is displayed. See **Figure 4-11.** If the Select button is not active (depressed), pick it to activate it. Then, select the two short lines and the arc on the sketch. When selected, the lines are displayed in cyan. Next, pick the Mirror Line button in the Mirror dialog box. Select the line you just constructed on the sketch. Finally, pick the Apply button in the Mirror dialog box. The curves are mirrored. Pick Done to close the Mirror dialog box.

Zoom in on the mirrored objects. Using the Show Constraints tool, show the constraints on the mirrored curves, as shown in **Figure 4-12.** The Mirror tool applies the symmetric constraint to the curves that are mirrored. The constraint can be applied manually to other curves by picking the Symmetric button in the constraint flyout in the 2D Sketch Panel. Notice in the constraint bar that the short lines are not connected (coincident) to Line AB or tangent to the arc. The coincident constraints indicated in the constraint bars are between the short lines and the arc.

Figure 4-11.
The **Mirror** dialog box.

Figure 4-12.
The constraints are displayed for the mirrored feature.

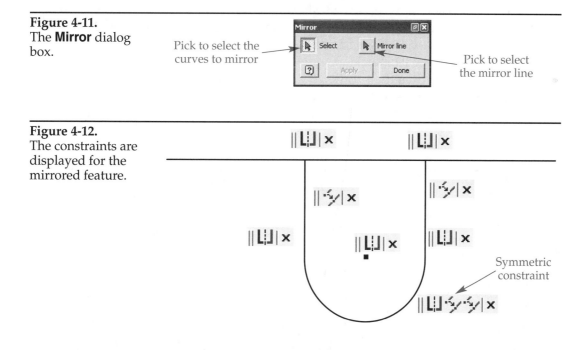

The Revolve Tool

The **Revolve** tool creates cylindrical parts by revolving a profile about an axis. The axis must be a straight line that, if extended indefinitely, would never cross the profile. The axis can touch the profile, such as an axis line that is tangent to a circle. This method produces a cylindrical model without a hole.

The style of the axis can be Normal, Construction, or Centerline. The advantage of using a centerline is that when you place dimensions in a sketch, they are placed as diametric dimensions, even though the complete diameter is not shown. *Diametric dimensions* measure the diameter of circular geometry. Place these dimensions by picking the geometry and the centerline. Be aware that if you pick the end point of the centerline, the dimension will be radial, not diametric.

Open Example-04-03.ipt from the CD at the back of this book. Edit the sketch by right-clicking on Sketch1 in the **Browser** and selecting **Edit Sketch** from the pop-up menu. Then, pick the **Look At** button on the **Inventor Standard** toolbar and select any line. Pick the **Zoom All** button. See **Figure 4-13.** Line BD is a construction line. Point B is the origin of the coordinate system. The top of the sketch is made of two separate lines, Lines JD and DK.

First, select and delete the dimension. Then, apply an equal constraint to Lines JD and DK, and to Lines EF and GH. This will center the sketch horizontally. If you do not first delete the dimension, the equal constraint on Lines JD and DK will over constrain the sketch.

Now, place a dimension between Line EF and the centerline (Line AC). Notice how the dimension is placed as a diameter dimension. Also, the dimension does not measure from the centerline to Line EF. Rather, it measures from a point on the opposite side of the centerline to Line EF. This is a diametric dimension. Edit the dimension and enter a value of 4.5".

Place a dimension between Line LM and the centerline. Edit this dimension and enter a value of 2.5". Also, place a dimension between Line JD and the centerline. This is the dimension you deleted earlier. The fully dimensioned sketch is shown in **Figure 4-14.** Display the isometric view and "finish" the sketch.

Now, you can revolve the profile about the centerline to complete the part. Pick the **Revolve** button in the **Part Features** panel or press the [R] key. The **Revolve** dialog box is displayed, **Figure 4-15.** As there is only one unambiguous, unconsumed sketch in the file, it is selected as the profile. The area of the profile is highlighted in cyan.

Figure 4-13.
This sketch will be used to create a symmetrical, circular part. Notice the centerline and the diametric dimension.

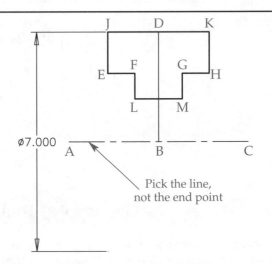

Figure 4-14.
The fully dimensioned sketch that will be revolved.

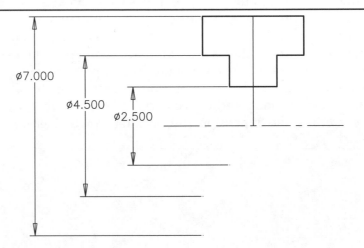

Figure 4-15.
The **Revolve** dialog box.

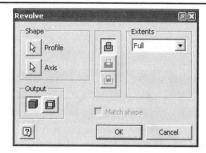

Also, the centerline in the sketch was created in the Centerline style. Since there is only one line created in this style, Inventor assumes this line will be the axis of rotation and selects it. The centerline is also displayed in cyan. The preview of the revolution is displayed in green.

If you have multiple profiles in the sketch, pick the **Profile** button in the **Shape** area of the **Revolve** dialog box. Then, select the profile in the graphics area. Like the **Extrude** tool, you can select multiple profiles. If there is more than one possible axis of revolution, you need to pick the **Axis** button in the **Shape** area of the **Revolve** dialog box. Then, pick the axis in the graphics area.

Once the profile and axis are selected, pick the **OK** button in the **Revolve** dialog box. This will create a solid model by revolving the profile through 360°. See **Figure 4-16.** There are other options with the **Revolve** tool that will be discussed later in the book.

Figure 4-16.
The sketch shown in **Figure 4-14** is revolved to create this part.

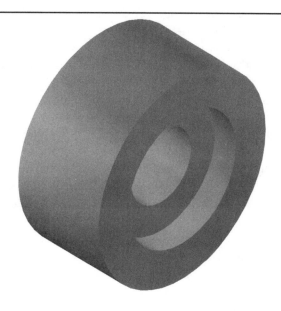

The **Inventor Precise Input** Toolbar

For the most part, you will not use the **Inventor Precise Input** toolbar to help construct geometry. Often, it requires more work than it saves. Remember the rules:

- Sketch the size reasonably close.
- Sketch so the constraints are exact.
- Dimension the objects precisely.

However, it can be very useful for drawing a rectangle, especially if the rectangle is the first curve sketched. This section provides an in-depth look at the **Inventor Precise Input** toolbar. You may want to quickly read through this section and then come back to it later when you know more about Inventor and applications where precise input may be useful.

Figure 4-17.
The default settings for the **Inventor Precise Input** toolbar. Note: A drawing tool, such as the **Line** tool, must be active to enable the toolbar options.

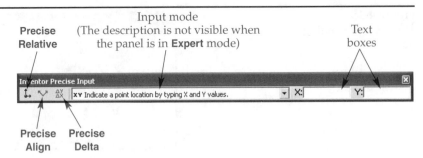

Precise Relative

Input mode
(The description is not visible when the panel is in **Expert** mode)

Text boxes

Precise Align   Precise Delta

The **Inventor Precise Input** toolbar has three tools, a drop-down list, and two text boxes, **Figure 4-17.** The **Precise Align** tool, the second button from the left, does not work in Sketch mode. Therefore, it is not discussed in this section.

The toolbar has four "modes" or types of input. Use the drop-down list on the toolbar to select the mode. The drop-down list indicates the mode in bold and provides a brief description. The four modes are:
- **XY.** Allows you to locate a point by providing an XY coordinate using the toolbar.
- **X∠.** Allows you to locate a point by providing its X coordinate and the angle from the X axis.
- **Y∠.** Allows you to locate a point by providing its Y coordinate and the angle from the X axis.
- **d∠.** Allows you to locate a point by providing the distance from the origin or the coordinate system and the angle from X axis.

The two text boxes on the right end of the toolbar change depending on which input mode you select, **Figure 4-18.**

You can input numbers or equations in the text boxes. However, the equation used to locate a point is not saved. Rather, the equation determines a numeric value for the location. The same rules that apply to equations in dimensions apply here. You can use the d0 variables in the equation. Remember, the units must be correct.

To provide an example of using the **Inventor Precise Input** toolbar, start a new English Standard (in).ipt file. Then, display the **Inventor Precise Input** toolbar by right-clicking on a blank space on the **Inventor Standard** toolbar. Select **Inventor Precise Input** from the pop-up menu. The toolbar appears floating by default. You can dock it or move it around the screen as you work. Now, use the toolbar to draw a line:
1. Pick the **Line** button in the **2D Sketch Panel**. The options in the **Inventor Precise Input** toolbar become available.
2. Notice that there is a blinking, vertical bar in the **X:** text box in the toolbar. This indicates that you can type a value or equation. Type 0 in the **X:** text box.
3. Press the [Tab] key or move the cursor and pick in the **Y:** text box. Then, type 0 in the text box.

Figure 4-18.
There are four input modes available for the **Inventor Precise Input** toolbar. A—**XY**. B—**X∠**. C—**Y∠**. D—**d∠**.

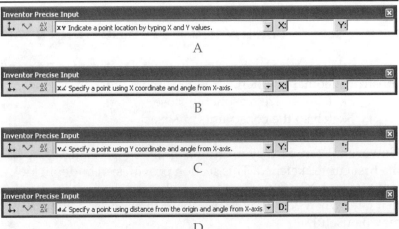

A

B

C

D

4. To place a point at the coordinates entered (0,0), press the [Enter] key.

A green line now appears from the first point (0,0) to the cursor location, just as if you had picked the first point with the cursor. Continue by entering the second point:

5. The blinking, vertical bar should be in the **X:** text box. If not, pick in the text box. Then, type 0.

6. Press the [Tab] key or pick in the **Y:** text box. Type 4.25 and press [Enter] to place the second point of the line.

7. The **Line** tool remains active to draw more line segments. Press the [Esc] key to end the tool.

You have constructed a vertical line from 0,0 to 0,4.25. You can verify this by placing a dimension on the line. Also, if you show constraints on the line, it is constrained vertical.

In the **Inventor Precise Input** toolbar, select the **X∠** mode from the drop-down list after picking the **Line** button. In order to enable the options on the toolbar, you must first activate the **Line** tool or other drawing tool. Then, draw another line from the origin:

1. Pick the first endpoint of the line at the origin (0,0).

2. Using the **Inventor Precise Input** toolbar, set the second endpoint by typing 3 in the **X:** text box. Press the [Tab] key or pick in the **°:** text box and enter 30. The angle is measured in degrees from the X axis.

3. Press [Enter] to create the second endpoint and draw the line.

4. Press [Esc] to end the **Line** tool. Zoom extents if necessary to see both lines.

A second line is drawn from the origin to a point three inches from the origin on the X axis and an at an angle of 30°. However, it is important to note that the *length* of the line is not three inches. This can be verified with dimensions:

1. Pick the **Fix** button in the constraint flyout in the **2D Sketch Panel**. Apply a fix constraint to both lines.

2. Draw a horizontal line of any length from the origin. Use the Construction style.

3. Pick the **General Dimension** button in the **2D Sketch Panel**. To place an angular dimension, pick the construction line and then the angled line. Drag the dimension to the desired location. Notice that the dimension value is 30°.

| **NOTE** | Depending on how Inventor placed constraints on the lines, you may get a message indicating the sketch will be over constrained. If this happens, pick the **Accept** button to place a reference dimension. Inventor calls reference dimensions *driven dimensions.* |

---

4. Place a dimension on the vertical line.

5. Now, dimension the horizontal distance between the endpoints of the angled line. Pick the line and move the cursor straight up or down. Then, drag the dimension to the desired location. The dimension value should be 3.000.

6. Finally, dimension the length of the angled line. Pick the line, move the cursor off the line, and then pick the line again. A dimension should appear that is aligned with the line. Drag the dimension to the desired location. You will receive a warning; pick the **Accept** button to place a driven (reference) dimension. Notice that the dimension value is 3.464, which is the length of the line.

Your sketch should look like **Figure 4-19.** When using the **X∠** input mode, it is important to remember that the X value is not the length of the line. Also, certain entries are not valid. For example, if you enter 3 for the X value and attempt to enter 90 for the angle, the 90 appears in red because Inventor cannot solve this "equation." An angle of 90° places the point on the Y axis (X = 0), yet you also told Inventor that the X value is 3, not 0. Hence, the error.

You can use relative input with the **Inventor Precise Input** toolbar. With relative input, the coordinates (or angle) are based on the last point, not the origin of the coordinate system.

1. Activate the **Line** tool. Make sure the style is set to Normal. Pick the right-hand endpoint of the inclined line as the first endpoint of the new line.

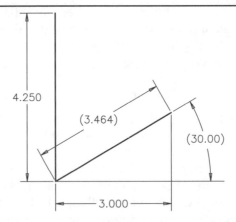

Figure 4-19.
The sketch after using the **Inventor Precise Input** toolbar to draw two lines.

2. On the **Inventor Precise Input** toolbar, pick the **Precise Delta** button. The button is depressed to indicate that the information entered will be relative to the last point (the first endpoint of the new line). Notice that a CSI appears at the first endpoint of the new line. Also, notice that the **Precise Relative** button on the toolbar is on (depressed).
3. Set the input mode to **XY**.
4. Type 0 in the **X:** text box and 1.75 in the **Y:** text box.
5. Press the [Enter] key to place the second point and draw the line, **Figure 4-20.**

A vertical line segment is added to the sketch. The second endpoint of the new line is 0 units on the X axis and 1.75 on the Y axis from the first point, not from the origin of the coordinate system. Notice that the **Precise Delta** button remains active (depressed). Also, the CSI has moved to the second endpoint of the new line. This is because the second endpoint, once drawn, becomes the "last" point. With the **Line** tool and **Precise Delta** button active, continue sketching:

6. Type the equation d2/–2 in the **X:** text box and 0 in the **Y:** text box. Dimension **d2** is the horizontal dimension between the endpoints of the angled line. Press [Enter] to draw the line.
7. Type 0 in the **X:** text box and .768 in the **Y:** text box. Press [Enter] to draw the line.
8. Pick the **Precise Delta** button on the **Inventor Precise Toolbar** to exit relative mode. The button is no longer depressed.
9. Type 0 in the **X:** text box and 4.25 in the **Y:** text box. Press [Enter] to draw the line.
10. Press [Esc] to end the **Line** tool.
11. Place dimensions as shown in **Figure 4-21** (not including the circle dimensions).

Notice how an equation and dimension name can be used to locate the midpoint of the angled line. The short vertical line has the same X value as the midpoint of the angled line. The negative value is used to indicate the line is to the left. Also, notice how after the **Precise Delta** button is turned off that the coordinates are based on the origin of the coordinate system, not the last point.

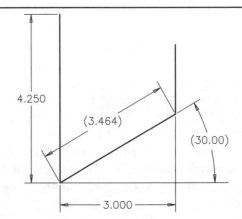

Figure 4-20.
Using relative input to add a vertical line to the sketch.

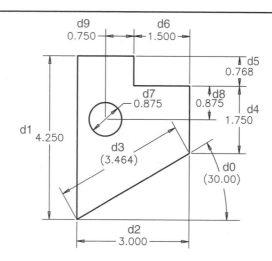

Figure 4-21.
The completed and dimensioned sketch. Make note of the dimension names indicated here. Adjust the procedure in the text based on the dimension names in your sketch.

When using relative entry in the previous example, a CSI is temporarily placed at the point from which the entry is measured. You can also manually place a temporary CSI and enter coordinates relative to it.

1. Pick the **Center point circle** button in the **2D Sketch Panel**.
2. On the **Inventor Precise Input** toolbar, pick the **Precise Relative** button. Then, pick the lower left corner of the cutout to locate the origin. This location will remain the origin until the button is turned off.
3. Select the **XY** input mode.
4. Type d6/–2 in the **X:** text box and d4/–2 in the **Y:** text box. Press [Enter] to locate the center of the circle as shown in **Figure 4-21.**
5. Type d6/–2 in the **X:** text box and d4/–4 in the **Y:** text box. Press [Enter] to create the circle, which has a diameter of .875.
6. Dimension the diameter of the circle. Also, locate the center of the circle with vertical and horizontal dimensions.

Chapter Test

Answer the following questions on a separate sheet of paper.
1. A sketch containing islands is considered a(n) _____ profile.
2. An island is also called a(n) _____.
3. If d3 = 2.5 mm and d2 = 1.5 mm, what is the solution calculated by Inventor for the equation d3/d2?
4. How does Inventor indicate an invalid or unacceptable equation?
5. For an item, such as a constant, that has no units, how does Inventor indicate the "unit" in an equation?
6. List one advantage of using the Construction style to draw construction lines.
7. Once a curve is drawn, how can you change its style?
8. What is the function of the sketch **Mirror** tool?
9. What is the function of the **Revolve** tool?
10. What relationship must the axis of revolution have to the profile when using the **Revolve** tool?
11. When using the **Revolve** tool, what is the advantage of drawing the axis of revolution in the Centerline style?
12. How can you select multiple profiles for extrusion?
13. How can you select multiple profiles for revolution?
14. When using the **Inventor Precise Input** toolbar, how long is an equation entered in the **X:** text box stored once the point is entered (placed)?
15. When the **Precise Delta** button on the **Inventor Precise Input** toolbar is on, from where are the coordinates you enter measured?

Chapter Exercises

Exercise 4-1. Start a new English Standard (in).ipt file. Construct the link shown below. Use construction lines, dimensions, and constraints on the sketch. Extrude the profile into a solid part. Do not create a title block drawing.

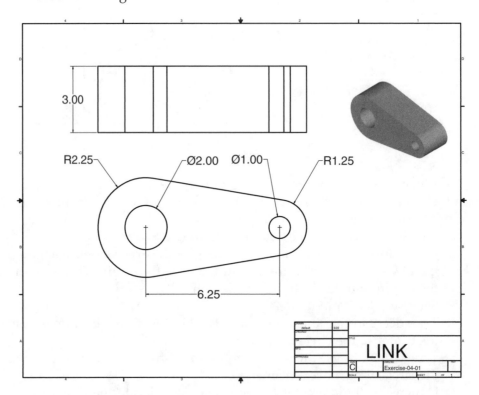

Exercise 4-2. Start a new metric Standard (mm).ipt file. Construct the valve shown below. Use dimensions and constraints on the sketch. Revolve the profile 360° into a solid part. Do not create a title block drawing.

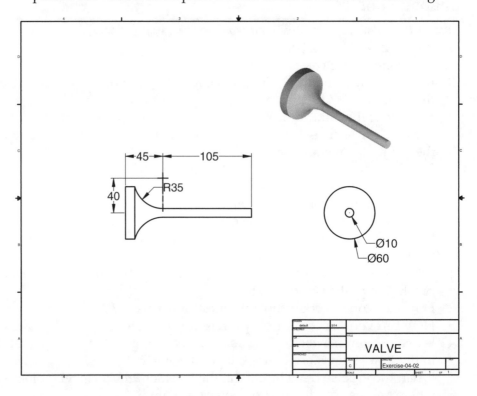

Exercise 4-3. Start a new English Standard (in).ipt file. Construct the stamping shown below. Use a horizontal construction line from the midpoint of the left side to locate the slot. Extrude the profile into a solid part. Do not create a title block drawing.

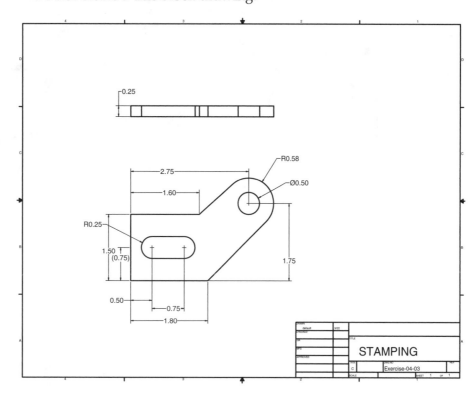

Exercise 4-4. Start a new English Standard (in).ipt file. Construct the mirror part shown below. Draw vertical and diagonal construction lines to locate the arcs. Draw a horizontal centerline from the midpoints of the sides to mirror the 1.200″ wide feature. Extrude the profile 1″ into a solid part. Do not create a title block drawing.

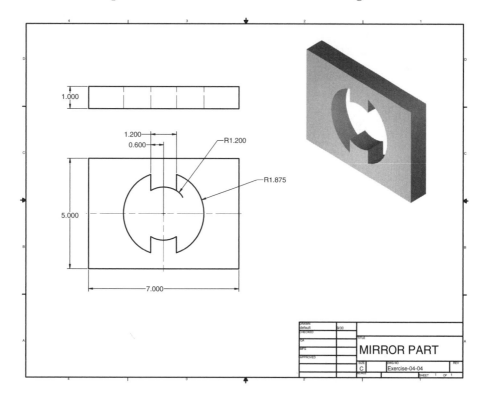

Exercise 4-5. Start a new metric Standard (mm).ipt file. Construct the cap shown below. Sketch half of the profile shown. Place the four diameter dimensions by applying the dimensions using the centerline. Revolve the profile 360° into a solid part. Do not create a title block drawing.

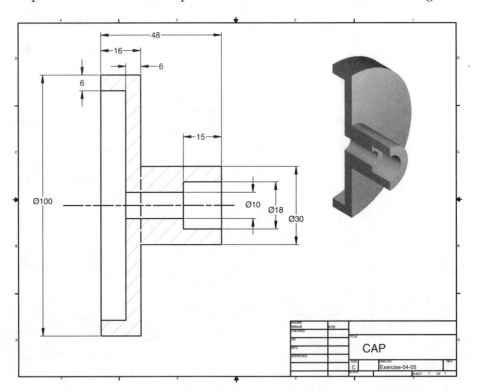

Exercise 4-6. Start a new English Standard (in).ipt file. Construct the wheel shown below. Start by sketching the profile as shown. Include the centerline and two construction circles. Use the circles to locate the vertical lines. Once the dimensions are placed, open the **Parameters** dialog box. Change the dimension names and enter equations as shown in the chart below. Refer to the default dimension names shown on the sketch below. Revolve the profile into a solid part. Now, show the dimensions on the part, edit the OD_WHEEL dimension to 24", and update the part.

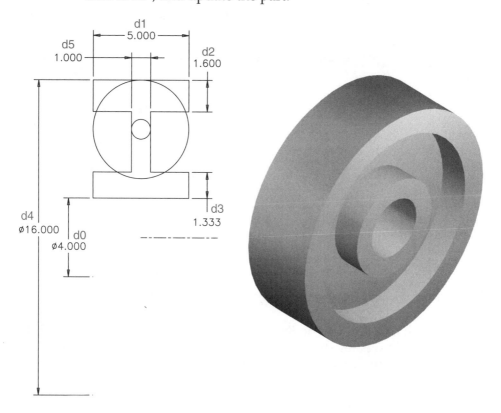

Dimension Name		Equation
Old	**New**	
d0	ID_WHEEL	OD_WHEEL/4ul
d1	WIDE_RIM	OD_WHEEL/3.200ul
d2	THICK_RIM	OD_WHEEL/10ul
d3	THICK_HUB	OD_WHEEL/12ul
d4	OD_WHEEL	*(current value)*
d5	THICK_WEB	OD_WHEEL/16ul

Exercise 4-7. Start a new metric Standard (mm).ipt file. Construct the blanked cam plate shown below. All arcs are tangent to connected arcs/lines. Use construction lines as needed. Extrude the profile into a solid part. Do not create a title block drawing.

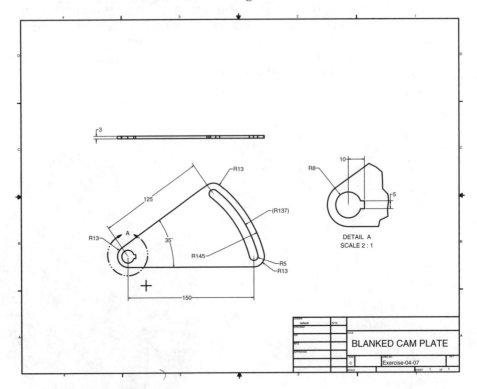

Exercise 4-8. Start a new metric Standard (mm).ipt file. Construct the pawl shown below. Some dimensions are missing; approximate those. Extrude the profile 6.6 mm into a solid part.

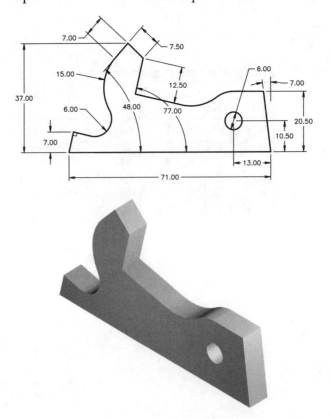

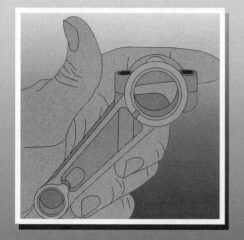

Secondary Sketches and Work Planes

Objectives

After completing this chapter, you will be able to:

✳ Explain how to create secondary sketch features.
✳ Explain and use the five termination features in the **Extrude** tool.
✳ Project silhouette curves onto a sketch plane.
✳ Use the **Cut** and **Intersect** options of the **Extrude** tool.
✳ Display, adjust, and use the three fundamental work planes of a sketch.
✳ Create, display, and use additional work planes.

User's Files

*The following is a list of files that you will need to work through this chapter. These files can be found on the **User's Files** CD included with this text.*

Examples	Practice	Exercises
Example-05-01.ipt	Practice-05-01.ipt	Exercise-05-01.ipt
Example-05-02.ipt	Practice-05-02.ipt	Exercise-05-03.ipt
Example-05-03.ipt	Practice-05-03.ipt	Exercise-05-06.ipt
Example-05-04.ipt		Exercise-05-07.ipt
Example-05-05.ipt		
Example-05-06.ipt		
Example-05-07.ipt		
Example-05-08.ipt		

Creating Secondary Sketch Planes and Features

The parts you created in *Chapter 3* and *Chapter 4* were all based on one primary sketch and profile. In this chapter, you will look at how to create additional, secondary sketches on the faces of the part and on work planes. Open the file Example-05-01.ipt. This part has five planar (flat) faces that can be used as sketch planes. It also has one curved face that cannot be used as a sketch plane. See **Figure 5-1**.

Figure 5-1.
This part has five planar faces that can be used as sketch planes. The curved face cannot be used as a sketch plane.

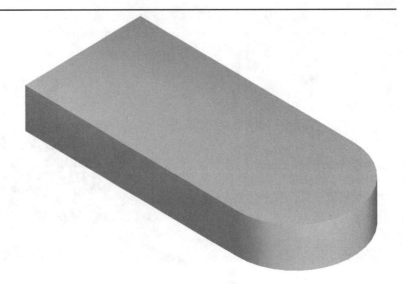

In this section, you will create a boss on the top face. First, to provide a visual reference, you will change the color of the top face. To do this, right-click on the top face. Then, select **Properties** from the pop-up menu. The **Face Properties** dialog box is displayed, **Figure 5-2.** Select Green (Flat) from the drop-down list and pick the **OK** button. The face is now displayed in green. You can change the color of any face using this procedure.

The procedure for creating the boss is to select the planar face on which to sketch, create the sketch, and extrude the profile. This basic procedure can be used to create a new feature on any planar face. The existing part edges can be used as part of the sketch or as reference geometry for dimensions and constraints.

1. Pick the **Sketch** button on the **Inventor Standard** toolbar or press the [S] key to enter Sketch mode. A pencil and paper icon appears next to the cursor in the graphics window. This indicates that you need to select a sketch plane, which is also indicated on the status bar at the bottom of the Inventor screen. You must select a plane on which to sketch before Sketch mode is entered.
2. Move the cursor over any face on the part. The edges of the face are highlighted in red. If you pause momentarily, the cursor changes to the **Select Other** tool. This tool allows you to choose between adjacent faces by picking the left or right arrow. To close the tool, simply move the cursor off of the tool.
3. Move the cursor over the green face and pick to select it.

The edges of the face and the center point of the arc are displayed in black. The grid is also displayed. Notice that the grid is on the selected face. This indicates that the current coordinate system coincides with the selected face. However, you do not know where the origin of the coordinate system is located. You can place a CSI at the origin as follows.

1. Select **Application Options...** from the **Tools** pull-down menu. The **Options** dialog box is displayed.
2. Pick the **Sketch** tab.
3. In the **Display** area of the **Sketch** tab, check the **Coordinate System Indicator** check box. (If it is checked, leave it checked.)

Figure 5-2.
Changing the color of a face using the **Face Properties** dialog box.

4. Pick the **OK** button or press [Enter] to close the dialog box and apply the setting. Do not pick the Windows close button (the small X) as this cancels the settings.

A CSI now appears at the origin of the current coordinate system. If the CSI is not immediately displayed, pick any of the edges on the part to display the CSI. The coordinate system determines horizontal and vertical directions in relation to the part. Later, you will learn how to reposition the coordinate system and why you would want to do so. Now, draw the sketch and extrude it into the boss:

1. Using the **Center point circle** tool, sketch a circle anywhere on the grid. For this example, do not pick the center of the circle coincident to the center of the arc.
2. Apply a concentric constraint between the circle and the arc.
3. Dimension the circle's diameter. Then, edit the dimension to a value of 25 mm.
4. "Finish" the sketch. You have now added an unconsumed profile (sketch) to the part.
5. Pick the **Extrude** button in the **Part Features** panel. Since the edges of the green face and the circle form an area that can be selected as a profile, this is an ambiguous sketch. A profile is not automatically selected; you must select the profile. With the **Profile** button active (depressed) in the **Extrude** dialog box, select the interior of the circle as the profile. The area is highlighted in cyan.
6. In the **Extrude** dialog box, make sure the **Join** button is on (depressed). Then, set the distance to 22 mm. Set the extrusion direction upward by picking the left-hand button below the distance drop-down list. Also, the output should be set to a solid. See **Figure 5-3A.**
7. Pick the **OK** button to extrude the feature, **Figure 5-3B.** Notice that the extrusion is given a different name in the **Browser**, such as Extrusion2, rather than being added to the previous extrusion.

You can rename the components of the part in the **Browser**. For example, Extrusion2 is really not very descriptive. However, Boss is more meaningful as a description. To rename the boss extrusion:

1. Right-click on the extrusion name in the **Browser** and select **Properties…** from the pop-up menu. The **Feature Properties** dialog box is displayed. See **Figure 5-4.**
2. In the **Name** text box, type Boss.
3. Just like individual faces, individual features can be assigned unique colors. This is usually done for emphasis or clarity. In the **Feature Color Style** drop-down list, select Chrome.
4. Pick the **OK** button. The feature is now displayed in simulated chrome. Also, the name of the extrusion in the **Browser** is Boss.

Figure 5-3.
A—The settings for extruding the profile. B—The extrusion is added to the part.

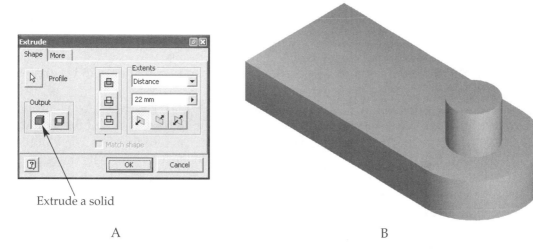

Extrude a solid

A   B

Figure 5-4.
Changing the name
and color of a
feature using the
Feature Properties
dialog box.

Feature
name

Feature
color

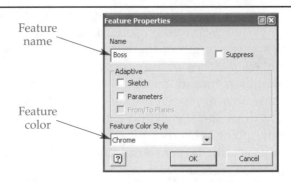

Operation and Extents Options of the Extrude Tool

There are three operation options in the **Extrude** dialog box—**Join**, **Cut**, and **Intersect**. **Join** adds the extrusion to the existing part. **Cut** removes the volume of the extrusion from the existing part. **Intersect** creates a new part based on the overlapping volume of the extrusion and the existing part.

There are also five extents options in the **Extrude** dialog box—**Distance**, **To Next**, **To**, **From To**, and **All**. These are used to specify how far the profile extrudes. **Distance** specifies an exact length for the extrusion, which is what you have used so far. **To Next** specifies that the extrusion will extend to the next surface of the part in the extrusion direction. **To** is used to pick a surface to which the extrusion will extend. **From To** allows you to pick a surface on the part from which the extrusion starts and another surface to which the extrusion extends. **All** specifies that the extrusion goes all the way through the part, including through all features.

In this next operation, you will use the **All** and **Cut** options. If you have not yet created the boss as described in the previous section, do so now. Then, continue as follows.

1. Pick the **Sketch** button on the **Inventor Standard** toolbar. Then, select the face on the top of the boss as the sketch plane.
2. Pick the **Look At** button on the **Inventor Standard** toolbar. Select the face on top of the boss to get a top view. This step is not necessary, but it may make locating the new sketch easier.
3. Sketch a circle. Then, constrain it concentric to the boss. Place a diameter dimension on the circle and edit the dimension to a value of 10 mm.
4. Right-click in the graphics window. Select **Isometric View** from the pop-up menu.
5. "Finish" the sketch. Then, pick the **Extrude** button in the **Part Features** panel. There are now several profiles, so select the small circle as the profile to extrude.
6. In the **Extrude** dialog box, pick the **Cut** button, which is the middle button in the center of the dialog box. Also, select **All** from the extents drop-down list. See **Figure 5-5A**.
7. Enter a taper angle of 10° in the **More** tab of the **Extrude** dialog box. See **Figure 5-5B**. A positive taper angle makes the feature get larger as it extrudes and a negative angle makes it smaller.
8. Pick the **OK** button to extrude the profile. The finished part is shown in **Figure 5-6** with half of the part removed to show the interior extrusion.

Now, open the file Example-05-02.ipt. This is a U-shaped part, **Figure 5-7.** You will add a cutout (hole) to one leg of the part, but not the other. The **To Next** option of the **Extrude** tool allows you to do this by limiting the extent of the extrusion to the next face that the extrusion encounters. The "to next" face can be either flat or curved. To create the cutout:

1. Pick the **Sketch** button and select the top face of the part as the sketch plane.
2. Draw a construction line from the midpoint of the edge. If it does not already exist, constrain the construction line perpendicular to the edge. Dimension the line as shown in **Figure 5-8**.

Figure 5-5.
A—The settings for extruding the profile. B—Entering a taper angle for the resulting extrusion.

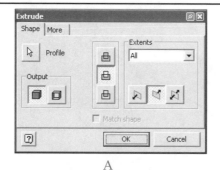
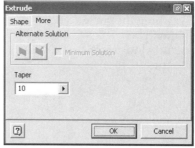

A   B

Figure 5-6.
The extrusion is subtracted from the part. The sketch was on the top face of the boss, which is why a positive taper angle results in the bottom of the cutout being larger than the top. Note: The part is shown sliced through the middle for illustration.

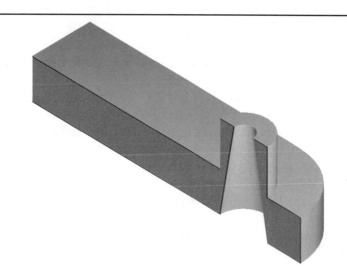

Figure 5-7.
A cutout (hole) will be added to one leg of this U-shaped part.

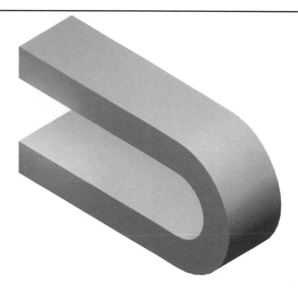

3. Draw a circle. Constrain the center of the circle coincidental to the endpoint of the construction line. Dimension the circle as shown in **Figure 5-8.**

CAUTION

The circle should not be drawn as construction style. If it is, then the circle cannot be extruded.

Chapter 5   Secondary Sketches and Work Planes   115

Figure 5-8.
Sketching the circle that will be extruded to create the cutout.

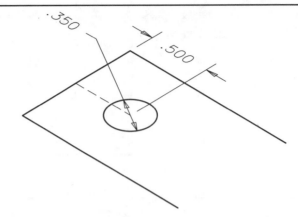

4. "Finish" the sketch. Then, pick the **Extrude** button in the **Part Features** panel. Select the circle as the profile.
5. In the **Extrude** dialog box, pick the **Cut** button. Also, select To Next from the **Extents** drop-down list. See **Figure 5-9.**
6. Pick the **Terminator** button in the **Extrude** dialog box. The tooltip for this button is Select body to end the feature construction. Then, pick anywhere on the part.
7. Pick the **OK** button to create the extrusion. The updated part is shown in **Figure 5-10.** Notice how the cutout (hole) does not pass through both legs of the part. You may want to use the **Rotate** tool to rotate the view and better see the feature.

Figure 5-9.
The settings for extruding the profile to create a cutout (hole) in one leg only.

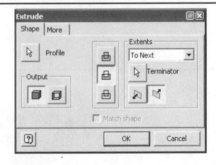

Figure 5-10.
The extrusion is subtracted from one leg only.

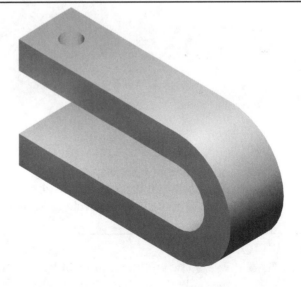

Now, open Example-05-03.ipt. This part is similar to the U-shaped part you have been working with. However, one leg has a small lip projecting to the interior of the part. See **Figure 5-11**. In this example, you will add a pin to the leg that extends only as far as the lip. The **To** option of the **Extrude** tool allows you to do this. The circle profile that you will extrude is already drawn. Continue as follows.

1. Pick the **Extrude** button and select the circle as the profile.
2. In the **Extrude** dialog box, pick the **Join** button. Also, select To in the **Extents** drop-down list.
3. When you select **To** in the drop-down list, the button below the list is activated. The tooltip for this button is Select surface to end the feature creation. Pick the bottom of the lip feature. The color of this face has been changed to green in the file to help you select the correct face. Rotate the view if you like to see the face better.
4. Once you pick the face, it is highlighted in cyan. Also, a check box appears next to the button in the **Extrude** dialog box. The tooltip for this check box is Check to terminate feature on the extended face. When checked, the face is treated as an infinite plane. Because the circle does not actually intersect the green face, check this check box. If you do not, you will get an error.
5. Pick the **OK** button in the **Extrude** dialog box to create the new feature. See **Figure 5-12**.

Figure 5-11.
An extrusion that extends to the same plane as the lip will be added to this part.

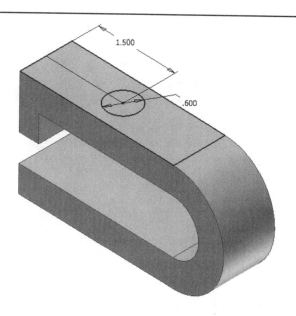

Figure 5-12.
The extrusion is added to the part. Notice how it extends to the same plane as the lip.

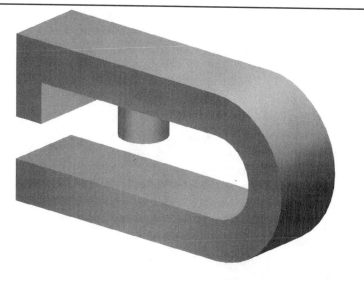

❑ Open the file Practice-05-01.ipt. Using the **Rotate** tool, rotate the view 180° about the horizontal axis. The outside face of the second leg is green and should be on top. Press [Esc] to exit the tool.

❑ Pick the **Sketch** button and select the green face as the sketch plane.

❑ Pick the **Look At** button and select the green face to display a top view.

❑ Pick the **Shaded Display** flyout on the **Inventor Standard** toolbar and select **Wireframe Display**. This allows you to see the existing circular cutout and use it for constraints.

❑ Sketch a rectangle around the hole.

❑ Apply tangent constraints between each side of the rectangle and the existing circle.

❑ "Finish" the sketch and display the previous view.

❑ Using the **Extrude** tool, extrude the square profile. Since this is an ambiguous profile, you will need to select the area of the circle and the square. Use the **Cut** and **To Next** options.

❑ The final result is shown below.

Projecting Silhouette Curves to the Sketch Plane

Open the Example-05-04.ipt file. **Figure 5-13A** shows the part in the isometric view as a solid with the curved surface in green. **Figure 5-13B** shows the same part in the front view as a wireframe. The line on the right side of the wireframe represents the extent of the curved surface. This type of line (curve) is called a *silhouette curve*. In reality, there is no edge there. The line is displayed to make the part "look right." The edges of the part can be used to define a sketch, but a special technique is necessary to use a silhouette curve. A silhouette curve must be projected onto the sketch plane. The **Project Geometry** tool is used to do this.

Now, you will construct a slot through the curved end of this part. To make visualization easier, stay in the isometric view, but change to a wireframe display. Then, continue as follows.

1. Pick the **Sketch** button. Then, select Face A as the sketch plane. Face A is indicated in **Figure 5-13.** Notice that a rectangle appears in black; this is the geometry that is on the current sketch plane. However, there is no line on the sketch plane that corresponds to the silhouette curve.

Figure 5-13.
Face A indicated here is used as the sketch plane. The silhouette curve representing the curved face will be projected onto the sketch plane.

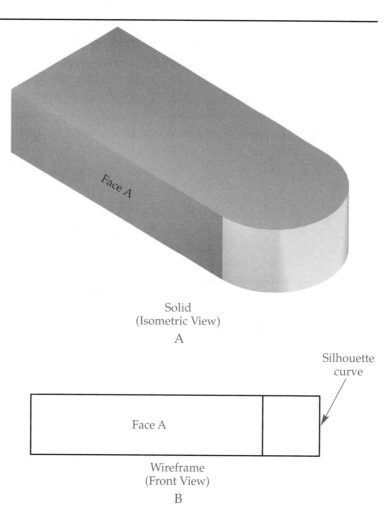

Solid
(Isometric View)

A

Silhouette
curve

Face A

Wireframe
(Front View)

B

2. Select the **Project Geometry** button in the **2D Sketch Panel**.
3. Move the cursor around the curved (green) face until you see a red vertical line at the midpoint of the curve. This line represents the silhouette curve. Pick the line and it is projected onto the sketch plane. See **Figure 5-14**.
4. Press the [Esc] key to exit the **Project Geometry** tool.
5. Sketch two construction circles, as shown in **Figure 5-15**. Constrain the large circle tangent to the top and bottom edges of the part. Constrain the small circle concentric (or coincidental) to the large circle. Dimension the small circle's diameter. Edit the dimension to a value of .2".
6. Sketch the three lines forming the profile of the slot. The two long lines should be constrained perpendicular to the projected line and tangent to the small circle. Use a dimension to constrain the location of the short line relative to the left edge of the part, as shown in **Figure 5-15**.
7. "Finish" the sketch.
8. Pick the **Extrude** button in the **Part Features** panel. Select the rectangular area formed by the projected line and the three lines you just constructed as the profile.
9. Select the **Cut** and **All** options in the **Extrude** dialog box. Then, pick the **OK** button to extrude the profile and remove its volume from the part. See **Figure 5-16**.

Figure 5-14.
Projecting the silhouette curve onto the sketch plane.

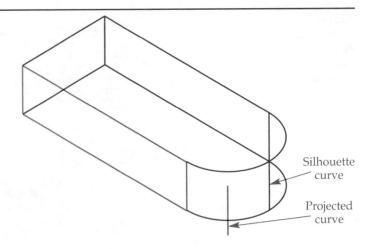

Silhouette curve

Projected curve

Figure 5-15.
Drawing the profile that will be extruded into the cutout. Notice how two construction circles are used to center the cutout profile.

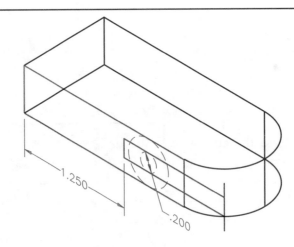

1.250

.200

Figure 5-16.
The extrusion is subtracted from the part to create the cutout (slot).

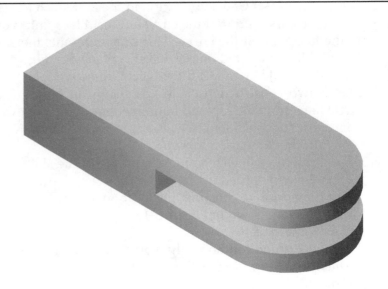

Default Work Planes and Midplane Construction

As you saw in the previous section, the planar faces of the part can be used to define sketch planes. In addition, you can use work planes to define the sketch plane. *Work planes* are construction geometry that can be used to help in creating sketches and features. There are three default work planes in every Inventor part file. These work planes are aligned with the three principle planes of the coordinate system—XY, XZ, and YZ. By default, the visibility of the default work planes is turned off, but they are listed in the **Browser**.

In this section, you will use one of the default work planes to construct the part shown in **Figure 5-17.** Only one work plane will be used. Start a new English Standard (in).ipt file and display the isometric view. In the **Browser**, expand the Origin branch of the part tree. This branch contains the three work planes, three work axes, and one center point. As you move the cursor over the names in the **Browser**, the objects are highlighted in the graphics area.

Although its visibility is turned off, the default sketch plane is the XY work plane. Right-click on XY Plane in the **Browser** to display the pop-up menu. Currently, the **Visibility** option does not have a check mark next to it. This means that the work plane is not visible. To turn the visibility of the work plane on, pick **Visibility** in the pop-up menu. Now, if you right-click on XY Plane to display the pop-up menu, there is a check mark next to the **Visibility** option. Also, the icon next to the name in the **Browser** is no longer grayed out.

Now, hold the cursor over the name in the **Browser**. The work plane is highlighted in the graphics window. Move the cursor into the graphics window and over the highlighted work plane. The highlight disappears until the cursor is over an edge of the work plane again. Pick the edge to select the work plane. What you have selected is an object representing an infinite plane; the edges are not true "edges."

You can move the work plane around in the graphics window. The work plane can even be resized. However, the **Auto-Resize** option must be turned off. Right-click on the work plane in the graphics window or its name in the **Browser**. Then, select the **Auto-Resize** option to uncheck it. Now, move the cursor to an edge of the work plane; the move cursor is displayed. Pick and drag the work plane to a new position. You can resize the work plane by moving the cursor to one of the four yellow circles at the corner of the plane. When the resize cursor is displayed, pick and drag to resize the work plane. This may help if the plane is displayed inside of a shaded part and, therefore, not visible. However, since the work plane is infinite, changing the size or location has no effect on the use of the plane, just its display.

Figure 5-17.
This part is created using a single work plane.

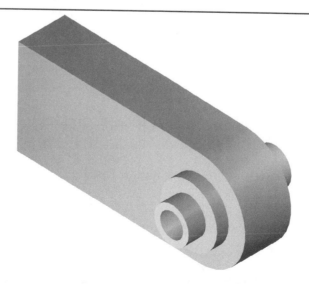

CAUTION

When moving a work plane, it may appear as if you are changing the elevation (Z value) of the plane. You are not.

Turning on the visibility of the center point can also be useful. The center point is at the origin of the coordinate system, which is the intersection of the XY, XZ, and YZ work planes. This point is fixed; it does not move when different sketch planes are active. When its visibility is turned on, the center point is a small diamond. Unlike a work plane, it is visible (when on) even when the cursor is not over it. To turn the visibility of the center point on, right-click on Center Point in the Origin branch in the **Browser**. Then, select **Visibility** in the pop-up menu. A check mark appears next to this option when the visibility of the center point is on.

To draw the part shown in **Figure 5-17,** turn on the visibility of the XY work plane. Make sure the **Auto-Resize** option is on for the work plane. Also, turn on the visibility of the center point. Then, continue as follows.

1. Sketch a 2" diameter circle at the origin.
2. Apply a fix constraint to the center of the circle. This will ensure the center of the circle is at the coordinate system origin for future sketches. Note: If you applied a dimension to the circle, delete it before applying the constraint. Otherwise, the fix constraint will over constrain the sketch.
3. Construct the three lines and place the two dimensions shown in **Figure 5-18.** Apply appropriate constraints. You can trim the inside portion of the circle, as shown in the figure, but this is not necessary.
4. "Finish" the sketch and display the isometric view.
5. Pick the **Extrude** button in the **Part Features** panel. In the **Extrude** dialog box, select Distance from the **Extents** drop-down list and set the distance to one inch. Also, pick the **Midplane** button, which is the right-hand button below the drop-down list. Notice that the preview in the graphics window shows the extrusion equally divided about the work plane. Pick the **OK** button to create the extrusion.
6. Pick the **Sketch** button on the **Inventor Standard** toolbar. Select the XY work plane as the sketch plane; pick an edge. This is why you turned on the visibility of the XY work plane. Otherwise, you would not be able to select it as the sketch plane. By using the XY work plane and the **Midplane** extrusion option, you can easily create the entire part.

Figure 5-18.
Sketching the first profile to be extruded.

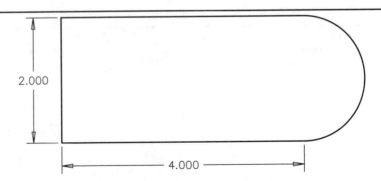

The sketch plane and grid pass through the part. Since the part is shaded, it would be hard to sketch within the boundary of the part. You can change to a wire-frame display. However, there is another method that allows you to keep the part shaded. Right-click in the graphics window to display the pop-up menu. Then, select **Slice Graphics** from the menu. See **Figure 5-19.** The portion of the part "above" the sketch plane (positive Z axis) is hidden. Now, you can sketch within the boundary of the part and see what you are doing. Continue creating the part as follows.

7. Using the **Center point circle** tool, draw a circle.
8. Apply a concentric constraint between the circle and the curved feature. To do this, select the arc at the "back" of the object. The arc representing the curved surface will be projected onto the current sketch plane.
9. Dimension the diameter and edit the value to 1.25".
10. "Finish" the sketch. The part is no longer displayed "sliced." Pick the **Extrude** button in the **Part Features** panel.
11. In the **Extrude** dialog box, pick the **Profile** button, if it is not already active. Then, select the circle as the profile. The circle is within the shaded part, but as you move the cursor over the circle, it is highlighted.
12. In the **Extrude** dialog box, pick the **Join** button. Also, select Distance from the **Extents** drop-down list and set the distance to 1.50". Pick the **Midplane** button and then pick **OK** to create the extrusion.
13. Using the same procedure, sketch a .75" diameter circle and extrude it 2.25".
14. Sketch a .50" circle and extrude it using the **Cut** and **All** options. Make sure the **Midplane** button is selected.

The part is complete and should look like Figure 5-17. To turn off the visibility of a work plane, right-click on it in the graphics window or on its name in the **Browser**. Then, select **Visibility** in the pop-up menu to turn off its visibility.

Figure 5-19.
Using the **Slice Graphics** option, the portion of the part that is above the current sketch plane (positive Z) is not displayed.

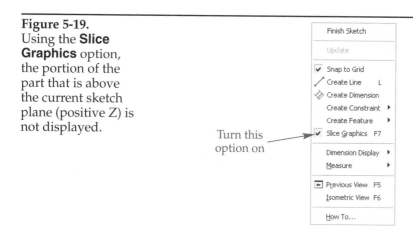

Turn this option on

Open the file Example-05-05.ipt. This is a revolved part to which you will add a cutout (hole) through one side of the outer shell, as shown in **Figure 5-20.** The **From To** option of the **Extrude** tool will be used in conjunction with the default work planes.

Expand the Origin branch of the part tree in the **Browser**. Right-click on X Axis and select **Visibility** in the pop-up menu. Similarly, turn on the visibility of the XZ work plane. Pick the **Sketch** button on the **Inventor Standard** toolbar. Then, pick the XZ work plane as the sketch plane. Display a "sliced" part and continue as follows.

1. Pick the **Project Geometry** button in the **2D Sketch Panel**. Pick the X axis and the large edge of the cylinder to project them onto the sketch plane. It is necessary to project the X axis as you cannot dimension to a work axis.
2. Change the style of the projected lines to Construction.
3. Draw a circle and dimension it as shown in **Figure 5-21.** Make sure it is drawn in Normal style.

Figure 5-20.
The cutout (hole) shown here will be added to the outer shell of this part.

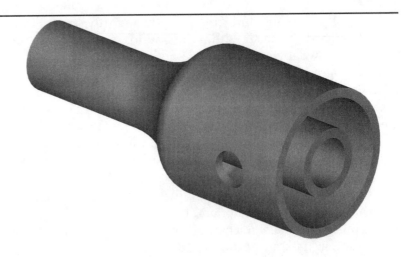

Figure 5-21.
Drawing the profile that will be extruded to create the cutout (hole).

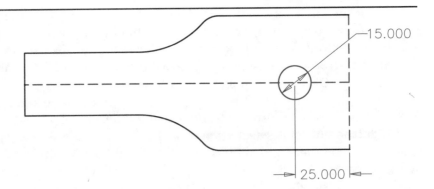

15.000

25.000

4. Apply a coincident constraint between the center of the circle and the projected X axis line. "Finish" the sketch.
5. Pick the **Extrude** button in the **Part Features** panel.
6. Select the circle as the profile.
7. In the **Extrude** dialog box, pick From To in the **Extents** drop-down list.

From To means that you have to pick two work planes or faces to define the extent of the extrusion. The extruded feature does not have to touch the plane on which you sketched the profile. Once you select this option, two buttons appear below the **Extents** drop-down list in the **Extrude** dialog box. Continue as follows:

8. Pick the top button below the **Extents** drop-down list. The tooltip for this button is Select surface to start the feature creation. Then, pick the outermost surface of the part. That surface is highlighted in an olive-drab color.
9. Once you select the first surface, the bottom button below the **Extents** drop-down list is activated. The tooltip for this button is Select surface to end the feature creation. With this button active, pick the inside surface of the outer shell of the part. The surface is highlighted in cyan.
10. Pick the **Cut** button in the **Extrude** dialog box. See **Figure 5-22.** Then, pick the **OK** button to extrude the profile.

The part is completed and should look like Figure 5-20. If the cutout (hole) is on the opposite side of the part, right-click on the extrusion name in the **Browser**. Select **Edit Feature** from the pop-up menu. This displays the **Extrude** dialog box. Then, pick the **More** tab. In the **Alternate Solution** area of this tab are two buttons, one of which is on (depressed). These buttons determine the direction of the extrusion. Pick the button that is not currently on. Then, pick the **OK** button to complete the edit. The cutout should now be on the opposite side of the part as before.

Figure 5-22.
The settings for
extruding the
profile.

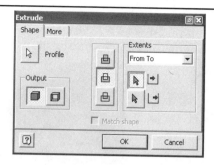

PRACTICE 5-2

❑ Open the file Practice-05-02.ipt. This is a stepped shaft constructed as a revolution. A keyway needs to be added to the shaft, as shown below.

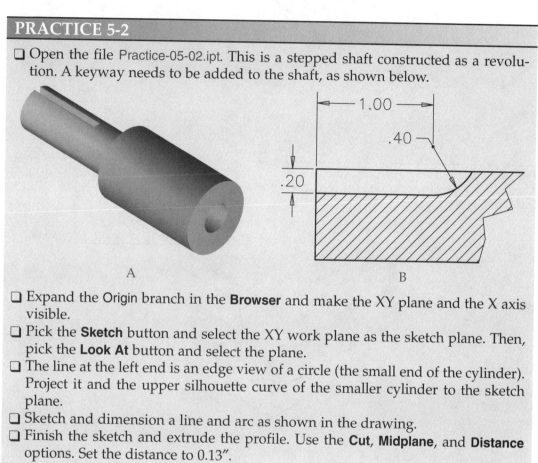

A   B

❑ Expand the Origin branch in the **Browser** and make the XY plane and the X axis visible.
❑ Pick the **Sketch** button and select the XY work plane as the sketch plane. Then, pick the **Look At** button and select the plane.
❑ The line at the left end is an edge view of a circle (the small end of the cylinder). Project it and the upper silhouette curve of the smaller cylinder to the sketch plane.
❑ Sketch and dimension a line and arc as shown in the drawing.
❑ Finish the sketch and extrude the profile. Use the **Cut**, **Midplane**, and **Distance** options. Set the distance to 0.13".

The Intersect Option of the Extrude Tool

The **Intersect** option of the **Extrude** tool creates a solid of the volume that is common to the part and the new extrusion. The effects of the three options—**Join**, **Cut**, and **Intersect**—are shown in **Figure 5-23.**

In this section, you will use the **Intersect** option to create a part that would be difficult to create any other way. Open Example-05-06.ipt. See **Figure 5-24.** The YZ and the XY work planes are visible. On each plane is a sketch. If you look at the **Browser**, the part tree shows two unconsumed sketches, as indicated by the pencil and paper symbol next to their names. An *unconsumed sketch* has not yet been used for an operation, such as extrude or revolve. A *consumed sketch* has been used for an operation. Consumed sketches are indicated by the same symbol being held by a hand.

1. Using the **Extrude** tool, extrude the left-hand sketch 300" to the right. The distance can also be entered as 25'. Notice that the symbol next to the name of the sketch in the **Browser** has changed to indicate a consumed sketch.

Figure 5-23.
A—Extruding this profile (shown in color) can result in one of three parts, depending on the option selected. B—**Join**. C—**Cut**. D—**Intersect**.

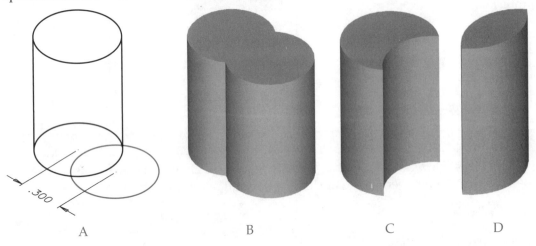

A   B   C   D

Figure 5-24.
These are two separate profiles (unconsumed sketches) that will be extruded using the **Intersect** option.

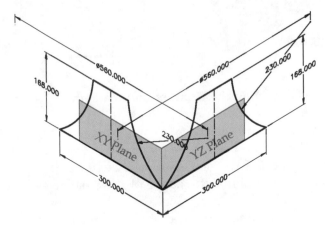

2. Pick the **Extrude** button again. Select the right-hand sketch as the profile. In the **Extrude** dialog box, select the **Distance** option and set the distance to 25′.
3. Pick the **Intersect** button in the **Extrude** dialog box. Also, pick the appropriate direction button so the preview shows the extrusion going into the existing part.
4. Pick the **OK** button to create the extrusion.
5. The result is the pagoda roof shown in **Figure 5-25.**
   The **Intersect** option can be very useful in converting AutoCAD 2D drawings into Inventor parts. This process is discussed in *Chapter 8.*

Creating and Using New Work Planes

New work planes can be created from part faces, edges, endpoints (vertices), and the default planes. The new planes are used to terminate extrusions, to provide angled and offset sketch planes, and as references for dimensions and constraints. The latter is very useful for the 3D constraints that you will use for assemblies in *Chapter 11.*

Figure 5-25.
This part is the result of extruding the profiles shown in **Figure 5-24** using the **Intersect** option.

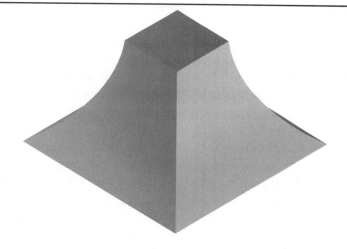

The **Work Plane** tool is used to create a new plane. The basic procedure is to activate the tool, which is located in the **Part Features** panel, and then pick some combination of objects or features on the part. The position and orientation of the new plane depend on which features you pick and how you pick them. Features of a part that can be selected include faces, lines, endpoints (vertices), existing work planes, work points, and work axes. The next sections present several different applications.

Offset from an Existing Face or Work Plane

You can create a new work plane that is offset a specific distance from a face or existing work plane. This feature is often used when creating a feature that will attach to a cylinder. The existing XY plane needs to be moved in the Z direction so a shape can be sketched. Since it is impossible to move the existing plane, a new one is created a set distance away from the old plane. Now, sketching the required shape is possible.

For example, the part in **Figure 5-26A** is the existing part. The finished part is shown in **Figure 5-26B**. To create the new feature, you will offset the YZ work plane. Open Example-05-07.ipt. Expand the Origin branch in the **Browser**. Right-click on YZ Plane and select **Visibility** from the pop-up menu to display the work plane. Then, continue as follows.

1. Pick the **Work Plane** button. Then, pick an edge of the YZ work plane and drag it to the right.
2. In the **Offset** dialog box that appears, enter a value of 100 mm. Then, press [Enter] or pick the check mark button in the dialog box.

Figure 5-26.
A—The original part. B—A feature is added to the part.

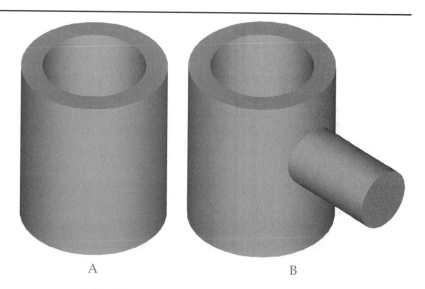

A

B

3. Pick the **Sketch** button on the **Inventor Standard** toolbar. Pick the new work plane as the sketch plane.
4. With the **Project Geometry** tool, project the top and bottom of the part onto the sketch plane. Change the style of the two projected lines to Construction.
5. Draw a construction line between the midpoints of the two projected lines.
6. Draw a circle in Normal style. Dimension the diameter and edit the value to 35 mm.
7. Apply a coincident constraint between the center of the circle and the construction line you drew in Step 5.
8. Add a dimension between the projected top edge and the center of the circle. Edit the dimension to a value of 40 mm. See **Figure 5-27.**
9. "Finish" the sketch and pick the **Extrude** button in the **Part Features** panel.
10. Select the circle as the profile.
11. In the **Extrude** dialog box, select the **Join** and **To Next** options. Pick the **Terminator** button and select the outside surface of the cylinder.
12. Pick the **OK** button to create the extrusion. The new feature is added to the part.

Figure 5-27.
Sketching the profile that will be extruded into the feature shown in **Figure 5-26B.**

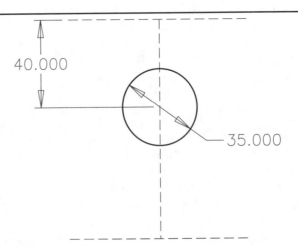

Offset through a Point

Instead of entering an offset distance, you can select a point through which the offset work plane will pass. Using the **Work Plane** tool, pick the interior of an existing planar face or the edge of an existing work plane. Then, select an endpoint of any edge on the part. The new work plane is created parallel to the selected face or plane and through the selected point.

Angled from a Face or Existing Work Plane

You can create a new work plane at an angle to a face or existing work plane. The new plane will pass through an edge on the part.

For example, look at the existing and finished parts in **Figure 5-28.** Notice how the added feature, the pin, is not extruded perpendicular to the face. In order to do this, you need to create a work plane at an angle to the face. Open Example-05-08.ipt. Then, continue as follows.

1. Pick the **Work Plane** button in the **Part Features** panel.
2. Pick Face A as indicated in **Figure 5-28.**
3. Pick Edge B as indicated in **Figure 5-28.**
4. In the **Angle** dialog box that is displayed, enter an angle of 30°. The feature you are adding is angled 30° from perpendicular to the face. Then, press [Enter] or pick the check mark button to create the new work plane.

Figure 5-28.
A—The existing part. B—A feature is added to the part.

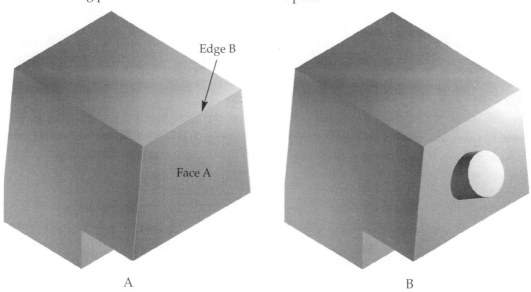

Edge B

Face A

A   B

5. Pick the **Sketch** button on the **Inventor Standard** toolbar. Pick the new work plane as the sketch plane.
6. With the **Project Geometry** tool, project the four edges of Face A onto the sketch plane. This can be done in one step by moving the cursor until all four edges are highlighted in red and then picking.
7. Draw a construction line from the midpoint of the top projected line to the midpoint of the bottom projected line.
8. Draw a circle with its center at the midpoint of the construction line, as shown in **Figure 5-29.** Dimension the circle's diameter to 30 mm.
9. "Finish" the sketch and pick the **Extrude** button in the **Part Features** panel.
10. Select the circle as the profile. Also, in the **Extrude** dialog box, select the **Join** and **To** options. Then, pick the button below the **Extents** drop-down list and select Face A.
11. Pick the **OK** button to create the extrusion. The new feature is added to the part.

Figure 5-29.
Sketching the profile that will be extruded into the feature shown in **Figure 5-28B.**

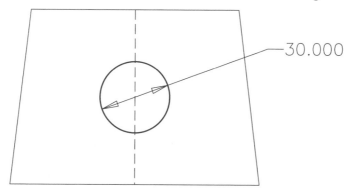

30.000

PRACTICE 5-3

- ❏ Open the file Practice-05-03.ipt.
- ❏ Offset a work plane 7.5" from the top face and make it the sketch plane.
- ❏ Project the two long edges of the part onto the sketch plane. Draw a construction line between the midpoints of the projected lines.
- ❏ "Finish" the sketch. Then, using the **Work Plane** tool, pick the offset work plane and then the construction line.
- ❏ Input –35° for the angle. See A below.
- ❏ Make the top face of the part the sketch plane. Then, draw two circles at the center of the part as shown in B below.
- ❏ "Finish" the sketch. Then, extrude the concentric circles as the profile using the **Join** and **To** options. Select the angled work plane as the "to" feature.
- ❏ Make the angled work plane the sketch plane.
- ❏ Construct a 7.5" by 10" rectangle centered on the circular feature. To center the feature, draw a construction line between the midpoints of opposite sides. Then, apply a coincident constraint between the midpoint of the construction line and the center of the circular feature.
- ❏ "Finish" the sketch and extrude the rectangle up 1.0". See C below.

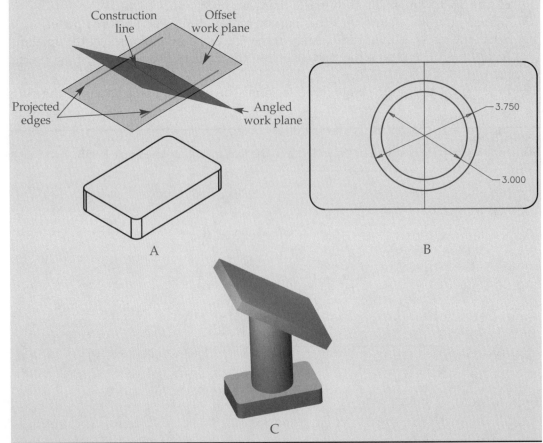

Perpendicular to a Line at Its Endpoint

To create a work plane at the end of, and perpendicular to, a line, pick the **Work Plane** button in the **Part Features** panel. Select the line and pick the endpoint at which you want the work plane. Using this method, you can create a work plane that is tangent to the curved surface of a cylinder. To do this, you must first sketch a line from the center of the cylinder to the circumference (at the intended point of tangency). See **Figure 5-30.** Then, using the **Work Plane** tool, select the line and its endpoint that lies on the cylinder's circumference.

Figure 5-30.
Creating a work plane tangent to a cylinder at a specific point.

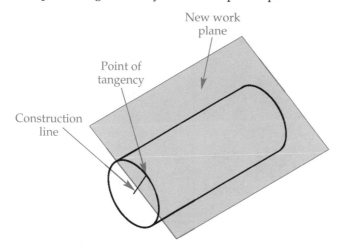

Parallel to a Face or Plane and Tangent to a Curved Face

A new work plane can be created parallel to an existing work plane or face and tangent to a curved surface. Using the **Work Plane** tool, select the planar face or work plane and the curved surface in either order. See **Figure 5-31.** Both faces must have their edge view in the same view.

Figure 5-31.
Creating a work plane parallel to a planar face and tangent to a curved face.

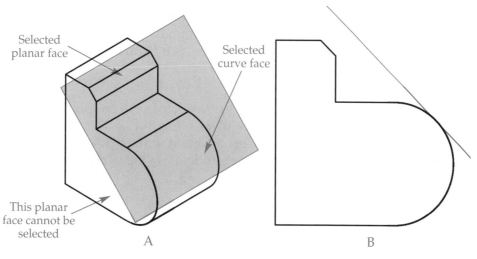

Tangent to a Curved Face through an Edge

You can create a new work plane that passes through an edge and is tangent to a curved surface. This is similar to the operation discussed in the last section. Using the **Work Plane** tool, select the curved face and the edge of a planar face in either order. See **Figure 5-32.** Note that the edge and the curved surface must appear as an edge view in the same view to create the tangent plane. The edge will appear as a point in an edge view and the curve as an arc or circle. Otherwise, the work plane will either not pass through the edge or not be tangent to the curved surface.

Figure 5-32.
Creating a work plane tangent to a curved face and through an edge.

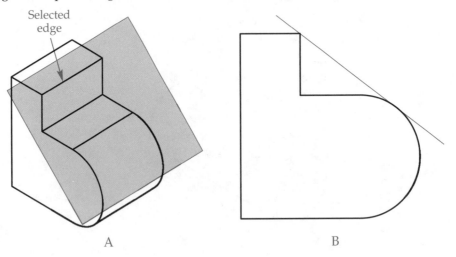

A

B

Through Two Parallel Work Axes

A *work axis* is another type of work feature. You cannot sketch on a work axis as you can a work plane. However, a work axis can be used to constrain features and create work planes. In essence, a work axis is a fixed construction line.

You can use a work axis to create a work plane. For example, you may have two cylinders and need a new work plane that passes through the center of each cylinder. To create a work axis, pick the **Work Axis** button in the **Part Features** panel. Then, pick each of the two cylinders. You will need to restart the tool after selecting the first cylinder. Work axes are displayed in a thin, yellow line similar to a construction line. Now, using the **Work Plane** tool, select the two axes in any order. A work plane is created that passes through the center of each cylinder. See **Figure 5-33.**

Perpendicular to an Axis and through a Point

A work plane can be created perpendicular to a work axis. An endpoint on a face edge is used to define the location of the work plane along the work axis. To create the work axis, use the **Work Axis** tool and select a curved surface or cylinder. Then, using the **Work Plane** tool, select the axis and the endpoint on the part in either order. See **Figure 5-34.**

Figure 5-33.
Creating a work plane through the center axes of two cylinders.

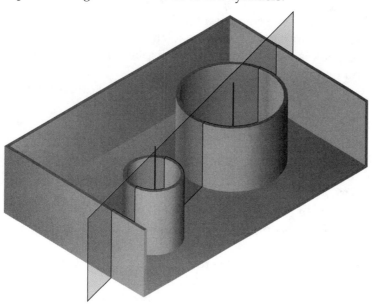

Figure 5-34.
Creating a work plane perpendicular to a work axis and through a point.

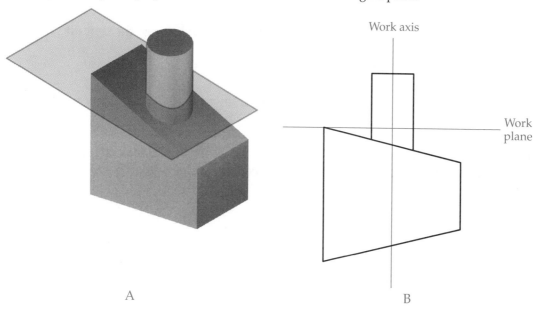

Work axis

Work
plane

A   B

Between Edges or Points

A work plane can be defined by three points that do not lie on the same line. By definition, any two points define a line. Therefore, a work plane can also be defined by an edge (line) and a point that is not on that edge. Finally, a work plane can be defined by two edges whose endpoints lie in the same plane. Using the **Work Plane** tool, select three points, an edge and a point, or two edges. The order of selection is not important. Note that selecting an edge and a point is the same as selecting a work axis and a point, as discussed in the previous section.

Chapter Test

Answer the following questions on a separate sheet of paper.

1. How can you change the color of a face?
2. Which faces of a part can be used as sketch planes?
3. Which dialog box is used to turn on the display of a CSI at the origin?
4. How do you open the dialog box in Question 3?
5. Suppose you have created an extrusion that has the default name Extrusion1. How can you change the name to Pin, Locking?
6. Explain the difference between the **Join, Cut**, and **Intersect** options of the **Extrude** tool.
7. Explain the **Distance** extents option of the **Extrude** tool.
8. Explain the **To Next** extents option of the **Extrude** tool.
9. Explain the **To** extents option of the **Extrude** tool.
10. Explain the **From To** extents option of the **Extrude** tool.
11. Explain the **All** extents option of the **Extrude** tool.
12. Define *silhouette curve.*
13. How do you project a silhouette curve onto the current sketch plane?
14. Define *work plane.*
15. How many default work planes are there? Name them.
16. Describe the relationship between the default work planes and the coordinate system.
17. Explain how to turn on the visibility of a default work plane.
18. How can you remove the display of the portion of a part that is above the current sketch plane?
19. Which tool is used to create a work plane that is *not* one of the default work planes?
20. Define *consumed sketch.*
21. Define *unconsumed sketch.*
22. Name four types of items that can be used, in part, to define a new work plane.
23. Give three applications where work planes may be used.
24. Define *work axis.*
25. Suppose you have drawn two cylinders. How can you create a sketch on a plane that passes through the center of each cylinder?

Chapter Exercises

Exercise 5-1. Open the file Exercise-05-01.ipt. Construct the boss as shown below. Use the appropriate constraints. Dimension the hole with the equation 0.8 * d1, where d1 is the 3.50″ width of the part. Notice that the hole passes through the entire part. Change the color of the extruded boss to Sea Green.

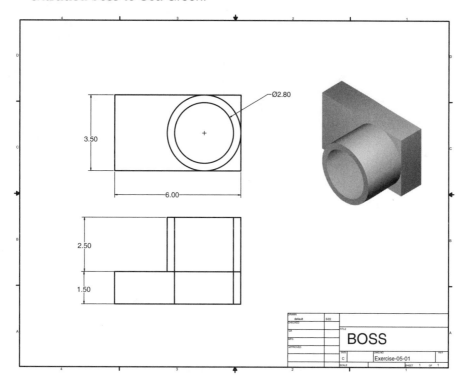

Exercise 5-2. Start a new English Standard (in).ipt file. Construct the bracket shown below. Use a construction line to locate the two small circles. Use the equal constraint to center the keyway and size the second small circle.

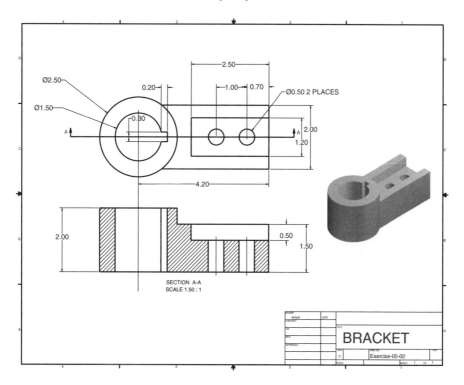

Exercise 5-3. Open the file Exercise-05-03.ipt. Construct three cutouts as shown below. To center the rectangular cutout, use horizontal construction lines and the equal constraint.

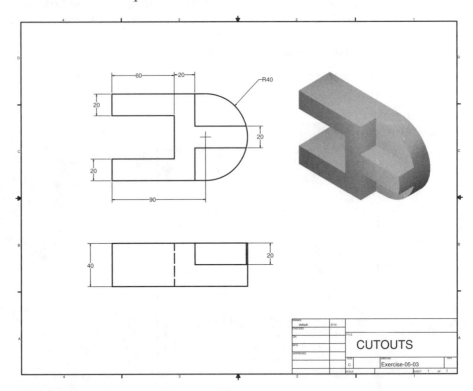

Exercise 5-4. Start a new metric Standard (mm).ipt file. Construct the tapered part shown below using three sketches. Fully constrain the three sketches with a total of two dimensions and tangent and equal constraints. Extrude the sketches, entering an appropriate taper angle to produce the 10° taper shown.

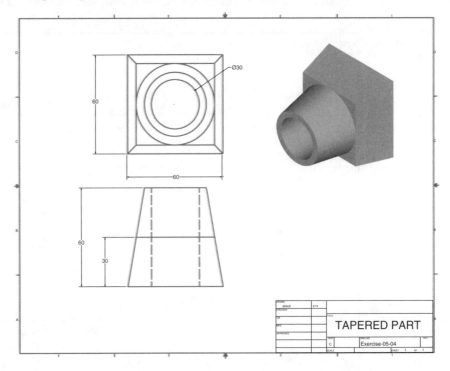

Exercise 5-5. Start a new English Standard (in).ipt file. Construct the mirror part shown below. Apply an equal constraint to the two horizontal lines on the top edge to center the top ear. Draw a horizontal centerline from the midpoints of the sides to mirror the ear to the bottom. Extrude the profile 1″ into a solid part.

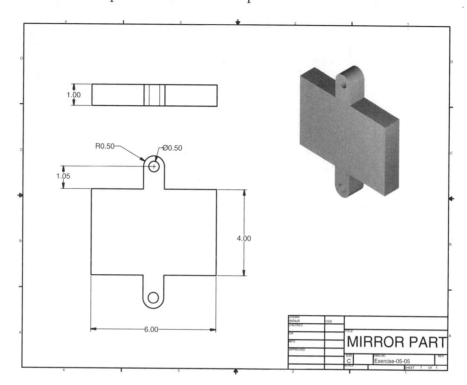

Exercise 5-6. Open the file Exercise-05-06.ipt. This is a hose nozzle stem. Construct the 5 mm diameter hole through the part as shown below.

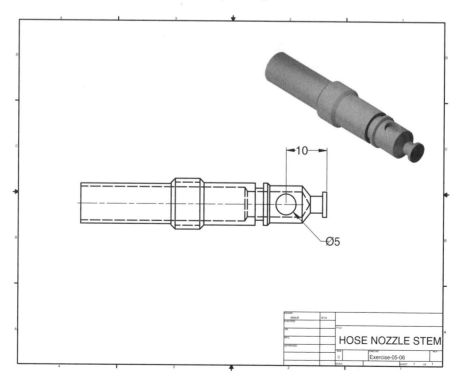

Exercise 5-7. Open the file Exercise-05-07.ipt. Construct the two holes and the shaft shown below.

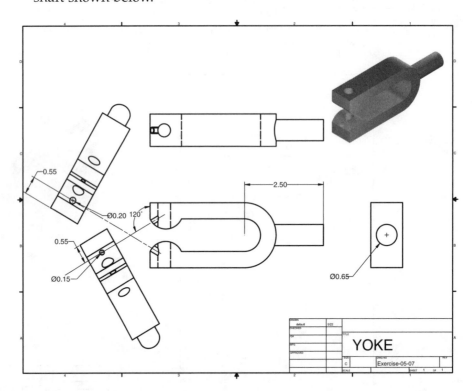

Exercise 5-8. Start a new English Standard (in).ipt file. Construct the brace shown below. Use work planes to define the extents of the first extrusion (the middle).

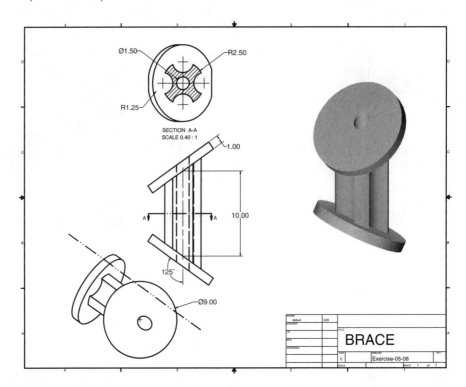

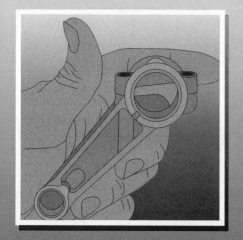

CHAPTER 6

Adding Features

Objectives

After completing this chapter, you will be able to:

* Add holes to a part.
* Add threads to a part.
* Fillet edges on a part.
* Chamfer edges on a part.
* Create rectangular and circular patterns.
* Mirror features.

User's Files

*The following is a list of files that you will need to work through this chapter. These files can be found on the **User's Files** CD included with this text.*

Examples		Exercises
Example-06-01.ipt	Example-06-10.ipt	Exercise-06-01.ipt
Example-06-02.ipt	Example-06-11.ipt	Exercise-06-02.ipt
Example-06-03.ipt	Example-06-12.ipt	Exercise-06-03.ipt
Example-06-04.ipt	Example-06-13.ipt	Exercise-06-04.ipt
Example-06-05.ipt	Example-06-14.ipt	Exercise-06-05.ipt
Example-06-06.ipt	Example-06-15.ipt	Exercise-06-06.ipt
Example-06-07.ipt	Example-06-16.ipt	Exercise-06-07.ipt
Example-06-08.ipt		Exercise-06-08.ipt
Example-06-09.ipt		Exercise-06-09.ipt

Adding Nonsketch Features to the Part

There are several features that can be added to a part without drawing a sketch profile. The tools for creating these features are found in the **Part Features** panel. For some, like the **Hole** tool, you will need to use a sketch plane. For the **Mirror Features** tool, a work plane is required. However, for other tools, like the **Fillet** tool, you need only select the edge or edges to be modified. In this chapter, seven of these feature tools are discussed—**Hole**, **Thread**, **Fillet**, **Chamfer**, **Mirror Feature**, **Rectangular Pattern**, and **Circular Pattern**. The rest of the feature tools are discussed in the next chapter.

So far, you have created holes as cutouts by extruding circles. However, the **Hole** tool has several advantages over extruded circles. Refer to **Figure 6-1** and the list below.

- The holes can be drilled, counterbored, or countersunk.
- Thread sizes and specifications can be easily applied.
- Threads will be displayed in the shaded image.
- Threaded holes will display correctly in the part drawings.
- "Hole Notes" can be automatically applied to the part drawings.
- "Hole Tables" can be created in the part drawings.

Figure 6-1.
These holes were created using the **Hole** tool. The part drawing appears correctly and tables can be created when done in this manner.

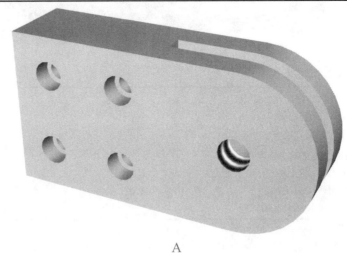

A

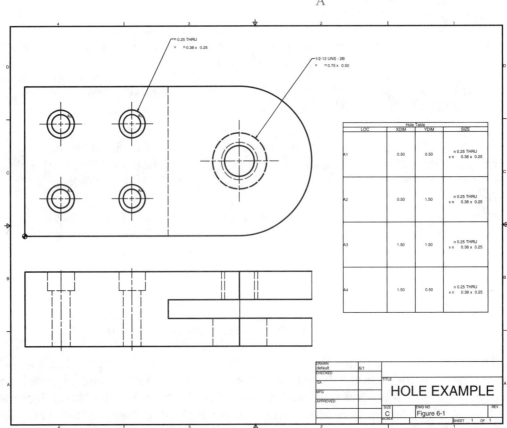

B

To use the **Hole** tool, you must draw special points, called *hole centers,* on a sketch plane. The **Point, Hole Center** tool is used to create these special points. The **Hole** tool is then used to specify the size and type of hole created on the hole center. By drawing multiple hole centers, multiple holes can be created in one session of the **Hole** tool. All the holes created in one tool session have the same size and characteristics.

The **Point, Hole Center** tool can create hole centers or sketch points; this choice is controlled by the next-to-last tool on the **Standard** toolbar called the **Hole Center** button. When this button is not selected, sketch points are created from the **Point, Hole Center** tool. When the **Hole Center** button is depressed, cross-hair shaped hole centers are created.

Open Example-06-01.ipt. Pick the **Sketch** button on the **Inventor Standard** toolbar and select the large, front face as the sketch plane. Next, pick the **Point, Hole Center** button in the **2D Sketch Panel**. Place two points anywhere on the face by picking at two locations. Apply a vertical constraint between the hole centers. This will keep them in a line parallel to the Y axis. Now, dimension the hole centers as shown in **Figure 6-2**. Then, "finish" the sketch.

In the **Part Features** panel, pick the **Hole** button or press the [H] key to activate the **Hole** tool. The hole centers in the sketch are automatically selected and the **Holes** dialog box is displayed. Also, a preview of the default settings is displayed at each hole. To remove a hole center from the operation, pick the **Centers** button in the **Holes** dialog box. Then, hold down the [Shift] key and pick the hole center to deselect it. Using the same procedure, you can add a hole center to the selection set.

Figure 6-2.
Placing hole centers
to create holes.

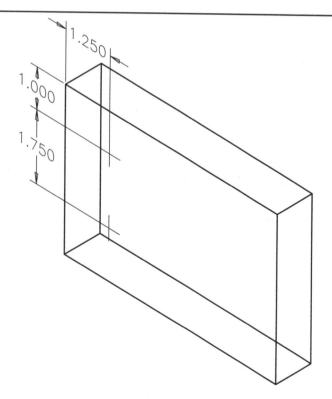

Figure 6-3.
The **Holes** dialog box is used to create holes.

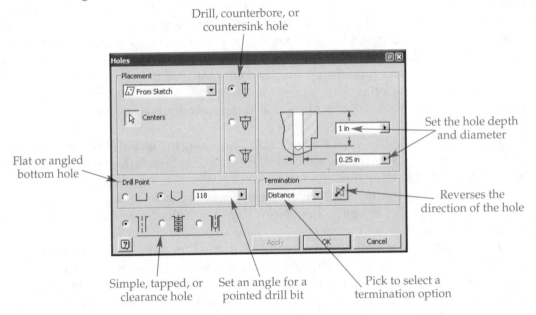

There are three options in the **Termination** drop-down list—**Distance**, **Through All**, and **To**, **Figure 6-3**. These are similar to the extents options for extruded profiles. **Distance** specifies a specific hole depth. The value is set by changing the dimension in the input box next to the image tile. The diameter is set by changing the dimension in the second input box. For this example, select the **Distance** option and change both the depth and diameter to .50".

Holes have a direction. The direction can be reversed or "flipped" by picking the button next to the **Termination** drop-down list. If the hole is "outside" of the part, an error is generated by Inventor. Holes must be to the interior of a part. There is no midplane option.

The settings in the **Drill Point** area of the **Holes** dialog box control how the bottom of the hole is created. You can set the angle of a pointed drill bit. The angle does *not* show in the image tile, but is shown correctly on the part when the hole is created. You can also select a flat-bottom hole, as is created by an end mill. For this example, pick the **Angle** radio button and use the default angle value.

Now, pick the **OK** button in the **Holes** dialog box to create the two holes. Using the **Rotate** tool, rotate the view to see the holes. Notice how they do not pass through the part. Change to a wireframe display. Then, rotate the view so you can see the side of the holes. Notice how the bottom is tapered according to the angle set in the **Holes** dialog box.

Threaded Holes

Adding threads to the holes can be done when the holes are created or later by editing the feature. Redisplay the isometric view. Also, shade the view. Now, right-click on the hole feature in the **Browser**. Select **Edit Feature** from the pop-up menu. The **Holes** dialog box is displayed.

Select **Through All** from the **Termination** drop-down list. This will change the holes so they pass completely through the part. Notice how the setting is reflected in the image tile. The three choices at the bottom of the dialog box are: **Simple Hole**, **Tapped Hole**, and **Clearance Hole**. Select **Tapped Hole** and the **Threads** area will appear in the dialog box. See **Figure 6-4**.

Learning Inventor

Figure 6-4.
Adding threads to a hole.

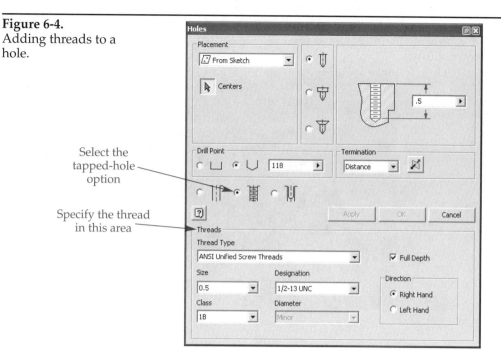

Select the tapped-hole option

Specify the thread in this area

Next, check the **Full Depth** check box. This check box specifies that the hole is threaded to its full depth. When unchecked, a dimension input box appears next to the image tile that specifies the depth of thread. The value can be changed in this box.

In the **Thread Type** drop-down list, select ANSI Unified Screw Threads. Since the hole was created at 0.50" diameter, the **Size** drop-down list displays 0.5 and the **Designation** drop-down list displays the standard pitches for ½" ANSI standards. Select ½-20 UNF for a fine thread. Leave the **Class** of the thread at 1B and pick **OK**.

All of the data required to properly size threads are located in a spreadsheet file called Thread.xls. Accessing and modifying this spreadsheet is beyond the scope of this book however, there is information in Inventor's help application that may be useful. The only reason to modify this spreadsheet is if a custom thread is required.

A representation of the threads appears in each hole. This representation is a bitmap, or picture, of the correct thread. See **Figure 6-5**. The bitmap display is based on data in the Thread.xls spreadsheet. The bitmap is a clever technique for displaying threads and has little effect on the file size. Showing actual cut threads (geometry) in the part can be done, as discussed in *Chapter 13*, but it results in very large files.

Clearance Holes

The third hole choice below the **Drill Point** area is **Clearance Hole**. See **Figure 6-6**. The size of this hole is determined by the fastener that will be located in the hole. Edit the same hole feature that you modified in the thread section. Once the **Holes** dialog box is visible, pick the radio button next to the clearance hole icon. The area below this button will change from **Threads** to **Fastener**. Most clearance holes are made to allow a fastener to go through a part and thread into an adjacent part, therefore, change the **Termination** to **Through All**. Now, set the **Standard** drop-down list to Ansi Unified Screw Threads, the **Fastener Type** to Socket Head Cap Screw, the **Size** to ½ inch, and the **Fit** to Normal. The hole diameter in the upper-right area of the **Holes** dialog box automatically changed to 0.531 inches. This is the proper hole size to accommodate the selected fastener.

Figure 6-5.
Threads are represented with a picture. Creating the correct model geometry increases the file size tremendously.

Figure 6-6.
Clearance holes are large enough for a fastener to pass through.

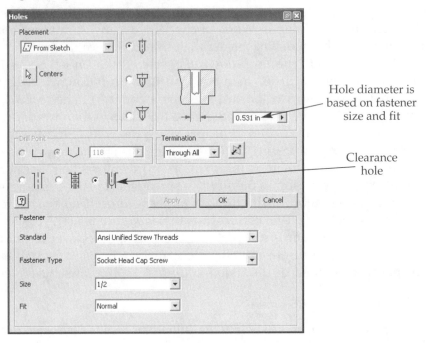

Hole diameter is based on fastener size and fit

Clearance hole

Counterbored and Countersunk Holes

You have seen drilled holes that are simple, threaded, and clearance. These three hole types can also be applied to counterbored and countersunk holes. Edit the same hole feature that you modified in the clearance holes section. At the top of the **Holes** dialog box is a column of three hole types; **Drilled**, **Counterbore**, and **Countersink**. Pick the button next to the **Counterbore** icon and review the dimensions on the right side of the dialog box. See **Figure 6-7**. These dimensions are calculated automatically based on the fastener information at the bottom of the dialog box. Change the **Fastener Type** to Hex Head Bolt and notice the counterbore dimensions change to make room for the wider bolt head.

Figure 6-7.
A counterbore clearance hole usually allows the head of the fastener to sit below the part surface.

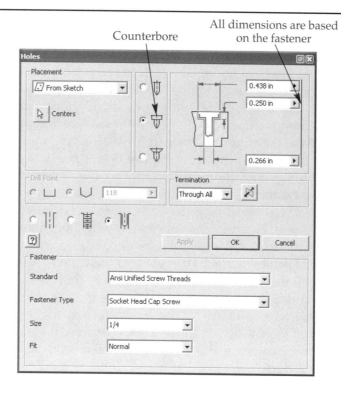

Figure 6-8.
Inventor assumes that a flat head or oval head fastener will be used if you are drawing a countersink hole.

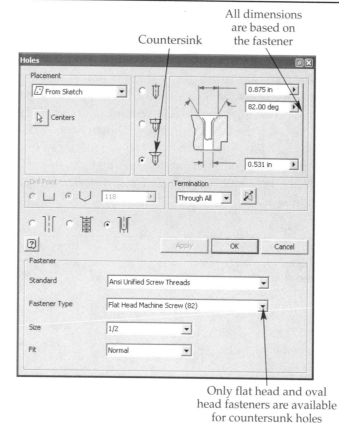

Pick the button next to the **Countersink** icon and review the dimensions. Now look at the fastener information at the bottom of the dialog box. See **Figure 6-8**. The **Fastener Type** changed to either a flat or an oval head style because these are the fasteners that are used with a countersunk hole.

The Thread Tool

A thread can be applied to cylindrical features, not just holes. The features can be external, such as a cylinder. The features can also be internal, such as cutouts created by extruding circles or revolving rectangular shapes. The **Thread** tool is used to do this.

Open Example-06-02.ipt, which is a small metric part. In the **Part Features** panel, pick the **Thread** button to activate the **Thread** tool and open the **Thread Feature** dialog box. With the **Face** button active in the dialog box, move the cursor over the outside diameter at Point A, shown in **Figure 6-9.** A preview of the thread appears on the surface. Notice that the thread is applied over the entire length of that surface. In the **Thread Feature** dialog box, uncheck the **Full Length** check box. Also, type 5 in the **Length** text box. Now, pick the **Face** button and move the cursor over the outside diameter again. The preview appears on only a portion of the face starting at the end nearest to the cursor. The length is 5 mm, as you entered in the dialog box. Now, in the **Thread Feature** dialog box, type 10 in the **Offset** text box. Pick the **Face** button and again move the cursor over the face. Notice that the thread preview begins 10 mm from the end nearest to the cursor. Change the offset to 0 and the length to 15. Then, pick the surface at Point A. The entire face is highlighted in an olive-drab color. Finally, pick the **OK** button in the **Thread Feature** dialog box to create the threads.

You can change the thread specification as you create the thread. However, since you have already created the threads, right-click on the thread name in the **Browser** and select **Edit Feature** from the pop-up menu. The **Thread Feature** dialog box is displayed. Pick the **Specification** tab, **Figure 6-10.** In this tab, there are options for setting the thread type, nominal size, pitch, class, and left- or right-hand threads. Use the drop-down lists to make the appropriate settings. Use the radio buttons to choose either left- or right-hand threads. However, be careful when changing the nominal size. This is a 12 mm diameter shaft and there are only two sizes of thread that can be cut on this shaft—12 mm and 14 mm. If you select any other size and pick the **OK** button, an error message is displayed, **Figure 6-11.** Pick the **Edit** button in the warning dialog box and change the thread specification as needed. When done changing the specifications, pick the **OK** button in the **Thread Feature** dialog box to update the threads.

Now, rotate the view of the part to see the large diameter bore. Place 8 mm deep threads on the bore that starts at the outside face. In the **Specification** tab of the **Thread Feature** dialog box, select specifications of your choice. Then, pick the **OK** button to place the threads.

Figure 6-9.
Threads will be
added to the shaft
on this object.

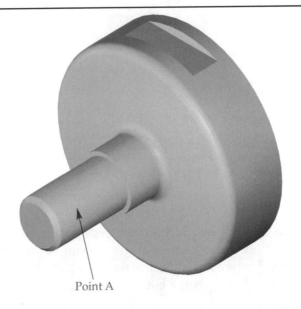

Point A

Figure 6-10.
Specifying threads using the **Thread** tool.

Select the type

Select the size

Select the class

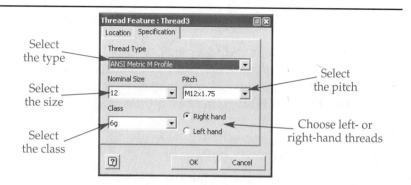

Select the pitch

Choose left- or right-hand threads

Figure 6-11.
An error is generated if you select a thread size that is not valid for the hole or shaft diameter.

The bitmap representations of the threads approximate the coarseness of the thread. Left- and right-hand threads are also represented by the bitmaps. However, the threads will be properly displayed in the part drawing in the correct ANSI standard. See **Figure 6-12.** Creating part drawings is discussed in *Chapter 8*.

Figure 6-12.
Threads are represented in Inventor as a bitmap, however, the part drawing conforms to standards for representing threads.

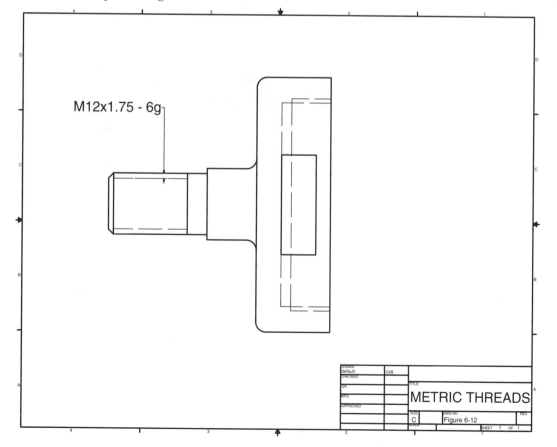

The **Fillet** tool puts a radius on any or all edges of a part, including intersections of part features. A radius on an inside edge adds material to the part and is called a *fillet.* A radius on an outside corner removes material and is called a *round.* See **Figure 6-13.** The terminology may be a little confusing as the **Fillet** tool creates both fillets and rounds.

Applying Fillets and Rounds

Open Example-06-03.ipt. In the **Part Features** panel, pick the **Fillet** button to activate the **Fillet** tool and display the **Fillet** dialog box, **Figure 6-14.** The **Fillet** tool allows you to set several radii in the **Fillet** dialog box. You can change the fillet radius of an existing setting by selecting the value in the **Radius** column, typing the new value, and pressing [Enter]. To set a new fillet radius, pick the Click here to add entry. A new row is added and you can set the fillet radius. In this way, you can apply a variety of fillets and rounds with the same session of the **Fillet** tool. In this example, there is currently a .125" fillet radius setting. Change this setting to .25". Then, add another fillet radius setting and change the fillet radius to .5".

Notice that the **Edge** radio button in the **Select mode** area is on. This mode allows you to select a single edge. Also, notice in the **Edges** column that the two fillet settings all list 0 Selected. Currently, nothing is selected for any of the fillet settings. Pick the 0 Selected in the **Edges** column for the .5" fillet setting. Then, pick the four short edges of the rectangular base. See **Figure 6-15.** As you select edges, red arcs representing the round that will be applied are displayed at each edge.

Figure 6-13.
Fillets and rounds.

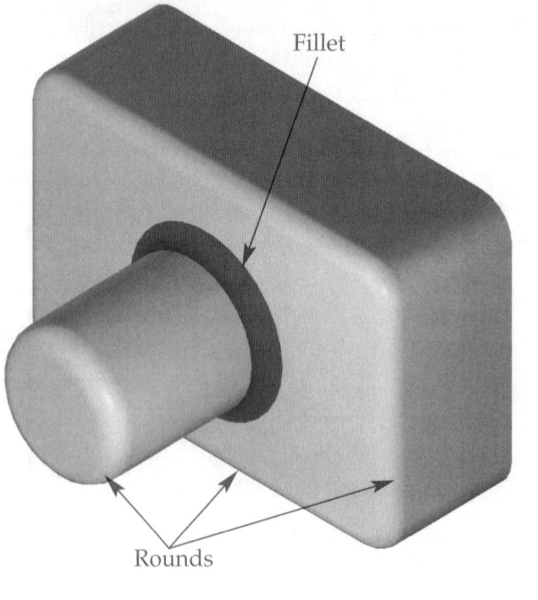

Figure 6-14.
Making settings for fillets and rounds.

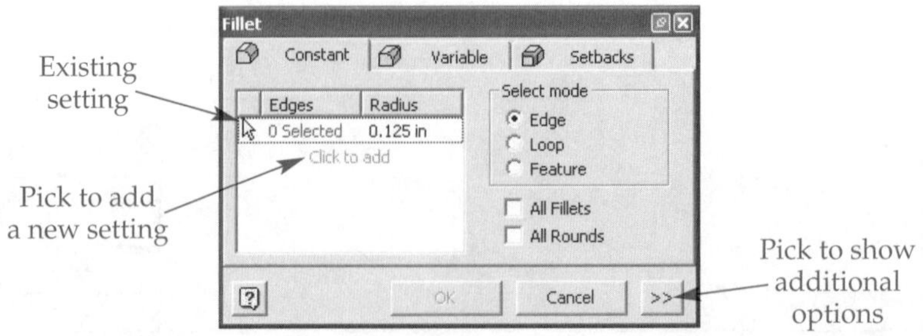

Figure 6-15.
A—Pick this edge for the .25″ rounds. B—The completed part. C—The fillet dialog box.

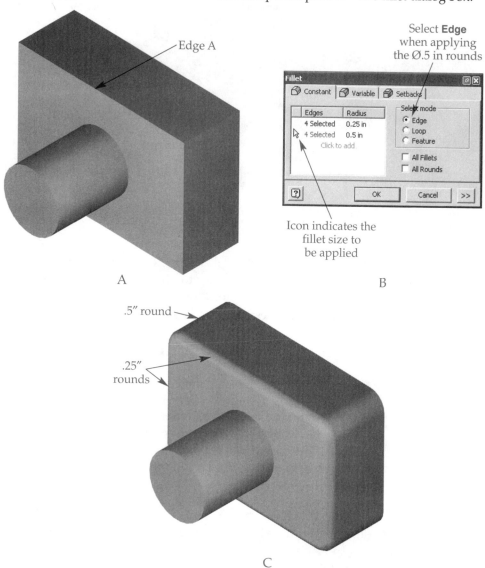

Now, select the .25″ fillet radius setting in the **Fillet** dialog box and pick the **Loop** radio button in the select mode area. Hover the cursor over Edge A as indicated in **Figure 6-15A**. Notice that as you move the cursor, two different loops of edges are highlighted. Pick the edge when the larger front face is highlighted. Select the 0.5″ radius setting and change the **Select mode** to **Edge**. Pick the four short edges of the part; try to pick the hidden edge without rotating the part. After the edges are picked, look at the dialog box and verify the number of edges selected, **Figure 6-15B**. Pick **OK** and compare your part with **Figure 6-15C**.

To see how a "loop" selection around sharp corners works, edit the fillet feature you just created. In the **Fillet** dialog box, pick the 0.5″ radius. Now, hold down the [Shift] key and pick the four short edges of the part. Pick **OK** to see the result.

The **Feature** radio button in the **Fillet** dialog box allows you to select a feature. Fillets and rounds will be applied to all edges on the feature, but not at intersections with other features. For example, open the **Fillet** dialog box. Add a 1″ radius fillet setting. Pick the **Feature** radio button in the **Select mode** area. Move the cursor over the part. When the cursor is over a feature, the edges of the feature that will be filleted or rounded are highlighted in red. Move the cursor over the post; only the circular edge on the left end is highlighted. This is because the edge on the other end is an intersection between features. Pick the post to select the feature. Then, pick the **OK**

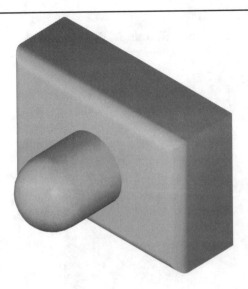

Figure 6-16.
A bullet nose round is applied to the shaft.

button in the **Fillet** dialog box to create the round. See **Figure 6-16.** Notice how the result is a spherical end on the cylinder. By setting the fillet radius equal to the radius of the cylinder, you can create a rounded pin. However, the fillet radius cannot be greater than the radius of the cylinder or an error is generated.

PROFESSIONAL TIP You can change the display color of fillets and rounds for emphasis. This is done by right-clicking on the fillet name in the **Browser** and selecting **Properties** from the pop-up menu. Then, change the color in the **Feature Properties** dialog box. You can also rename the fillet in this dialog box.

Filleting and Rounding All Edges

You can have Inventor automatically apply fillets and rounds to all edges of the part. Open Example-06-04.ipt. Activate the **Fillet** tool. In the **Fillet** dialog box, change the fillet radius setting to .10″. Then, check the **All Fillets** check box. There are four inside edges on the part, so these are automatically selected. The fillets are also previewed in red on the part. Pick the **OK** button to place the fillets. See **Figure 6-17A.** Repeat the operation, this time checking the **All Rounds** check box. There are 33 external edges on the part, so these are automatically selected. The rounds are previewed on the part. Pick the **OK** button to place the rounds. See **Figure 6-17B.**

Now, there are two separate fillet operations listed in the **Browser.** Also, notice how all edges have been filleted and rounded, including the edges of the holes. However, holes are not normally filleted or rounded. This can be easily corrected by moving the fillet operations above the hole operations in the **Browser.** Pick one of the fillet names in the **Browser,** hold down the [Ctrl] key, and pick the other name. Both names are now selected. Drag the names up so they are above the Hole1 and Hole2 features in the part tree and release. See **Figure 6-18A.** Now, the two fillet operations are not applied to the holes, **Figure 6-18B.**

It is possible to check both **All Fillets** and **All Rounds** in the same operation. However, in this example, an error will result. The error message that is displayed refers to the inability of Inventor to perform operations with mixed convexity at a vertex. The two corners where the base meets the web will have both a concave fillet and convex rounds applied. Inventor cannot calculate the result in one operation, which is why fillets and rounds were applied in two steps for this example.

Figure 6-17.
A—Fillets are added using the **All Fillets** check box. B—Rounds are added using the **All Rounds** check box.

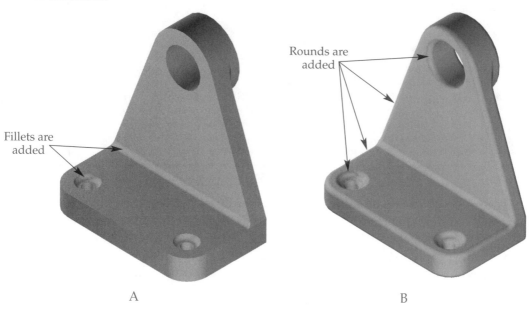

A   B

Figure 6-18.
A—Moving the fillets and rounds above the holes in the part tree. B—The holes are no longer filleted or rounded.

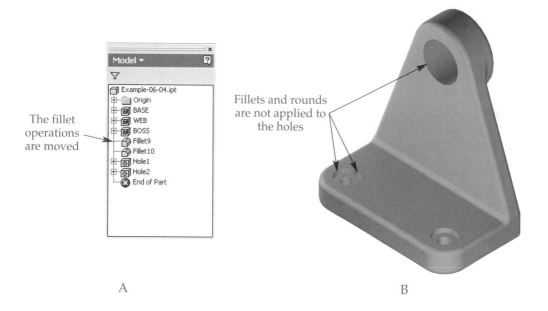

A   B

More Fillet Tool Options

You may have noticed the **>>** button at the bottom of the **Fillet** dialog box. Refer to **Figure 6-14.** Picking this button displays an expanded dialog box. See **Figure 6-19.** There are four options in the expanded dialog box—**Roll along sharp edges**, **Rolling ball where possible**, **Automatic Edge Chain**, and **Preserve All Features**.

The **Automatic Edge Chain** check box is normally checked. This results in all edges in a chain being selected by picking any edge in the chain. When unchecked, you can pick individual edges in the chain.

Figure 6-19.
The expanded **Fillet** dialog box.

Pick to hide the additional options

Additional options

When the **Roll along sharp edges** check box is unchecked, a constant fillet radius is applied along the edge, even if this means extending adjacent faces. For example, open Example-06-05.ipt. Put a .75″ fillet on the intersection where the cylinder meets the base. See **Figure 6-20.** Notice how Face C is extended to keep the constant radius. This is obvious in the end view. Now, edit the fillet feature and check the **Roll along sharp edges** check box. Notice how the fillet radius is now varied to keep the edge along Face C a straight line. See **Figure 6-21.** This is done by varying the radius of the fillet.

The **Rolling ball where possible** check box controls the intersection of inside edges. Open Example-06-06.ipt. Place a .75″ radius fillet on the three visible edges inside the box. Make sure this option is checked. The result of the operation at the inside corner looks as if a ball end mill machined the cavity. See **Figure 6-22A.** Now, edit the fillet feature and uncheck the option. The result of the operation at the inside corner is now a smooth blend. See **Figure 6-22B.**

When the **Preserve All Features** check box is checked, some features, such as holes, are retained when they intersect the fillet. Open Example-06-07.ipt. Place a 1″ radius round (fillet) on the short edge of the top by the hole and the post. Make sure the **Preserve All Features** check box is unchecked. When the round is applied, the hole and post disappear. See **Figure 6-23A.** In some cases, you will get an error message. Now, edit the fillet feature and check the **Preserve All Features** check box. When you pick **OK** to update the feature, the hole reappears, but not the post. See **Figure 6-23B.**

Figure 6-20.
A—Face C is extended when the **Roll along sharp edges** check box is unchecked. B—The end view of the part clearly shows how the face is extended.

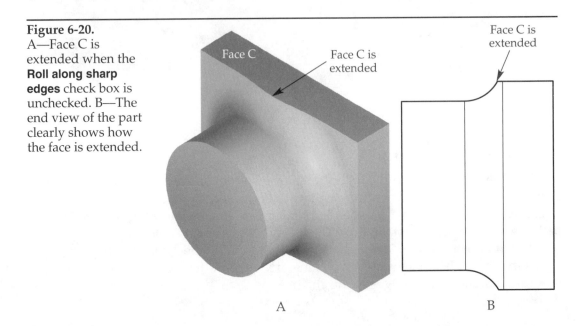

Face C

Face C is extended

Face C is extended

A

B

Figure 6-21.
A—Face C is not extended when the **Roll along sharp edges** check box is checked. B—The end view of the part clearly shows how the face is not extended.

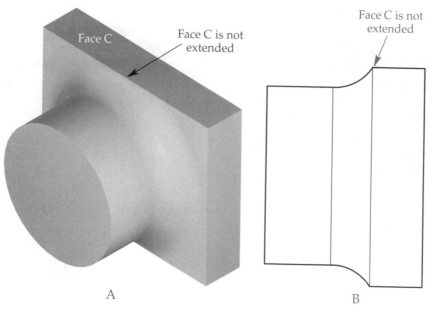

A B

Figure 6-22.
A—The intersection of these three fillets appears as if a ball end mill was used to create the cavity. B—Unchecking the **Rolling ball where possible** check box smoothes the intersection.

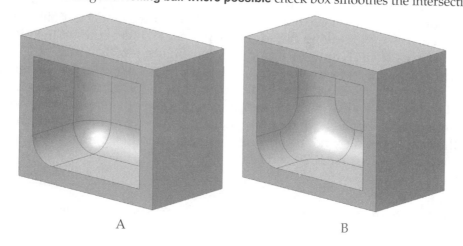

A B

Variable-Radius Fillets and Setbacks

In some applications, the radius of a fillet must vary over the length of the fillet. This is called a *variable-radius fillet.* Open Example-06-08.ipt Activate the **Fillet** tool and pick the **Variable** tab in the **Fillet** dialog box. Then, pick the long, front edge of the top face. The two endpoints are automatically selected as the Start and End points in the **Variable** tab. The point highlighted in the **Point** list in the tab is also highlighted on the part. The radius of the fillet at the highlighted point is given in the **Radius** text box in the tab. You can change the value by picking in the text box and typing a new value. Do not press [Enter], because this will apply the fillet. You will apply the fillet later. Set the radius of the Start point to .1″ and the End point to 1.0″.

Figure 6-23.
A—When the **Preserve All Features** check box is unchecked, a fillet or round may remove some features. B—When the check box is checked, some features are retained. However, some features may still be removed by the fillet or round.

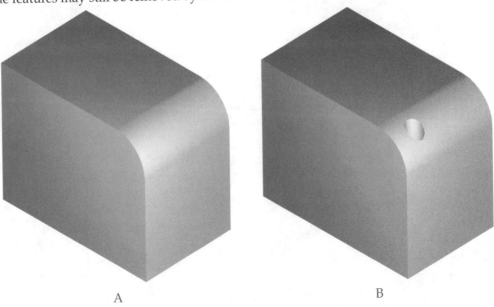

A   B

This setting creates a smooth, variable-radius round (fillet) between the two endpoints. See **Figure 6-24A.** However, you can have multiple points along the length of the fillet, each with a different radius. To add points between the endpoints, simply move the cursor along the edge and pick where you want the points. For this example, add two points along the edge. Change the radius values of the two points so the fillet goes from .1, to .2, to .5, and to 1. See **Figure 6-24B.**

Once any points between the Start and End points are picked, you can change their location along the fillet. Highlight the point in the **Point** list. Then, pick in the **Position** text box and type a value. The value in this text box represents a percentage along the length of the fillet, where 1.00 equals 100%. You can move a point anywhere along the length, but it cannot be outside of the Start and End points, nor can it share the same position as either (1.00 and 0.00).

Figure 6-24.
The radii are given at the transitions. A—This is a variable-radius round (fillet). B—This variable-radius round has multiple transitions.

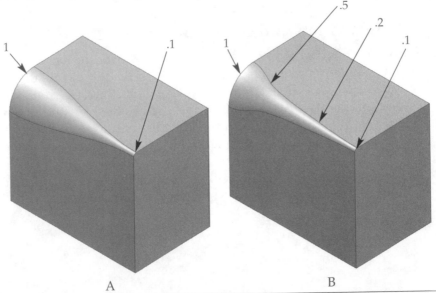

A   B

A *setback* is the distance from the corner where three fillets or rounds meet. It is measured from the point where the edges would meet to form a square corner. See **Figure 6-25.** By increasing the setback, the corner is smoothed by blending the fillets together. Open Example-06-09.ipt. The block has a .5" fillet with no setback on three intersecting edges. See **Figure 6-26A.** Edit the fillet feature and select the **Setback** tab in the **Fillet** dialog box. As you move the cursor over the part, a yellow circle will appear on valid vertices. Since there is only one vertex where fillets intersect, there is only one valid vertex. Pick the corner where the three fillets intersect. In the **Setback** tab, three edges are listed in the **Edge** column, each with a default setback setting of .25". Pick in the **Setback** column for each edge and change the value to 2". See **Figure 6-27.** Then, pick the **OK** button to update the rounds (fillets). The corner is smoothed, as shown in **Figure 6-26B.**

Figure 6-25.
The setback is the distance from the theoretical sharp corner to the radius. A—No setback. B—1.00 setback.

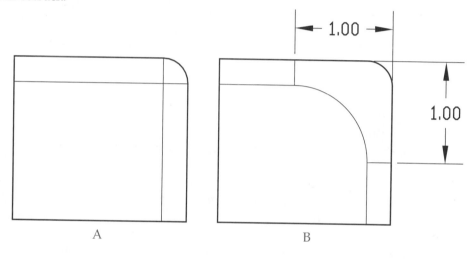

A   B

Figure 6-26.
A—The intersection of these three rounds is created without setbacks. B—Setbacks allow you to smooth the corner.

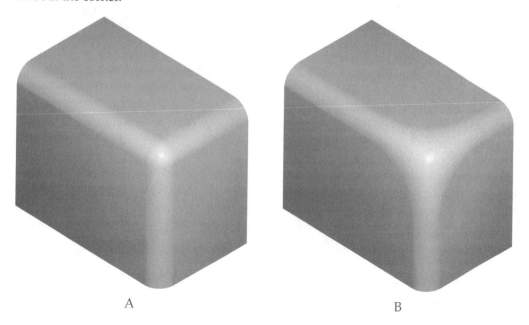

A   B

Figure 6-27.
Adding setbacks to the intersection of three fillets or rounds.

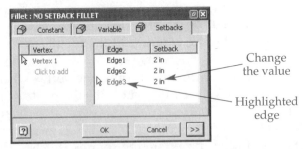

Change the value

Highlighted edge

Adding Chamfers

A *chamfer* is a bevel—a flat sloping face—on the edge between two intersecting faces of the part. The faces can be flat or curved. Like fillets and rounds, chamfers add material to inside edges and remove material from outside edges. **Figure 6-28** shows chamfers applied to three types of edges.

To create a chamfer, pick the **Chamfer** button in the **Part Features** panel. The **Chamfer** dialog box is displayed. There are three methods for defining the chamfer—two equal distances, a distance and an angle, and two unequal distances. The three buttons on the left side of the **Chamfer** dialog box allow you to choose between these methods. See **Figure 6-29.**

- **Distance button.** The same distance from the edge is used on each face. This creates a 45° chamfer. With this option, specify the distance and pick the edge to chamfer.
- **Distance and Angle button.** A chamfer is created a specified distance from the edge on one face and at a specified angle to the second face. With this option, specify the distance and angle, then pick the face from which to measure the angle and the edge to chamfer. The distance is applied to the selected face.
- **Two Distances button.** Allows a different distance from the edge on each face. With this option, specify the two distances and pick the edge. Inventor automatically chooses the faces to which the distances are applied, but you can flip the order.

Figure 6-28.
A—The unchamfered part. B—Three different chamfers are applied to the part.

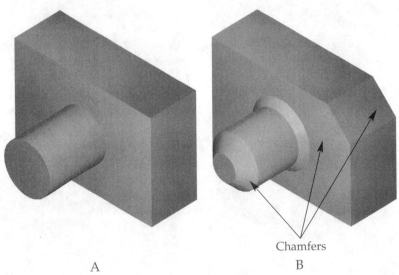

Chamfers

A   B

Figure 6-29.
There are three methods for defining a chamfer. A—**Distance**. B—**Distance and Angle**.
C—**Two Distances**.

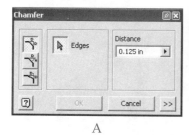

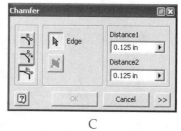

A	B	C

Open Example-06-10.ipt. The small knob has a circular pattern creating a fine knurl. This feature has been suppressed to reduce the time needed to create the chamfers. Pick the **Chamfer** button in the **Part Features** panel. Then, continue as follows.

1. In the **Chamfer** dialog box, pick the **Distance** button.
2. Pick in the **Distance** text box and change the distance to .5".
3. Pick the **Edges** button. Then, pick the edge at the end of the large cylinder (not the intersection with the base). A preview of the chamfer appears on the part in red.
4. In the **Chamfer** dialog box, pick the **OK** button to place the chamfer.
5. Pick the **Chamfer** button in the **Part Features** panel. Note: You can apply multiple chamfers of the same type in one session of the **Chamfer** tool, but you cannot apply multiple chamfers of different types.
6. In the **Chamfer** dialog box, pick the **Distance and Angle** button.
7. Set the distance to .3" and the angle to 30°.
8. Pick the **Face** button in the **Chamfer** dialog box. Then, pick the face on the base that intersects the large cylinder.
9. The **Edge** button in the **Chamfer** dialog box should be automatically selected. If not, pick it. Then, pick the intersecting edge between the cylinder and the base. A preview of the chamfer appears on the part in red.
10. In the **Chamfer** dialog box, pick the **OK** button to place the chamfer.
11. Pick the **Chamfer** button in the **Part Features** panel.
12. In the **Chamfer** dialog box, pick the **Two Distances** button.
13. Change the value in the **Distance1** text box to .5" and the value in the **Distance2** text box to .75".
14. Pick the **Edge** button in the **Chamfer** dialog box and then pick the short edge just above the small cylinder.
15. In the **Chamfer** dialog box, pick the flip button so that the .75" distance is on the vertical face.
16. Pick the **OK** button to place the chamfer. Note how the knob extends back into the chamfer. See **Figure 6-30A**.

Now, edit the last chamfer by right-clicking on its name in the **Browser** and selecting **Properties...** from the pop-up menu. In the **Chamfer** dialog box, change the .75" distance to 1.5". This will extend the chamfer down below the knob. Pick the **OK** button to update the feature. The knob disappears. See **Figure 6-30B**.

When using the **Chamfer** tool, you can select a chain or loop of edges. However, unlike fillets, they cannot have sharp corners. Also, the corner radii must be tangent to the straight edges. Pick the **>>** button in the **Chamfer** dialog box to show the expanded options. There are two buttons in the **Edge Chain** area that determine if chains or single edges are selected. Chain is the default. See **Figure 6-31**.

Open Example-06-11.ipt. See **Figure 6-32A**. Face A has fillets on each corner; in other words, the face has no sharp corners. Face B, on the other hand, has three sharp corners and one smooth corner. Activate the **Chamfer** tool and expand the dialog box. Make sure the chain button is on. The tooltip for this button is All tangentially connected

Figure 6-30.
A—The chamfer is placed and the knob is retained. B—Increasing the chamfer distance on the vertical face removes the knob.

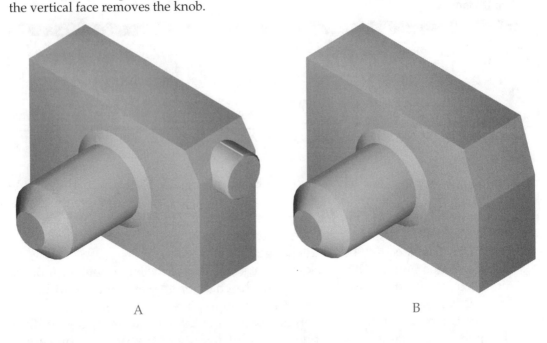

A   B

Figure 6-31.
You can determine if selecting an edge selects the entire chain or just the edge.

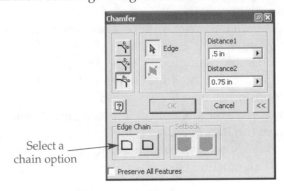

Select a chain option

edges. Then, pick the **Distance** button and set the distance to 5 mm. Pick the **Edges** button in the dialog box and move the cursor over Face A and Face B. Notice how all edges of Face A are highlighted in red. However, as you move the cursor over Face B, the two edges that are connected by the fillet, and the other two edges, are highlighted independently. In either case, if the single edge button in the **Edge Chain** area of the **Chamfer** dialog box is on, only a single edge is selected. Pick the highlighted edges on Face A. Also, pick the two edges on Face B that are connected by the fillet. Then, pick the **OK** button to place the chamfers. See **Figure 6-32B.**

Rectangular and Circular Patterns

Patterns, or arrays, of features or sketches can be created in Inventor. The process is very similar for both features and sketches. There are two basic types of patterns—rectangular and circular. A *rectangular pattern* is an arrangement in rows and columns. A *circular pattern* is an arrangement about a center point or axis. Feature patterns are discussed first. Once you understand rectangular and circular patterns for features, you will be able to create sketch patterns with ease.

Figure 6-32.
A—The edge of Face A does not have any sharp corners; the edge of Face B does. B—The chain for a chamfer stops at sharp corners, which results in the chamfer on Face B.

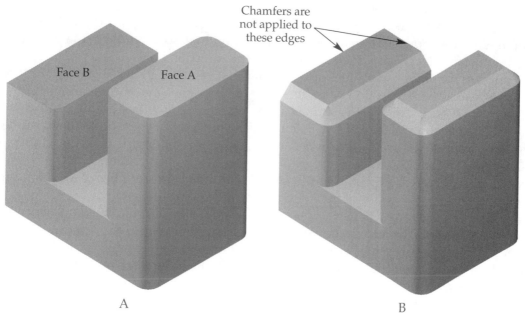

Feature Patterns

Open Example-06-12.ipt. This is a metric part with a circular sketch that has been extruded through the part to create a cutout (hole). The color of the cutout feature was changed to blue. In the **Part Features** panel, pick the **Rectangular Pattern** button. The **Rectangular Pattern** dialog box is opened. See **Figure 6-33.**

With the **Feature** button in the dialog box active, select the blue extrusion as the feature. You can select it in the **Browser** or on the part. With the feature selected, you need to define the pattern by setting directions and values for the columns and rows.

Pick the arrow button in the **Direction 1** area of the dialog box. Then, pick near the B end of Edge AB. A green arrow appears on the part indicating the direction of the pattern. Next, enter 4 in the top text box in the **Direction 1** area. The tooltip for this text box is Column Count = and the current value. Also, enter 25 mm for the spacing in the middle text box. The tooltip for this text box is Column Spacing = and the current value.

In the **Direction 2** area, pick the arrow button. Then, pick near the C end of Edge AC. The green arrow on the part that indicates the direction of the pattern should point toward C. If not, pick the **Path** flip button in the **Direction 2** area. Next, enter 3 for the row count and 20 mm for the spacing.

Figure 6-33.
Defining a rectangular pattern.

As you make changes in the **Rectangular Pattern** dialog box, a preview of the pattern is shown in green on the part. Once the pattern is defined, pick the **OK** button in the dialog box. The selected feature is arrayed in the defined pattern.

If you expand the Rectangular Pattern in the **Browser**, each of the holes you created is listed as an "Occurrence". Pick the fourth one down and then right-click on it. The pop-up menu gives you a choice of suppressing individual occurrences in the pattern.

You can select edges, axes, or paths for the directions. The edges can be straight or curved; the paths can be lines, arcs, or splines. Even though the **Rectangular Pattern** tool is used, these selections can generate patterns that are not rectangular. Open Example-06-13.ipt. This part is similar to the one in the previous example. However, there is an unconsumed sketch. The curve is a spline that is part of the unconsumed sketch. Repeat the procedure in the previous example, except pick the curve for **Direction 1**, enter 5 for the number of columns, and enter 10 mm for the column spacing. Also, select Edge AB for **Direction 2**, enter 4 for the number of columns, and enter 20 mm for the column spacing. See **Figure 6-34** for the result.

To create a circular pattern, you must select an axis of rotation about which the feature is arrayed. This axis can be a work axis or a circular feature. Open Example-06-14.ipt. The ear contains an extrusion, hole, and two fillets. All features that comprise the ear will be arrayed about the center of the main shaft. Pick the **Circular Pattern** button in the **Part Features** panel. The **Circular Pattern** dialog box is displayed, **Figure 6-35**. With the **Features** button active, select all four features that comprise the ear. Next, pick the **Rotation Axis** button in the dialog box. Since the three components of the main shaft share the same centerline, you can select any one of these features to define the axis. A preview of the array appears in green on the part. In the **Circular Pattern** dialog box, enter 3 in the left-hand text box in the **Placement** area to create three items. Also, enter 360 in the right-hand text box to indicate the pattern will be rotated around the entire circumference. Finally, pick the **OK** button to create the pattern. See **Figure 6-36** for the result.

Pattern the Entire Part

It is possible to pattern the entire part. Open the part Example-06-15.ipt. Display this part in wireframe to allow the work axis at the bottom of the part to be visible. This axis will be used for the circular pattern. Open the **Circular Pattern** dialog box and pick the **Pattern the entire solid** button. See **Figure 6-37**. The **Rotation Axis** button is automatically depressed allowing you to pick the work axis. Now, in the **Placement** area, set the **Occurrence Count** to 3 and the **Occurrence Angle** to 360 degrees. Pick **OK** to complete the pattern.

Figure 6-34.
Using a curved line as one path creates a different pattern.

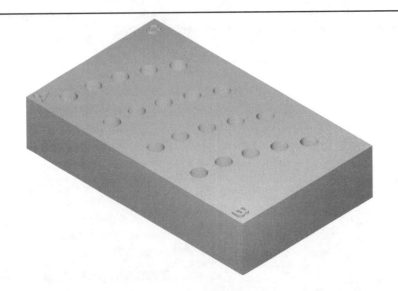

Figure 6-35.
Defining a circular
pattern.

Figure 6-36.
One ear on this part
was first created. A
circular pattern was
then created to
place the other two
ears.

Figure 6-37.
Expand the **Circular
Pattern** dialog box
to modify the
positioning method.

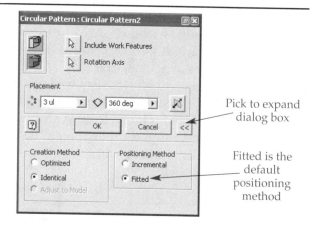

Pick to expand
dialog box

Fitted is the
default
positioning
method

Edit the circular pattern you just created. Expand the **Circular Pattern** dialog box by picking the **More >>** button. Within the **Positioning Method** area are two options, **Incremental** and **Fitted**. The default option, **Fitted**, spaces the occurrences of the solid equally throughout the occurrence angle, 360 degrees. Pick the **Incremental** option and change the **Occurrence angle** to 90 degrees. There are still 3 occurrences, but they are not spaced *within* the 90 degree angle. Instead, the solids are 90 degrees apart.

Sketch Patterns

You can also create patterns of sketches. Open Example-06-16.ipt. This part is a hinge plate with an unconsumed sketch. Edit the unconsumed sketch. Then, continue as follows.
1. Activate the **Rectangular Pattern** tool.
2. Pick all four edges of the samll rectangle as the geometry.
3. For **Direction 1**, pick the edge that passes through the rectangle. Flip the direction if needed so the green arrow points to the right.
4. Set the number of occurrences to 4 and leave the distance as 1".
5. Do not make any settings for **Direction 2**. In this way, only one row is created.
6. Pick the **OK** button to create the pattern.

Figure 6-38.
A rectangle was
sketched, and then a
rectangular pattern
was created. The
four profiles were
then extruded to cut
the part.

7. Finish the sketch.
8. Pick the **Extrude** button in the **Part Features** panel. Extrude all four rectangles in the pattern using the **All** and **Cut** options.
9. The result is shown in **Figure 6-38.**

The Mirror Feature Tool

The **Mirror Feature** tool mirrors features about a work plane, straight edge on the part, or a planar face. It performs much the same function as the sketch **Mirror** tool. Open Example-06-17.ipt. This part is similar to the one used to create a circular pattern, except a work plane and work axis have been added.

Pick the **Mirror Feature** button in the **Part Features** panel. The **Mirror Pattern** dialog box is displayed, **Figure 6-39.** With the **Features** button active, select all four features that comprise the ear. Next, pick the **Mirror Plane** button in the dialog box. Select the work plane. Finally, pick the **OK** button in the dialog box to mirror the features. See **Figure 6-40.**

Figure 6-39.
The **Mirror Pattern**
dialog box.

Figure 6-40.
All features in one
ear were mirrored
to create the second
ear.

Chapter Test

Answer the following questions on a separate sheet of paper.

1. List four advantages of creating holes using the **Hole** tool over extruding circles.
2. What type of object must be drawn to use the **Hole** tool?
3. Which tool is used to create the object in *Question 2* and where is it located?
4. What are the three termination options for a hole?
5. What are the two options for the bottom of a hole?
6. What is the purpose of a counterbore?
7. Which tool is used to create a counterbore?
8. How can you place threads in a cutout (hole) created by extruding a circle?
9. Define *fillet*.
10. Define *round*.
11. Which tool is used to create a fillet?
12. Which tool is used to create a round?
13. How can you smooth the corner where three rounds intersect?
14. What is a variable-radius fillet?
15. Define *chamfer*.
16. What are the three methods for defining a chamfer?
17. What is the difference between a loop for a fillet and a chamfer in Inventor?
18. What are the two basic types of patterns that can be applied to features and sketches?
19. Briefly, how can you create a pattern that has one row and four columns?
20. Which tool is used to mirror selected features of a part?

Chapter Exercises

Exercise 6-1.   Open Exercise-06-01.ipt. This is a sheet metal part; this type of part is discussed in *Chapter 18*. To display the **Part Features** panel, pick the **Sheetmetal Features** title in the **Panel Bar** to display a drop-down list. Then, pick **Part Features** in the list. The tools you have been using in this chapter are now visible.
- Create a .375″ diameter tapped hole through the end with the weld nut. Set the thread to 3/8-16 UNC and full depth.
- Create a .385″ diameter clearance hole through the other rounded end.
- Drill a .5″ hole through the back face. Center the hole.

Exercise 6-2. Open Exercise-06-02.ipt. Create four clearance holes on the corners of the base for M8 socket head cap screws. Center the holes in the fillets and counterbore to a depth so the head is recessed flush. Create a 23 mm clearance hole through the top surface concentric to the large radius. Create an M22 × 1.5 hole tapped full depth ANSI Metric M Profile through the base.

Exercise 6-3. Open the part Exercise-06-03.ipt. The bottom faces of this casting are to be machined flat. Place .20″ fillets and rounds on all the other cast edges. Do not fillet or round the holes. Also, place a .25″ × .25″ chamfer on the bottom of the large bored hole.

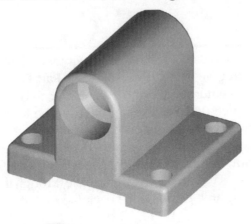

Exercise 6-4. Open Exercise-06-04.ipt. Place a 15 mm round on the vertical edges of the base. Place 5 mm fillets and rounds on all other edges. You will need to do this in more than one step.

Exercise 6-5. Open the part Exercise-06-05.ipt. Place a round on the end of the pin to create a spherical end. Also, chamfer the edges of the base opposite the pin with a 2″ × 2″ chamfer.

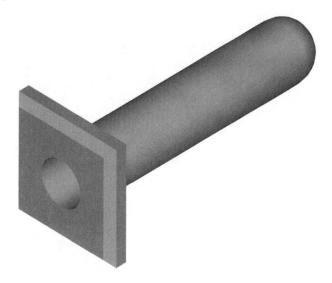

Exercise 6-6. Open Exercise-06-06.ipt. This part is the preliminary stage in the design of a bracket for a windshield wiper arm. Put a variable-radius round on both curved top edges. The rounds are 10 mm at the left end, 3 mm at the center, and 8 mm at the right end.

Exercise 6-7. Open Exercise-06-07.ipt. Complete the part as shown in the drawing below. Create a circular pattern to place the 5 mm holes.

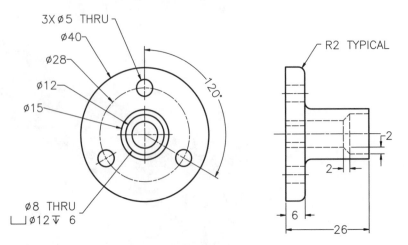

Exercise 6-8. Open Exercise-06-08.ipt. This is a rod for a caulking gun. Put the 15 notches in the rod as shown in the drawing below. The file has a work plane created to help you. Note the small fillet at the bottom of each notch.

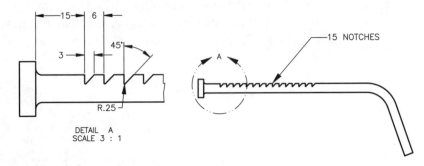

Exercise 6-9. Open Exercise-06-09.ipt. Finish the part as shown below by mirroring the features.

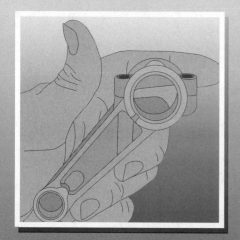

Adding More Features

Objectives

After completing this chapter, you will be able to:

* Create shelled parts.
* Add ribs and webs to parts.
* Create embossed and engraved parts.
* Add decals to parts.
* Create face drafts on parts.
* Create splits.

User's Files

*The following is a list of files that you will need to work through this chapter. These files can be found on the **User's Files** CD included with this text.*

Examples

Bricks.bmp
Caution.bmp
Example-07-01.ipt
Example-07-02.ipt
Example-07-03.ipt
Example-07-04.ipt
Example-07-05.ipt
Example-07-06.ipt
Example-07-07.ipt
Example-07-08.ipt
Example-07-09.ipt
Example-07-10.ipt
Example-07-11.ipt

Example-07-12.ipt
Example-07-13.ipt
Example-07-14.ipt
Example-07-15.ipt
Example-07-16.ipt
Example-07-17.ipt
Example-07-18.ipt
Example-07-19.ipt
G-W Logo.bmp

Practice

Practice-07-01.ipt

Exercises

Exercise-07-01.ipt
Exercise-07-02.ipt
Exercise-07-04.ipt
Exercise-07-05.ipt
Exercise-07-08.ipt
Exercise-07-09.ipt
Exercise-07-10.ipt
Exercise-07-11.ipt
Knurl Pattern.bmp
Serial Number.xls

The Shell Tool

The **Shell** tool turns solid parts into "boxes" by hollowing out a solid part. The wall thickness of the shell and any openings are specified. The result of the operation is typical of a cast metal or molded plastic part. Part faces, both flat and curved, can be removed by the shell operation to create open boxes.

Open Example-07-01.ipt. This part has eight faces, **Figure 7-1.** Four of these faces will be removed during the shell operation. First, pick the **Shell** button on the **Part Features** panel. This opens the **Shell** dialog box, **Figure 7-2.** Make sure the **Remove Faces** button in the dialog box is on (depressed). Then, move the cursor over the part. As you move the cursor, the face that will be selected is outlined in red. Move the cursor over Face A, as indicated in **Figure 7-1,** and pick to select it. The face is highlighted in cyan. Then, in the **Shell** dialog box, pick the **OK** button to accept the default shell thickness of 0.1". A cavity is created inside the part. See **Figure 7-3A.** Notice how Face A is removed to create an open part. If it was not removed during the operation, the cavity would be completely enclosed.

Multiple faces can be removed while using the **Shell** tool. This can be done in the initial operation or by editing the feature. To edit the feature, right-click on the shell name, Shell1, in the **Browser** and select **Edit Feature** from the pop-up menu. In the **Shell** dialog box, pick the **Remove Faces** button to turn it on. Then, select Face B on the part. Notice how selecting Face B also selects the two tangent faces. Pick the **OK** button in the dialog box to update the feature. See **Figure 7-3B.**

Figure 7-1.
This part will be shelled. The faces indicated here will be removed in various combinations to produce different results.

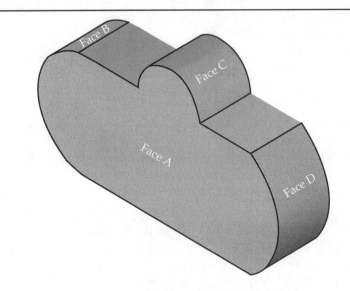

Figure 7-2.
Making settings for the shell operation.

Wall thickness

Direction options

Figure 7-3.
The shell operation with various faces removed. A—Face A. B—Faces A and B. C—Faces A and C. D—Faces A and D.

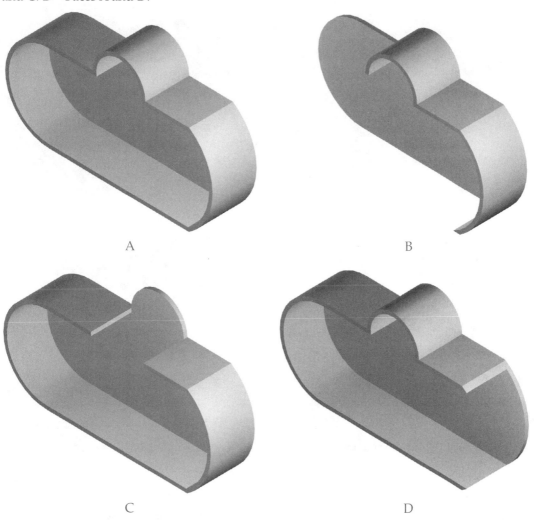

A

B

C

D

When editing the feature, you can also remove faces from the exclusion. In other words, add the face back onto the part. Edit the feature again and pick the **Remove Faces** button. The faces currently removed by the operation are highlighted in cyan. Hold down the [Shift] key and pick Face B to deselect it. Then, release the [Shift] key and pick Face C to select it. Pick the **OK** button in the dialog box to update the feature. See **Figure 7-3C.** Edit the feature again, add Face C, and remove Faces A and D. See **Figure 7-3D.** Notice the sharp chamfers created by removing Face D.

The **Shell** tool has three options that determine how the cavity wall is created in relation to the base part—**Inside, Outside,** and **Both.** The **Inside** option keeps the outer shape of the part the same size. In other words, the cavity wall is to the inside of the part. The **Outside** option adds material to the outside of the part to create the cavity wall. The cavity has the dimensions of the original part. The **Both** option splits the difference; half of the wall thickness is to the inside and half to the outside. Open Example-07-02.ipt. See **Figure 7-4A.** Activate the **Shell** tool and remove the large, front face. Also, set the wall thickness by picking in the **Thickness** text box and entering .5″. Make sure the **Inside** button is on and pick the **OK** button to create the shell, **Figure 7-4B.** Edit the feature and select the **Outside** button. Notice how this option makes the hole smaller. See **Figure 7-4C.** Edit the feature again and select the **Both** button. See **Figure 7-4D.**

Figure 7-4.
A—The part before shelling. B—The shelled part with the **Inside** option selected. C—The shelled part with the **Outside** option selected. D—The shelled part with the **Both** option selected.

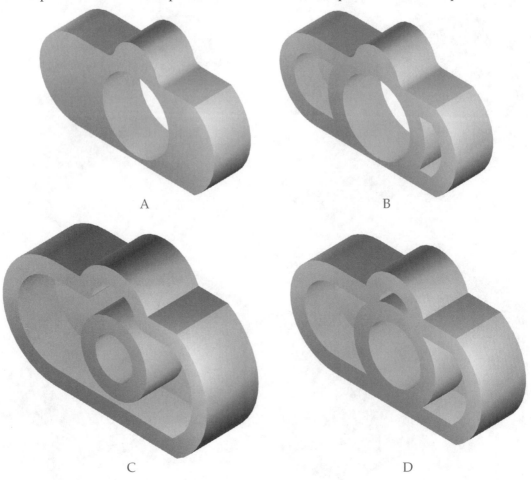

A

B

C

D

By default, the same wall thickness is applied to all faces. However, you can apply different wall thicknesses to selected faces. In the **Shell** dialog box, pick the **>>** button to expand the dialog box. See **Figure 7-5A.** The **Unique face thickness** area is used in much the same way as setting up multiple, different-radius fillets. Open Example-07-03.ipt. This simple box was shelled with a thickness of .1" and the front face was removed. Edit the shell feature and expand the dialog box. Pick the entry **Click to Add** to add a setting. Then, pick in the **Thickness** column and set the value to .3". Pick in the **Selected** column and then select the top surface of the part. Finally, pick the **OK** button in the dialog box to update the part. The result is shown in **Figure 7-5B.** You can add multiple settings; each setting can have multiple faces if needed.

Open Example-07-04.ipt. This part was built with two extruded circles and then shelled with a 2 mm wall thickness. The side and bottom face were removed in the operation. See **Figure 7-6A.** Sketch a 70 mm diameter circle on the top of the part and extrude it 50 mm high. Shell the new feature to the inside with a thickness of 10 mm and the top face removed. The result is shown in **Figure 7-6B.**

Figure 7-5.
A—The expanded **Shell** dialog box. B—The wall thickness at the top of the part is different than for the other walls.

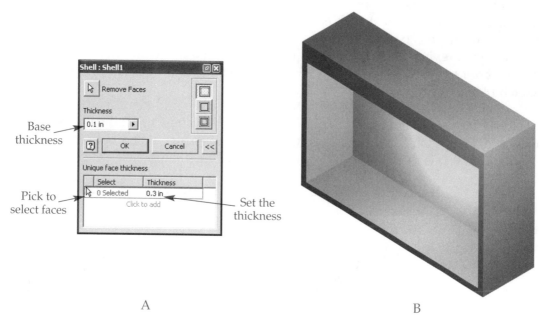

Base thickness

Pick to select faces

Set the thickness

A

B

Figure 7-6.
A—The original shelled part (shown in section). B—An extruded feature is added to the part (shown in section) and a different shell operation performed.

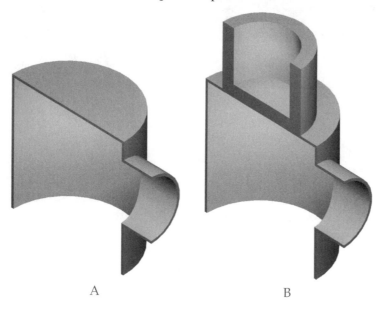

A

B

CAUTION

If you try to move the extrusion you just created above the first shell operation in the **Browser**, you will get an error message.

The Rib Tool

Often, features are added to help support or strengthen the part. These features are called ribs and webs. A *web* is a thin support feature with an opening or space between it and the part; only the ends contact the part. A *rib* is similarly a thin support feature, however, it is closed instead of open. See **Figure 7-7.** The **Rib** tool creates both ribs and webs of a specified thickness from a very simple sketch on a sketch plane. Often, the sketch is a single line.

Open Example-07-05.ipt. There is a work plane in the middle of the part. Pick the **Sketch** button and select the work plane as the sketch plane. Project the Edges AB and CD onto the sketch plane, creating two points. Construct a line between the two points. This line is the profile for the rib (or web). Finish the sketch.

Figure 7-7.
A—A rib is added to the angled bracket. B—A web is added to the angled bracket.

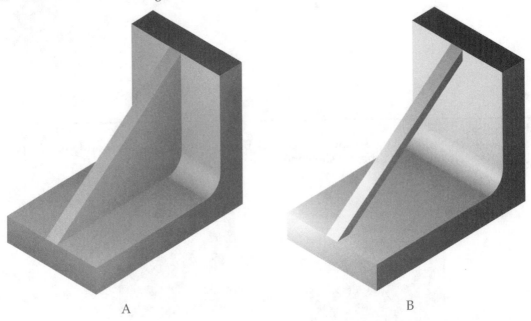

A   B

Pick the **Rib** button in the **Part Features** panel. This accesses the **Rib** dialog box. See **Figure 7-8A.** To select a profile, pick the **Profile** button in the **Shape** area and select the profile in the graphics window. However, since there is only one unambiguous sketch, it is automatically selected as the profile. Also, the **Direction** button in the **Shape** area is automatically turned on (depressed). There are four choices for direction of the rib or web. If you look straight down the line, so it is a point, the four directions are similar to the four cardinal directions of a compass. The direction for this rib is downward toward the part. This can best be seen in a side view of the part. Use the **Look At** tool and look at the side of the part. Then, move the cursor over the line until the green arrow points toward the part and pick. In this view, there will only be two options. Redisplay the isometric view. A preview of the rib appears on the part in green.

In the **Thickness** area of the dialog box, pick in the text box and enter .375″ as the thickness. Also, select the midplane button so that the thickness is applied equally on each side of the line. In the **Extents** area of the dialog box, make sure the **To Next** button is on. This button creates a rib. Then, pick the **OK** button to create the feature. See **Figure 7-8B.**

Figure 7-8.
A—Making settings for the rib operation. B—The rib is created.

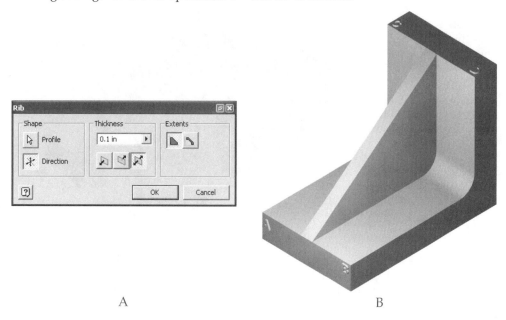

A

B

Now, edit the rib feature by right-clicking on its name in the **Browser** and selecting **Edit Feature** from the pop-up menu. In the **Extents** area of the **Rib** dialog box, pick the **Finite** button. A text box appears below the button. Pick in the text box and enter a value of 1″. Pick the **OK** button to update the part. The rib is now a web. See **Figure 7-9.** Notice that the 1″ dimension is measured perpendicular to the line.

The sketch curve (line) for a rib or web does not have to be drawn full length. By checking the **Extend Profile** check box in the **Rib** dialog box, the curve will be extended to the next faces. Open Example-07-06.ipt. The line was drawn with the top of the box as the sketch plane. The length of the line is arbitrary, but the line does pass through the center of the part. Activate the **Rib** tool, pick the line as the profile, and set the thickness to 2 mm. Pick the **Direction** button and set the direction down toward the part. Note that the **Extend Profile** box appears and is checked. Pick the **OK** button to create the rib. The rib extends from wall-to-wall. Also, notice that the rib is not created inside the hole. See **Figure 7-10.**

Figure 7-9.
The rib is changed into a web by editing the feature.

Figure 7-10.
A rib is added to the inside of the part. The rib does not pass through the hole.

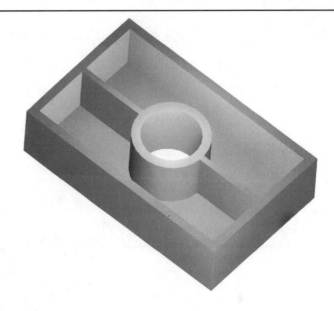

Edit this feature. Uncheck the **Extend Profile** check box and pick **OK**. The rib does not extend to the outside walls of the part. Edit the feature again and pick the **Finite** button. When the **Extend Profile** check box is not checked, the web does not extend to the outside walls of the part. When checked, the web extends to the outside walls. However, in both cases, notice how the web passes through the hole.

The **Rib** tool also accepts multiple lines as profiles. Open Example-07-07.ipt. Five lines are drawn through the centers of the six hole bosses. Note that the lines do not touch the walls of the part. Activate the **Rib** tool, select all five lines as the profile, set the thickness to .1″, and pick the **To Next** button. Pick the **Direction** button and set the direction in toward the part. Then, pick the **OK** button to create the ribs. See **Figure 7-11.**

You are not limited to straight lines as the rib or web profile. Circles, arcs, and splines can also be selected as profiles. **Figure 7-12** shows a part where the ribs were created with a circle and three lines. Open Example-07-08.ipt. Study this part to see how the profiles were created. Notice that there is only one rib operation listed in the **Browser**. Try to select and move the hole and circular pattern operations above the rib operation. The ribs pass through the holes. The reason you can move the hole and circular pattern above the rib operation, unlike in Example-07-06, is because the rib

Figure 7-11.
Five lines were used to create the ribs in this part in a single operation.

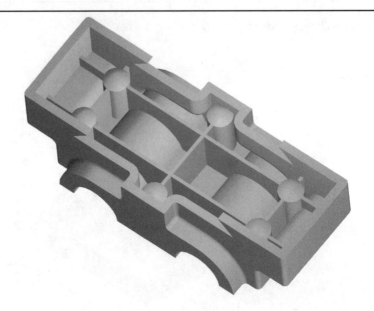

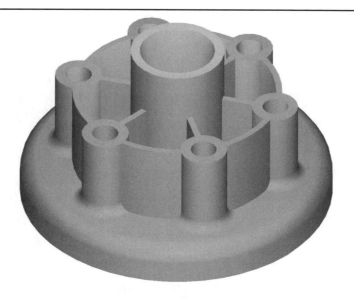

Figure 7-12.
A combination of lines and a circle were used to create the ribs on this part.

operation is not dependent on those two operations. In fact, you can move the rib operation anywhere in the part tree up to Extrusion2. The rib operation is dependent on Extrusion2 and all operations above it.

Moving the Coordinate System on a Sketch Plane

So far, you have changed location and orientation of the coordinate system by selecting work planes or part faces as sketch planes. The coordinate system for a sketch is called the *sketch coordinate system (SCS)*, which is represented by the icon called the *Coordinate System Indicator* or CSI. However, you can also change the coordinate system, including its orientation, using the **Edit Coordinate System** tool. The orientation of the coordinate system is important for some tools, such as the **Emboss** tool that will be discussed in the next section. The text used to emboss is always aligned with the X axis of the coordinate system. Since you can change the orientation of the coordinate system, you can change the orientation of the text.

Open Example-07-09.ipt. See **Figure 7-13.** The top of the part was picked as the sketch plane to create Sketch3. If the CSI is not displayed at the origin, turn it on in the **Sketch** tab of the **Options** dialog box, which is accessed by selecting **Application Options...** from the **Tools** pull-down menu. Right-click on its name in the **Browser** and select **Edit Sketch** from the pop-up menu. The origin of the SCS was automatically placed at the intersection of the straight and arc edges when the face was selected as the sketch plane. A line, circle, and text were then created with the appropriate sketch tools. The Y work axis was also made visible. You will learn how to create text in the next section.

PROFESSIONAL TIP

To see the location and orientation of the axes, expand the Origin in the **Browser**. Then pass the cursor over an axis and it will be highlighted in the graphics window.

The top of this part has eight vertices on the sketch plane. The origin of the SCS can be moved to any of these vertices or the center of curved face, which is also a point on the sketch plane. In addition, the X or Y axis can be aligned with any straight

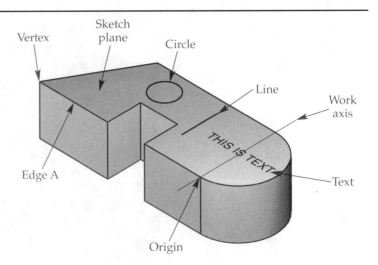

Figure 7-13.
The circle, line, and text on the top of this part were created using the default coordinate system.

Vertex

Sketch plane

Circle

Line

Work axis

THIS IS TEXT

Text

Edge A

Origin

edge of the part or with the work axis. The X axis is indicated by the red arrow on the CSI, the Y axis by the green arrow. The axes *cannot* be aligned with the sketched line. To align the Y axis of the SCS with Edge A, as indicated in **Figure 7-13,** do the following.

1. Pick the **Edit Coordinate System** button in the **2D Sketch Panel**. Notice how the CSI changes to long red and green arrows and a blue dot.
2. Move the cursor over the green Y axis arrow so it is highlighted in red. Pick the arrow; it is highlighted in cyan.
3. Move the cursor around the graphics window. Notice how the orientation of the long arrows change. Pick Edge A. The SCS is rotated so the Y axis aligns with the edge.
4. Right-click in the graphics window and select **Done** from the pop-up menu. Do *not* press [Esc]; this cancels the operation.

The SCS is oriented so the Y axis aligns with Edge A. Notice how the CSI has rotated to reflect the new SCS orientation.

Notice how the orientation of all the sketch geometry changed with the SCS. See **Figure 7-14.** Notice that the line moved, but did not rotate. This is because a perpendicular constraint is applied to the line and one of the edges. You can relocate the sketch geometry by dragging it around and rotate geometry using the **Rotate** tool. However, you cannot change the orientation of the text. Therefore, it is good practice to orient the SCS in such a way that text will be properly aligned before you construct any sketch geometry.

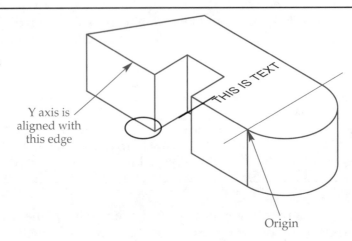

Figure 7-14.
When the orientation of the coordinate system is changed, the geometry also rotates. The line did not rotate because of a constraint applied to it.

Y axis is aligned with this edge

THIS IS TEXT

Origin

Now, the orientation of the SCS is changed. However, the origin remains in the same location. As mentioned earlier, you can move the origin to any of the eight vertices or the center of the arc. Move the origin of the new SCS to the center of the arc as follows.

1. Pick the **Edit Coordinate System** button in the **2D Sketch Panel**.
2. Move the cursor to the center of the SCS. When the blue dot is highlighted red, pick it.
3. Move the cursor around the graphics window. The X and Y axis arrows jump to the vertex closest to the cursor. Move the cursor over the arc and the arrows jump to the center of the arc.
4. When the X and Y axis arrows are at the center of the arc, pick to move the origin of the SCS.
5. Right-click in the graphics window and select **Done** from the pop-up menu. Remember, do not press [Esc].

The CSI is now displayed at the center of the arc, indicating that this is the origin of the SCS. However, notice that the orientation of the SCS has not changed. The Y axis is still aligned with Edge A.

The Text and Emboss Tools

To *emboss* text means to raise or project it above a face. To *engrave* text means to cut it into a face. The **Emboss** tool can both emboss and engrave text created by the sketch **Create Text** tool. The face on which text is embossed or engraved can be flat or curved.

Open Example-07-10.ipt. You will place text on the front face. Later, you will emboss this text. Pick the **Sketch** button on the **Inventor Standard** toolbar and pick the front face as the sketch plane. Then, pick the **Text** button in the **2D Sketch Panel**. Move the cursor near the top-left corner of the face and pick. The **Format Text** dialog box is displayed, **Figure 7-15**.

Figure 7-15.
Adding text to a sketch.

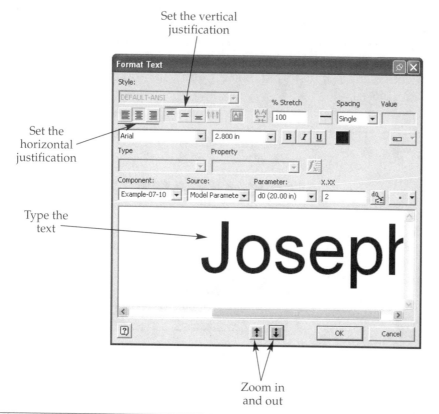

In the **Format Text** dialog box, pick in the large edit box and type your name. Then, using the cursor, highlight the text as you would in a text editor by picking and dragging over the text. With the text highlighted, pick the **Center Justification** and **Middle Justification** buttons at the top-left of the dialog box. You can change the font using the lower-left drop-down list. If Arial is not displayed, pick the drop-down list and select it. Finally, pick in the text box to the right of the font drop-down list and type 2.8 to set the text height at 2.8". When you press [Enter], the dialog box is closed and the text is created. Now, drag the text so your name is approximately centered on the face. Finally, finish the sketch.

Pick the **Emboss** button in the **Part Features** panel. The **Emboss** dialog box is displayed. With the **Profile** button on (depressed), move the cursor over the text in the graphics window. The entire front face is highlighted in red. Pause the cursor to display the **Select Other** tool, pick an arrow to highlight the text, and pick the center button to select the text. In the dialog box, pick in the **Depth** text box and change the value to .5". Notice the button below the **Depth** text box. This is the **Top Face Color** button. See **Figure 7-16.** Pick it to open the **Color** dialog box. Select **Gold Metallic** from the drop-down list in the **Color** dialog box and pick the **OK** button. Finally, pick the **OK** button in the **Emboss** dialog box to emboss the text. Notice how the top face is in gold instead of the part color.

Now, right-click on the name of the emboss operation in the **Browser** and select **Properties** from the pop-up menu. In the **Feature Color Style** drop-down list, select Black Chrome. You can also rename the feature if you like. Then, pick the **OK** button to update the feature. Notice how the color on the top face of the embossed text *and* the color on the text sides are replaced.

Open Example-07-11.ipt. This part has text created on a work plane offset from the curved face. The text is Arial font, bold, and .5" high. Pick the **Emboss** button in the **Part Features** panel. Then, select the text as the profile, set the height to .1", and pick the **Engrave from Face** button. Finally, pick the **OK** button to engrave the text. See **Figure 7-17.**

Notice how the text is engraved straight into the cylinder. The three letters are distorted at the beginning and end of the word. Edit the feature by right-clicking on its name in the **Browser** and selecting **Edit Feature** from the pop-up menu. In the **Emboss** dialog box, check the **Wrap to Face** check box. Then, pick the **Face** button so it is on (depressed) and select the cylindrical face. Pick the **OK** button in the dialog box to update the feature. The text is now wrapped on the face before being engraved. In this manner, the letters are not distorted. See **Figure 7-18.** You can only wrap text to planar or cylindrical faces.

Figure 7-16.
The dialog box used for the emboss operation.

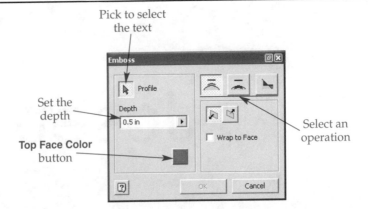

Pick to select the text

Set the depth

Top Face Color button

Select an operation

Figure 7-17.
A—Text is engraved in a curved feature, but distorted. B—In a top view, you can see that the lines are parallel, resulting in the distortion of the text.

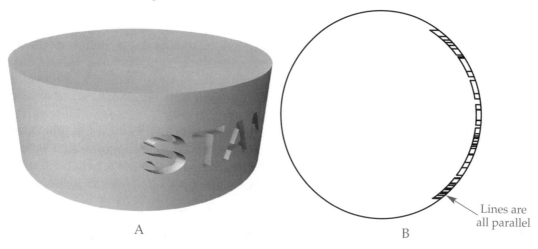

Lines are
all parallel

A

B

Figure 7-18.
A—By wrapping the text onto the curved surface, the distortion is eliminated.
B—In a top view, you can see that the lines are perpendicular to the curved face.

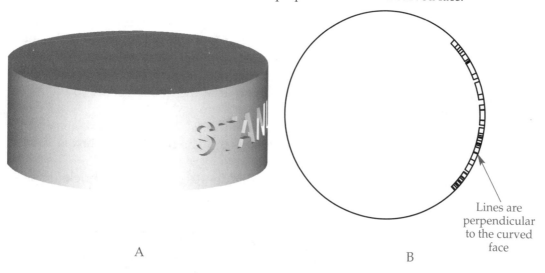

Lines are
perpendicular
to the curved
face

A

B

The Decal Tool

An image can be placed on a part, much like a label on a bottle. The "image" can be any picture, drawing, or text saved as a bitmap (BMP extension), Word file (DOC extension), or Excel file (XLS extension). Some applications include labels, logos, art, part information such as part numbers, or warranty data to parts. You can insert an image onto a sketch plane and position it with constraints or dimensions. The **Decal** tool can then be used to attach the image to a part face.

An important property of the image to be used is its aspect ratio. The *aspect ratio* is the image height divided by its width. The aspect ratio cannot be changed in Inventor. The aspect ratio of bitmaps can be modified in graphics programs, such as Adobe Photoshop. In order to keep the colors consistent when inserting, the bitmaps need to be relatively small; about 400 or less pixels along the long axis.

In many industries, a label has a part number and, therefore, a drawing. A drawing of a label can be easily created by placing the image on a very thin part and creating a decal. This serves several functions:

- The label can have a part number.
- The label can have a drawing.
- Most importantly, the drawing can be used in an assembly, which is discussed in *Chapter 11*.

The part on which the label is placed should have the same dimensions as the label. Also, the part should be the same shape; i.e., a box for a rectangular label or a cylinder for a round label.

Open Example-07-12.ipt. The two bitmap images you will use on this part are a photo of a brick wall, with an aspect ratio of .6, and a company logo, with a ratio of .71, **Figure 7-19.** To make this example easy, Extrusion1 has the same aspect ratio as the bricks and Extrusion2 has the same aspect ratio as the logo. First, add the brick image as follows.

1. Start a new sketch on the large front face.
2. Pick the **Insert Image** button in the **2D Sketch Panel** or select **Image...** from the **Insert** pull-down menu. The **Open** dialog is displayed. This is a standard Windows open dialog box.
3. Navigate to the Chapter 7 Examples folder, select the file Bricks.bmp, and pick **OK**
4. A small rectangle now appears attached to the cursor. This represents the bitmap image. Position the box at the upper-left corner of the part and pick.
5. Notice how the green box is still attached to the cursor. You can insert another instance of the bitmap. Since you only need one, press [Esc] to end the **Insert Image** tool.
6. Move the cursor to the lower-right corner of the image. When the corner is highlighted red, pick to select the corner. Then, drag the image corner to the lower-right corner of the part and release.
7. "Finish" the sketch.
8. Pick the **Decal** button in the **Part Features** panel. The **Decal** dialog box is opened, **Figure 7-20.**
9. With the **Image** button in the dialog box on (depressed), pick the brick image on the part.

Figure 7-19.
A—This image of bricks will be used as a decal. B—This company logo will be used as a decal.

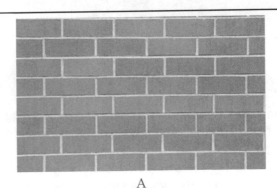

A

B

Figure 7-20.
Placing a decal on a part.

10. With the **Face** button in the dialog box on, pick the front face of the part. The face is highlighted in cyan.
11. In the **Decal** dialog box, pick the **OK** button to apply the decal. Notice that the decal is not applied to Extrusion2 because it was not selected as a face, which is the intended result. See **Figure 7-21.**

Figure 7-21.
The image of the bricks is applied to the part.

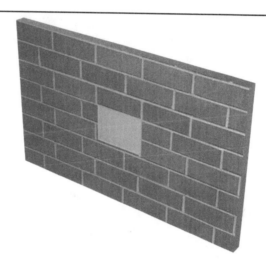

Using a similar process, apply the file G-W Logo.bmp to the front face of Extrusion2. See **Figure 7-22.** You may want to use the **Slice Graphics** option to improve visibility when resizing the image. A good way to locate the bitmap is to insert it off of the part, then use the coincident constraint on the corners of the bitmap and the face.

Like the Emboss tool, the Decal tool has a Wrap to Face option for wrapping an image onto a curved face. Open Example-07-13.ipt. This is a model of a firecracker that needs a caution label applied. The model has the Z axis visible and a work plane offset from the part. Create a new sketch on the work plane and project the Z axis onto it, as shown in **Figure 7-23.** Also, project the circular end of the part onto the sketch plane. These projection lines will be used to help align the label.

Notice the direction of Z axis (blue) on the coordinate system. This is the normal direction for the sketch plane. The destination face must point the same way or the decal will not be visible. Also, notice the orientation of the X and Y axes. The X axis of the image is aligned with the X axis of the sketch plane. The X axis of the label is along the bottom of the text. Therefore, the image will have to be rotated after it is inserted. You will use the Rotate tool to do this.

1. Pick the **Insert Image** button. Navigate to the Chapter 7 Examples folder and select the file Caution.bmp.
2. Place the image anywhere in the graphics window. Notice how the image is rotated 90° counterclockwise. Press [Esc] to end the tool.
3. Pick the **Rotate** tool in the **2D Sketch Panel**. The **Rotate** dialog box is opened.
4. With the **Select** button on in the **Rotate** dialog box, pick the inserted image; pick all four edges.
5. With the **Center Point** button on in the **Rotate** dialog box, pick any corner of the image.

Figure 7-22.
The company logo
is added to the part.

6. In the **Rotate** dialog box, enter –90 in the **Angle** text box. Then, pick the **Apply** button to rotate the image. Then, pick the **Done** button to close the dialog box. The image should be right reading.
7. Apply a coincident constraint between the midpoint of the left and right borders (short edges) of the image and the projected axis.
8. Dimension the sketch as shown in **Figure 7-23.** Notice how the image stayed proportional. Constraining the midpoints to the projected axis allowed this.
9. "Finish" the sketch.
10. Pick the **Decal** button in the **Part Features** panel. Select the label as the image and the large-diameter cylinder as the face. Also, check the **Wrap to Face** check box. Pick the **OK** button to place the label on the part, **Figure 7-24.**

Figure 7-23.
Adding a caution
label to a part.

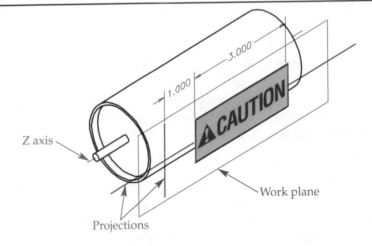

Figure 7-24.
The part with the
caution label
applied.

Figure 7-25.
A—When the **Chain Faces** check box is unchecked, the decal is cut off at the edge of the face. B—When **Chain Faces** is checked, the decal wraps around the edge of the face.

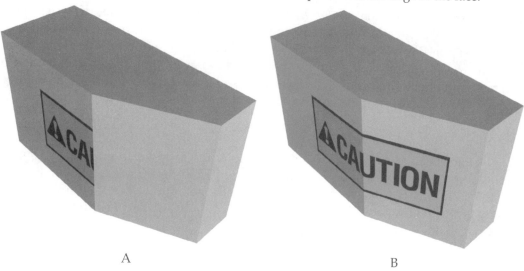

A   B

You can project a decal across adjacent faces using the **Chain Faces** option in the **Decal** dialog box. In **Figure 7-25A,** the **Chain Faces** check box was not checked and the decal is cut off. In **Figure 7-25B,** the check box was checked. Notice how the image is distorted as it is projected onto the adjacent face. This distortion is more pronounced the closer the face angle is to 90°, **Figure 7-26A.** This distortion can be eliminated by checking the **Wrap to Face** check box, **Figure 7-26B.** Open Example-07-14.ipt. Then, make the appropriate settings to create **Figure 7-25A.** Undo the operation and create **Figure 7-25B.** Start a new part file and create **Figure 7-26A** and **Figure 7-26B.** Use your own dimensions.

Figure 7-26.
A—When the **Chain Faces** check box is used to wrap a decal around an edge, distortion can occur. B—Checking the **Wrap to Face** check box in addition to the **Chain Faces** corrects the distortion.

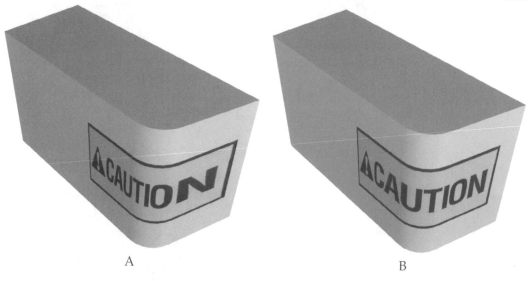

A   B

Some manufacturing processes, such as casting, forging, and injection molding, require a draft angle on the part. A *draft angle* is an angle on the sides of a part so that the part can be removed from the forming tool. Therefore, the forming tool itself must have corresponding angles on its cavity walls. The actual feature, the angled sides, is called the draft. A draft angle can be created on a part when a sketch is extruded or later using the **Face Draft** tool. This tool can also be used to put a unique draft angle on one or more faces, simplifying the construction process.

Fixed Edge

The **Face Draft** tool requires several selections and is best explained with a simple part. Open Example-07-15.ipt. This part has two extruded blocks. The first block, **Figure 7-27A,** has two faces; one green face and one yellow face. The draft will be applied to the yellow face. Pick the **Face Draft** button in the **Part Features** panel. The **Face Draft** dialog box is opened, **Figure 7-28.**

The draft angle is set in the **Draft Angle** text box. Typically, a draft angle is very low, 2° or 3°. For this example, a large angle is used to emphasize the effect of the operation. Pick in the text box and enter 20° as the draft angle.

Figure 7-27.
A—The pick point determines which edge is fixed and used to create the draft. The top edge is fixed. B—The resulting draft. C—Using a work plane to create the draft. D—The resulting draft.

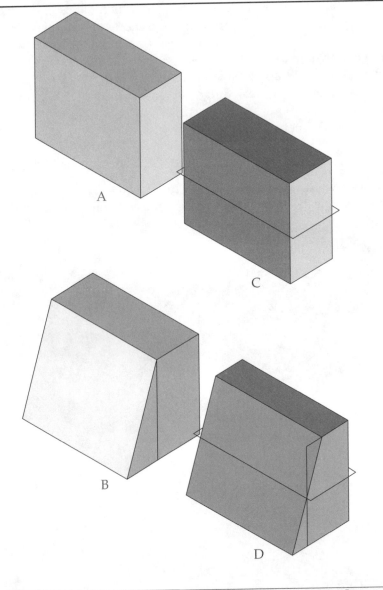

Figure 7-28.
The **Face Draft** dialog box is used to create a face draft.

Fixed Edge

Fixed Plane

Pick the **Fixed Edge** button at the far-left of the dialog box. Just to the right of this button are the two buttons. The **Flip pull direction** button is grayed out. Pick the **Pull direction** button so it is on (depressed), if it is not already. Select the green face on the part. Work planes can also be used to define the pull direction. An arrow should appear on the face pointing away from the part. The draft will be at the specified angle to this arrow. If you imagine a mold over the top of this part, the pull direction is the direction the mold is "pulled" to remove the molded part. You can flip the pull direction by picking the **Flip pull direction** button.

Now, pick the **Faces** button if it is not already on. You will select the yellow face, however, you have two choices. If you pick near the top edge, that edge remains fixed and the part gets larger. If you pick near the bottom edge, it remains fixed and the part gets smaller. You *cannot* change this by editing the feature later. As you move the cursor over the face, a preview indicates how the face draft will be applied. Pick the yellow face near the top edge. Then, pick the **OK** button in the dialog box to create the draft angle. See **Figure 7-27B**.

Fixed Plane

The second block, **Figure 7-27C**, will be drafted using a work plane to locate the draft. Pick the **Face Draft** button in the **Part Features** panel. Pick in the text box and enter 20° as the draft angle, if it is not already entered. This time, pick the **Fixed Plane** button. Notice that the label of the buttons to the right changed from **Pull Direction** to **Fixed Plane**. You will be selecting a plane as the starting point for the draft. Go ahead and select the work plane in the middle of the second block. An arrow should appear on the plane and the **Flip pull direction** button is no longer grayed-out. Now select the purple face to add the draft, then pick **OK** to close the dialog box. Selecting the blue face near the top or the bottom of the block did not matter. The work plane controlled the location and direction of the draft, unlike the previous method. The resulting draft is significantly different than the previous one. See **Figure 7-27D**. Compare the two methods to see where material was added and subtracted.

PROFESSIONAL TIP

Both methods of adding draft affected the entire face. If the face above or below the work plane needs to remain undrafted, the face has to be split. You will learn how to do this in the next section.

Other Face Drafts

Open Example-07-16.ipt. Two of the cavity's vertical edges are filleted; the other two are square. Activate the **Face Draft** tool. In the **Face Draft** dialog box, set the angle to 12° (for emphasis). Then, pick the **Pull direction** arrow button if it is not already on. Select the green face; the arrow should point up out of the cavity. Pick the **Faces** button in the dialog box and select the red face on the cavity wall near the top edge. When you select the face, notice how all faces connected by radii are selected, similar to the **Fillet** tool. The selection ends at the square corners. Finally, pick the **OK** button in the dialog box to create the face draft. See **Figure 7-29.** If you look at the top edge of the cavity, you can see that three of the walls have a face draft. The fourth wall, between the square corners, remains straight.

There is a second type of draft that can be created with the **Face Draft** tool—a shadow draft. This fills the shadow that a curved face projects on a flat face. Open Example-07-17.ipt. Imagine a light placed at a 50° angle to the vertical from the flat blue face, as shown in **Figure 7-30A.** The area that is not illuminated by the light, the shadow area, is filled by the **Face Draft** tool. The volume of the part is increased. See **Figure 7-30B** and **Figure 7-30C.** The operation is listed in the **Browser** as a TaperShadow*xx*.

Figure 7-29.
The face draft is applied to the three walls connected by radii.

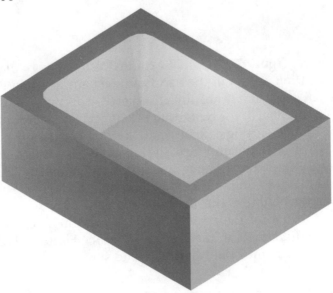

To place a shadow draft on this part, activate the **Face Draft** tool. In the **Face Draft** dialog box, set the angle to 50°. Then, pick the **Pull direction** arrow button and select the blue face on the part. The arrow should point up away from the part. In the dialog box, pick the **Faces** button and then select the yellow face on the part. Finally, pick the **OK** button to create the shadow draft. The name of the feature in the **Browser** is FaceDraft1.

Figure 7-30.
A—Imagining a light source casting a shadow on the part. B—The part before the face draft operation. C—The part with the shadow draft.

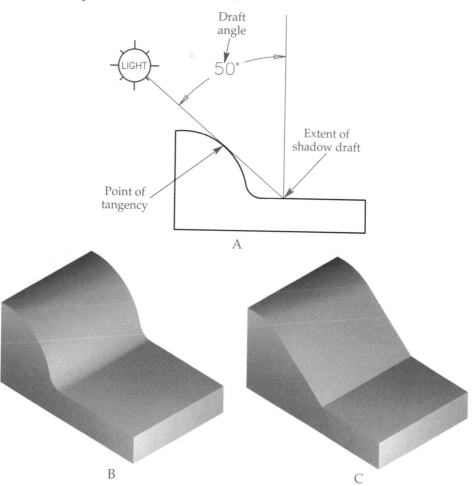

B

C

PRACTICE 7-1

❑ Open the part Practice-07-01.ipt.
❑ Put a 30° taper on the front face without creating another sketch. Use the **Face Draft** tool picking the blue face as the pull direction. Pick the red face, near the bottom edge, as the face to taper.
❑ The result is shown below.

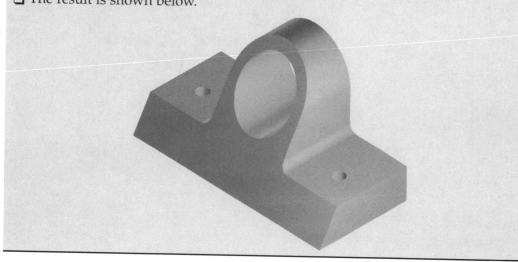

The Split Tool

The **Split** tool is used to divide a part or face based on selected geometry. The split can be along a face, work plane, or sketched curve. When used on a part, one side of the split is discarded. When used on faces, both sides of the split are retained. Also, when used on faces you can select multiple faces or apply the split to all the faces of the part. The examples in this section will demonstrate the principles of the **Split** tool. At the end of this section, a practical example is provided.

Using the Split Tool

Open Example-07-18.ipt. There is an unconsumed sketch on the front face consisting of two lines and two arcs. This sketch will be used to split the part. Pick the **Split** button in the **Part Features** panel. The **Split** dialog box is opened, **Figure 7-31.** Pick the **Split Part** button in the **Method** area. This applies the operation to the part, not faces. The outline of the part is highlighted with a cyan centerline. Now, pick the **Split Tool** button if it is not already on. Then, select the sketch in the graphics window. A preview of the split line appears in red. An arrow points in the direction of the side of the split to be removed. In the **Remove** area of the dialog box, pick one of the two buttons until the arrow points toward the top half. Then, pick the **OK** button in the dialog box to split the part. The top of the part is removed. See **Figure 7-32.**

Figure 7-31.
The dialog box used
for a split operation.
A—Part. B—Faces.

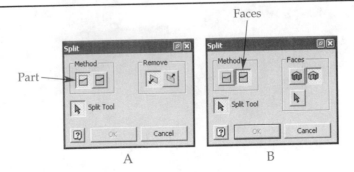

Figure 7-32.
This part was
created by splitting
a block.

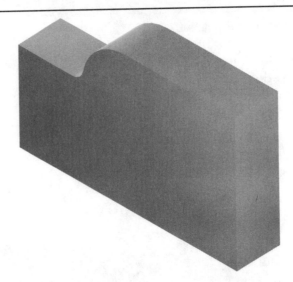

Undo the split operation. Then, activate the **Split** tool. This time, pick the **Split Face** in the **Method** area. With the **Split Tool** button on, pick the sketch on the part. Notice that a double arrow points perpendicular to the face. Also, the **Faces to Split** button in the **Faces** area is turned on. Select the front face on the part; it is highlighted in cyan. Finally, pick the **OK** button in the dialog box to split the face. The face is now divided into two parts. Change the color of each face by right-clicking on it and selecting **Properties** from the pop-up menu. In the **Face Properties** dialog box, select the new color. The end result is shown in **Figure 7-33.**

Splitting a Helical Gear

Open Example-07-19.ipt. This part is a gear blank with a sketch of a gear tooth. In this example, you will sweep the sketch along a curve, create a circular pattern, and trim off the excess. The end result is a helical gear. See **Figure 7-34.** Swept features are covered in detail in *Chapter 10*.

Figure 7-33.
The front face is split. Each face can be changed to a different color.

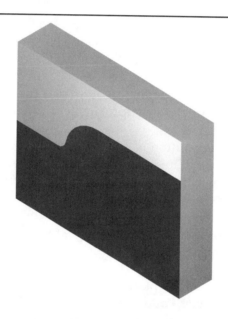

Figure 7-34.
A helical gear that will be created from a "blank."

Gear teeth generally have a profile based on an involute curve. This complex curve can be closely approximated by calculating and creating a number of points on the profile and then drawing a spline through the points. This process, though accurate, results in large files. To show gear teeth while keeping the file small, the spline can be approximated with an arc. This process has been used for many years by drafters to show gear teeth and was used to create the tooth profile in the file. For those interested, the gear being created has a diametral pitch of 5 and 18 teeth, which define the diameters and profile.

The visibility of the YZ plane and Y axis has been turned on. These will be used to create a helical line on the face of the cylinder to sweep the sketch. This line will have a 21° angle to the YZ plane. You need to create a new work plane before sketching the line.

1. Activate the **Work Plane** tool.
2. Pick the Y axis and then the YZ plane. Enter an angle of –21° in the **Angle** dialog box.
3. Activate the **Split** tool. In the **Split** dialog box, pick the **Split Face** button in the **Method** area. Then, pick the **Split Tool** button and select the new work plane. Then, select the outside curved face. Finally, pick the **OK** button to split the face. See **Figure 7-35.** Note: This actually splits the face in two places because the work plane intersects the face at the top and bottom of the part.
4. Start a new sketch on the work plane you created.
5. Use the **Project Geometry** tool to project the edge of the face split onto the sketch plane. This edge closely approximates the helical path needed to sweep the sketch.
6. Finish the sketch.
7. Pick the **Sweep** button in the **Part Features** panel. This tool is covered in detail in *Chapter 10*, but in this case, it is very easy to use.
8. Pick the **Profile** button in the **Sweep** dialog box and select the gear tooth sketch. Select an area as if you are extruding the profile. Then, pick the **Path** button in the dialog box and select the projection of the split face. Finally, pick the **OK** button to create the swept feature. A single tooth is created. See **Figure 7-36.**

A few more steps are required to finish this first tooth, create the pattern, and clean up the gear.

1. Using the **Fillet** tool, place a .04″ fillet on each side of the base of the tooth.
2. With the **Circular Pattern** tool, make a pattern of the tooth and the fillets as the features. Use the cylinder for the rotation axis, and a placement of 18 at 360°.
3. Create a sketch plane on the front of the gear. Draw a 4″ diameter circle at the gear center. In gear terminology, this dimension value is the pitch diameter plus twice the addendum.

Figure 7-35.
Splitting the curved face of the blank.

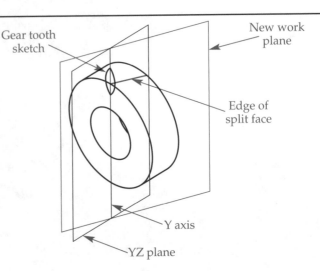

Gear tooth sketch

New work plane

Edge of split face

Y axis

YZ plane

Figure 7-36.
The profile is swept to create one tooth. The top of the tooth needs to be removed.

4. Finish the sketch.
5. Using the **Split** tool, remove the portion of the gear teeth that is outside of the circle. Refer to **Figure 7-34.**

The gear is almost complete. However, use the **Rotate** tool to rotate the view and look closely at the back of the gear. A small error is present at the intersection of the tooth fillets and the back face. You may need to zoom in to see the error. To correct the error, split the gear again using a new work plane.

1. Create a work plane offset from the back face a distance of –.1″. This will put the plane "in" the part.
2. Use the **Split** tool and this work plane to remove a thin section from the back face. Make sure the direction arrow is pointing out from the back of the gear.
3. Use the **Rotate** tool to rotate the view and verify that the error has been corrected.

Chapter Test

Answer the following questions on a separate sheet of paper.
1. What is the purpose of the **Shell** tool?
2. Why would faces be removed during the shell operation?
3. Which types of faces can be removed during a shell operation.
4. Give the three shell options for determining the position of the cavity wall in relation to the base part.
5. Define *web.*
6. Define *rib.*
7. Which tool is used to create a web? a rib?
8. What types of objects are used as the path by the tool(s) in Question 7?
9. What is a *sketch coordinate system (SCS)?*
10. Which tool is used to change an SCS?
11. Define *emboss.*
12. Define *engrave.*
13. Which tool is used to emboss text? to engrave text?
14. Which tool is used to create text in a sketch?
15. Define *aspect ratio.*
16. Which three types of "images" can be placed on a part using the **Decal** tool?
17. Define *draft angle.*
18. Which tool is used to create a draft angle on a part?
19. What is the purpose of the **Split** tool?
20. If the **Split** tool is used to split a part, what is true of one side of the split?

Chapter Exercises

Exercise 7-1. Open the part Exercise-07-01.ipt. Shell the part with a wall thickness of .375"; remove *all* of the bottom faces. The bottom of the part is opposite the two shaft journals. The finished part is shown below in a section view. In the **Browser**, drag the hole feature down below the shell and review the results.

Exercise 7-2. Open the part Exercise-07-02.ipt. Shell the part with a wall thickness of 2 mm on the long edges and 3 mm on the flat ends; remove the bottom (rectangular) face.

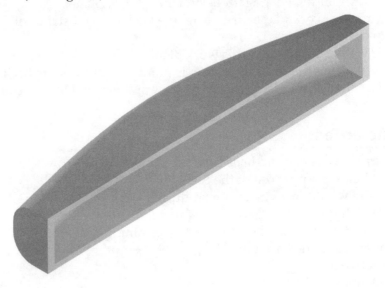

Exercise 7-3. Start a new metric Standard (mm).ipt file. Construct the ribbed collar shown below. Use the **Rib** tool to create one rib and then create a circular pattern. Use your own dimensions for fillets and rounds that are not dimensioned.

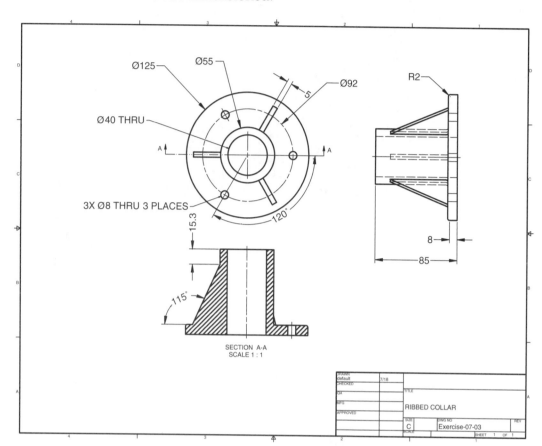

SECTION A-A
SCALE 1 : 1

RIBBED COLLAR
Exercise-07-03

Exercise 7-4. Open Exercise-07-04.ipt. This is the base for a computer keyboard. Create three .1″ thick ribs, as shown below. The long rib terminates on the midpoint of the sides. The distance between the intersection of each short rib and the end of the long rib is 1/4 of the length of the long rib. You can use dimensions and equations to set this spacing.

Exercise 7-5. Open Exercise-07-05.ipt. Create the text INVENTOR on the work plane. The text should be in Arial font, bold, and .5" high. Approximately center the text on the part. Then, emboss the text on the curved face with a height of about .2". Edit the emboss feature, check the **Wrap to Face** text box, and select the curved face as the face. What happens?

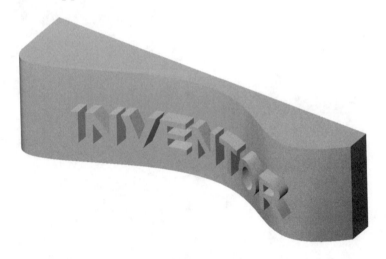

Exercise 7-6. Start a new metric Standard (mm).ipt file. Create a 100 mm × 30 mm × 15 mm block. Engrave text on the top face, as shown below, to a depth of 1 mm. The large text is 10 mm high and the small is 3.5 mm high. Approximate the text position, but use the orientations shown below.

Exercise 7-7. Creating a complex surface finish, such as a knurl, results in a large file. Instead, the finish can be represented with an image. Start a new English Standard (in).ipt file. Create a 5″ × 5″ × 1″ block. Using the **Decal** tool, place the image file Knurl Pattern.bmp on one of the large faces. Make sure the image completely covers the face.

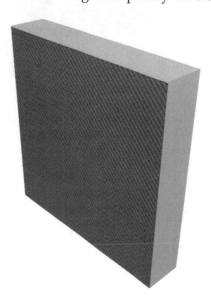

Exercise 7-8. Open the part Exercise-07-08.ipt. Using the **Decal** tool, place the Excel file Serial Number.xls as an "image" on the front faces, as shown below. Approximately center and size the label, but it should be "square" to the part.

Exercise 7-9. Open Exercise-07-09.ipt. Using the **Face Draft** tool, add the four tapers shown below. Each taper angle is 20°.

Exercise 7-10. Open Exercise-07-10.ipt. Create a 50° shadow draft. Select Face A as the pull direction. Select Face B as the face. Note that the draft is applied to both sides of the shaft journal, as shown below.

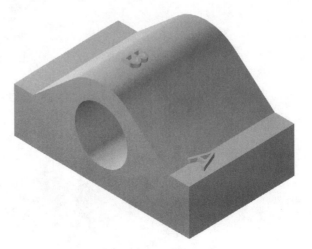

Exercise 7-11 Open Exercise-07-11.ipt. Split the flat, circular face on the end of the part into two faces. The edge between the two faces should pass through the center of the part. Then, use the **Face Draft** tool to create a 45° chisel point, as shown below. Remember, you can use work planes when splitting faces and creating drafts.

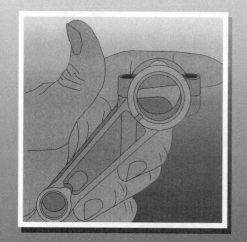

CHAPTER 8

Creating Drawings

Objectives

After completing this chapter, you will be able to:

✳ Create 2D drawings of your models.
✳ Understand the different views and how to create them.
✳ Specify paper size and border.
✳ Edit existing drawing views.
✳ Update the drawing views when the part changes.

User's Files

*The following is a list of files that you will need to work through this chapter. These files can be found on the **User's Files** CD included with this text.*

Examples	Practice	Exercises
Example-08-01.ipt	Practice-08-01.ipt	Exercise-08-01.ipt
Example-08-02.ipt		Exercise-08-02.ipt
Example-08-03.dwg		Exercise-08-03.ipt
		Exercise-08-04.ipt
		Exercise-08-05.ipt
		Exercise-08-06.ipt

Preparing to Create a Drawing Layout

Open Example-08-01.ipt and examine the part. Notice the XYZ coordinate system and the part's orientation to that coordinate system. This will help you to understand the creation of the drawing views. Pick **New** from the **File** pull-down menu. In the **Open** dialog box, double-click on Standard.idw template. A new Inventor drawing will be created with a default border and title block. See **Figure 8-1.**

This drawing consists of a sheet, a border, and a title block. The *sheet* is the size of the beige drawing area. It determines the size of the border and the title block. For example, change the size of the sheet and the border and title block change with it. The *border* is the two rectangles with the alphanumeric zone labels. The *title block* is

Figure 8-1.
Inventor file with a default border and title block. The shaded area is the sheet.

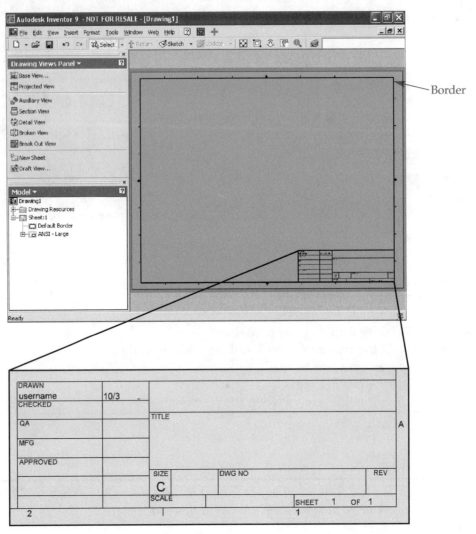

located in the lower-right corner of the border and consists of areas for the drawing title, date, designer's initials, sheet size, etc. Notice that this drawing's current sheet size is C. This is the default size.

With manual drafting, the drafter would go to the drawer and choose a sheet of vellum that was calculated to be large enough for their drawing views. Suppose a D-size sheet was needed. In Inventor, the **Browser** is used to change the sheet size to D. Find Sheet:1 in the **Browser** and right-click on it. Choose the **Edit Sheet...** option in the pop-up menu. This accesses the **Edit Sheet** dialog box. Use the **Size:** drop-down box to change the sheet size from C to D. In the **Name:** text box, change the name of the sheet to D-Size. See **Figure 8-2.** Pick **OK** to exit. In the graphics window, you should see the sheet get larger, the border expands to fill the new sheet, and the title block text updates to read D.

Figure 8-2.
The **Edit Sheet** dialog box is used to change the sheet size, rename the sheet, and change the sheet orientation.

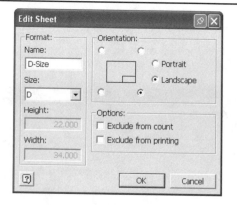

Creating the Drawing Views

As the Inventor 2D drawing layout is created, you are actually working on a *separate* drawing file. However, when you create the drawing views of the part then that part file is then linked to the drawing file. This means that any future changes made to the part are automatically updated in the drawing view. This also works the other way as well. Change a dimension on the drawing view and the part model updates to reflect the new values. This will be covered in more detail later in this chapter. Save this drawing as Example-08-01.idw. The IDW file extension is the default and does not need to be entered. Pick the **Window** pull-down menu and notice the two Inventor files currently open. See **Figure 8-3.** The IPT file is the part model for this example and the IDW file is the 2D drawing that we are creating.

Figure 8-3.
The **Window** pull-down is used to manage multiple open files.

PROFESSIONAL TIP
Keep in mind that more than one part file can be used to create drawing views in a single drawing layout. For example, if you were creating a sheet of details that contains views of many parts.

Creating a Base View

The next step is to create drawing views. Start by creating a *base view* from which all others will be generated. To create the base view, pick the **Base View...** button on the **Drawing Views Panel**, pick **Base View...** in the **Model Views** cascading menu of the **Insert** pull-down menu, or right-click on D-Size:1 in the **Browser** and choose **Base View** from the pop-up menu. This accesses the **Drawing View** dialog box. See **Figure 8-4.**

CAUTION

Follow the next step carefully because you can do things in the graphics window while the dialog box is still up on the screen.

Figure 8-4.
The **Drawing View** dialog box is used to create, scale, and label different views.

Pick to access
Options tab

New scale
value

Hidden
line

Hidden
line
removed

Shaded

Cursor and
image

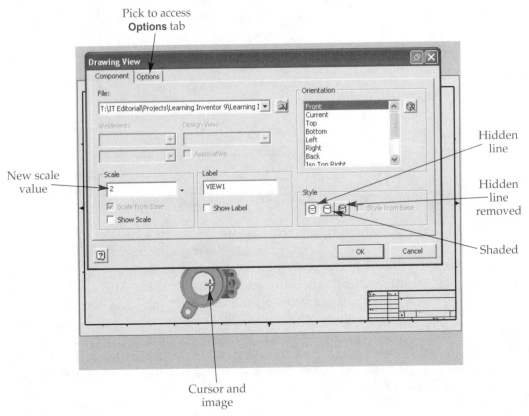

Before you change anything in the dialog box, move the cursor around the graphics window. Notice the cursor and image of the example part model in Figure 8-4. The image will be the base view if a point is selected on the sheet. Before you pick a point, look at the dialog box. If you inadvertently picked a point, just undo and start again.

The **File:** field shows the opened part model as the source for our 2D drawing layout. If no part file was open, then you can locate and specify another part file as the source. The scale of the drawing view is entered in the **Scale** area. Inventor calculates the scale based on the size of the model and the sheet size. You may accept it or override it with your own value. All other views will inherit this scale. Use a value of 2. A name for the view is entered in the **Label** area. Labels are usually used for section views or detail views. The **Orientation** area gives you the option of choosing any one of the standard orthographic drafting views for the first base view. Remember, all subsequently created views will be generated from the base view, so choosing the proper start is important. For this example, we will use Front. In the **Style** area, there are three buttons for the three states that a drawing view can be displayed in Inventor. The left button is for showing visible and hidden lines, the middle button is for showing visible lines only, and the right button is for showing a shaded object. The **Shaded** button is extremely useful for the final isometric view we will be doing in this example. It is also helpful for visualizing the final part's shape. The **Options** tab is only available for base views

Finally, pick a point in the upper left-hand corner of the drawing sheet. The front view of the part should be created and the dialog box closed automatically. The example part file was manipulated, so that it's "top" was really what Inventor calls the front view. This is derived from the orientation of the coordinate system when the part is started. This is not a problem, because all subsequent drawing views will be generated from this base view using standard orthographic projection. In the **Browser**, this will be listed as VIEW*n*. Rename it so it is more descriptive. Left-click twice (not double-click) on the name and rename it Top View.

PROFESSIONAL TIP

If you want to change the scale of the linetypes used in the drawing views, pick **Styles Editor...** in the **Format** pull-down menu. Double-click on **Default Standard (ANSI).** Change the value in the **Global Line Scale** text box and pick **Save** to see the changes immediately. Pick **Done** to exit the dialog box.

Now, project a view below the top view just created. This new view is normally referred to as a *front view* in drafting terminology. In the **Browser**, right-click on Top View to access the pop-up menu. Pick **Projected** from the **Create View** cascading menu. You should get a line originating at the top view that follows your cursor movements. Pick a point on the screen below the top view, right-click, choose **Create** from the pop-up menu. See **Figure 8-5.** Rename this view Front View. You can move the view by selecting its edge and dragging it to a new location. Although this view is constrained because it is projected from the base view, you can also use crossing and window methods to select the views.

Now as far as drawing views go, this one is pretty poor. Technically it does represent the *actual* front view of our part, but that is not a *descriptive* view of our part. You might have learned in drafting class that since space on the paper is limited, every view must do the work of two views. Instead of a standard front view, you are going to experiment with Inventor's section views. Move the front view by picking it in the graphics window. Now left-click on the view's edge, hold, and drag the front view down below the bottom border of the sheet. Leave it there because it will be used later in this chapter.

Figure 8-5.
The front view is placed below the base view, or the top view.

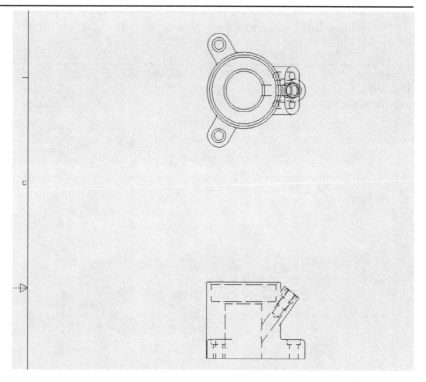

Creating Other Views

Inventor can be used to create a number of other views using the base view. These other views include full-sections, half-sections, aligned sections, isometric views, auxiliary views, detail views, broken views, and breakout views.

NOTE

Breakout views are used to remove covering geometry to reveal the inner geometry of a part. These will be covered in detail in *Chapter 17*.

Full Sections

In the **Browser**, right-click on Top View. In the pop-up menu, pick **Section** from the **Create View** cascading menu. Be careful because your mouse clicks need to be very subtle. Notice that there is now a red border around the top view in the graphics window. This will be important in the next steps. Follow the steps as listed to define two points of the section line:

1. Hover over the center of the part until you "receive" the center point, but do *not* click anything.
2. Move your cursor straight to the left until it is beyond the red border. Keep a straight projection line as you do this. A dotted line gives you visual feedback that you are keeping it straight. Once your cursor is at a point beyond the red border and the projection line is straight, pick that point. This point is the start of your section line.

Figure 8-6.
The cut line and the resulting section view.

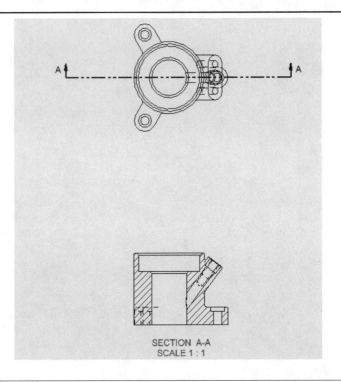

3. Move your cursor in a straight line to the right until it is beyond the red border on the right side of the view. Again pick a point.
4. Right-click and choose **Continue** in the pop-up menu. This accesses the **Section View** dialog box. Make sure A is shown in the **Label** text box, and not A-A or B.
5. Pick a point below the top view and the section view is created. See **Figure 8-6.**

This is a full section view of the part. If care was not taken when defining the cutting line, the section view can be redefined by clicking and holding the cutting line, and then dragging it to a new location. Create a new section view and do not select the center of the part this time—just sketch a straight line through the part. Then if you click and drag the cutting line you will notice the section view is redefined. These views are fine, however, the views are not fully descriptive. Delete all the section views just created.

Half-Section View

In the **Browser**, right-click on Top View and pick **Section** in the **Create View** cascading menu. Be very careful in the next steps because your clicks need to be precise. Again, notice that there is a red border around the top view. This will be important in the next steps. Follow the steps as listed to define three points of the section line:

1. Hover over the center of the part until you "receive" the center point, but do *not* click anything.
2. Move your cursor straight *down* until it is beyond the red border. Keep a straight projection line as you do this. A dotted line gives you visual feedback that you are keeping it straight. Once your cursor is at a point beyond the red border and the projection line is straight, pick that point. This point, Point 1, is the start of your half-section line. See **Figure 8-7.**
3. Move your cursor in a straight line back to the center of the part. A green dot will appear when you are over the center. Pick this point, Point 2.
4. Now, move the cursor to the right until it is beyond the red border on the right side of the view. Once your cursor is at a point beyond the red border and the projection line is straight, pick that point, Point 3. These are the three points needed.

Figure 8-7.
The cut line and the resulting half-section view.

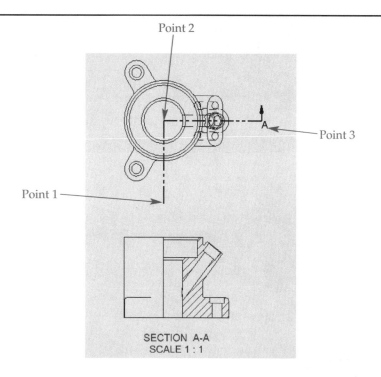

SECTION A-A
SCALE 1 : 1

5. Right-click and choose **Continue**. In the **Section View** dialog box, make sure that the section label is A. Pick a point below the top view.

If you look closely, the mounting foot is not depicted very well in this new section view. The next section will cover how to correct this.

Aligned Section View

The aligned section view will best describe the type of part in this example. The part has mounting feet that are not lying in the same plane. In the **Browser**, right-click on Top View and pick **Section** in the **Create View** cascading menu. Be very careful in the next steps because your clicks need to be precise. Again, notice that there is a red border around the top view. The section line is created using the following steps:

1. Hover over the far right quadrant of the mounting foot until you have "acquired" the point. See **Figure 8-8A.** Do not pick this point.
2. Move your cursor to the right until it is beyond the red border. Keep a straight projection line as you do this. A dotted line gives you visual feedback that you are keeping it straight. Once your cursor is at a point beyond the red border and the projection line is straight, pick that point. This point, Point 1, is the start of your half-section line. See **Figure 8-8B.**

Move your cursor in a straight line back to the center of the part. A green dot will appear when you are over the center. Pick this point, Point 2. See **Figure 8-8C.**

4. Then move to the center of the upper-left mounting foot and hover to acquire that point. Continue in the same direction until you are outside the red border. Pick this point, Point 3. See **Figure 8-8D.**
5. Right-click and choose **Continue**. In the **Section View** dialog box, make sure that the section label is B. Pick a point below the first section view.

Figure 8-8.
Creating an aligned section view. A—Hover over the edge of the mounting foot. B—Pick a point to the right of the cut line. C—Pick the center of the part. D—Pick the final point beyond the upper mounting foot.

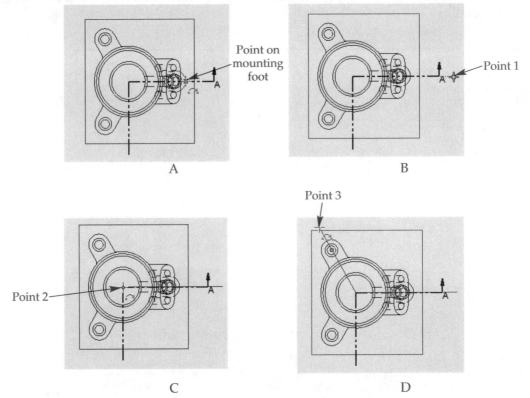

Figure 8-9.
From top to bottom.
The top view, the
half section, and the
aligned section
view.

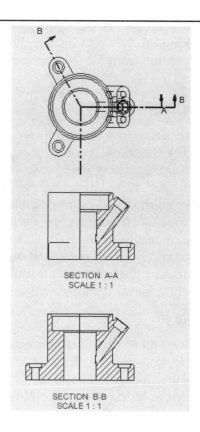

The finished section views are shown in **Figure 8-9.** If you need to move the views to arrange them on the sheet, simply pick their edges and drag them into position. Notice that the projected views are constrained to the base view from which they are derived. Move the base view and all other views follow. It is clear that the aligned section view is the best to depict the front view of our example part. The mounting feet are shown according to drafting convention for a cylindrical part with features that do not lie in the same plane. Try moving the cutting lines for these views in the top view. You will find that you will not be able to, because of the way they were defined. As the points were picked, Inventor inferred constraints. They are in essence "nailed down" to the part's features. To move the cutting lines, you would have to edit the cutting lines, and show and remove the constraints.

CAUTION

Be careful renaming the views in the **Browser**, because it will affect the label shown under the section views. For example, if you rename the view to Section A-A in the **Browser**, then the label under the view would read SECTION Section A-A-Section A-A.

Section Views with Depth

You can also specify the depth, or thickness, of the *sectioned slice*. To do this, create a simple full section view of the base view. Before placing the view, in **Section View** dialog box pick Distance in the **Section Depth** area and change the value to .50. As the view is placed, in addition to the cutting line notice there is now a thick black line in the base view. This represents the .50 offset. The resulting section view depicts the slice of the part between the cutting line and the new thick black line.

Place the view and observe the results. You can change the section depth at any time by right-clicking on the view on the sheet or in the **Browser** and choosing the **Edit Section Depth** option from the pop-up menu. You also can convert it back to a full depth section view.

Section Views Based on Sketch Geometry

For all of the section views created thus far, the cutting line has been specified *on the fly* before placing the view. Another method uses sketch geometry that already exists as the basis for the cutting line. Ensure that the Top View is selected by clicking once in it. Once the top view is highlighted, pick **Sketch** on the **Inventor Standard** toolbar to enter Sketch mode. The **Drawing Sketch Panel** and the sketch tools will be available. Use the **Line** tool to draw a line across the base view. The line does not need be centered on the centerline of the part. Pick **Return** to exit Sketch mode. Right-click on the sketched line and choose **Create Section View** from the pop-up menu. Move the resulting cutting line in the base view by picking and dragging one of the green endpoints. The section view changes to reflect the new position of the cutting line.

Isometric Views

Inventor allows you to create isometric views automatically. If you have ever drawn an isometric view with AutoCAD or with a drafting board, you can really appreciate this feature. To create an isometric view we will use SECTION A-A, which is the half-section view. In the **Browser**, right-click on A:Example-08-01.ipt. This is the listing for the section. Pick **Projected** from the **Create View** cascading menu. A projection line will connect from SECTION A-A to the cursor. Pick a point to the upper right of SECTION A-A. Now, pick a second point to the lower right, thereby creating two isometric views. Right-click and choose **Create**. See **Figure 8-10**.

The two isometric views can be moved by dragging them. It is nice to keep them lined up vertically, so pick both of them at the same time by holding down the [Shift] key and dragging them both to the new position. You can also select them with a crossing or window selection method. You need to make some room for the auxiliary view coming up.

Figure 8-10.
Isometric views of
the part.

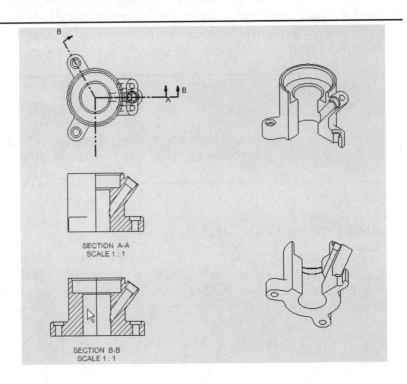

Right-click on one of the isometric views and choose **Edit View** from the pop-up menu. In the **Style** area of the **Drawing View** dialog box, uncheck **Style from Base** and pick the **Shaded** button. See **Figure 8-11**. Pick the **Options** tab and check the **Tangent Edges** and **Foreshortened** check boxes. Pick **OK** to exit and the isometric view is now a shaded rendering. Do the same for the other isometric. See **Figure 8-12**.

Auxiliary Views

An *auxiliary view* can be used to get a true representation of any inclined or oblique face on a part. The angled feature on our example requires an auxiliary view to show its inclined face so it can be dimensioned. SECTION A-A will be used to create the auxiliary view. You cannot use SECTION B-B, because it is not a true representation of our part. If you try to use SECTION B-B the software will not allow it. In the **Browser,** right-click on A:Example-08-01.ipt. Pick **Auxiliary** from the **Create View** cascading menu. This accesses the **Auxiliary View** dialog box and you are prompted to select a line. Leave the defaults in the dialog box and pick the line/edge shown in **Figure 8-13**. Move your cursor to the upper right and place the view where you can for now. Notice that the auxiliary is not exactly what we want. It was projected from a half-section view and is a true representation of that view. Now, delete this view.

Figure 8-11.
Changing the style of the view to **Shaded**.

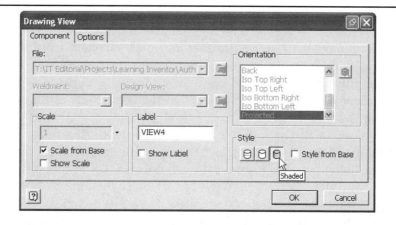

Figure 8-12.
The isometric views shown shaded.

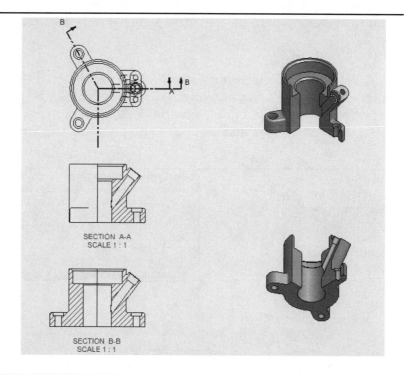

Figure 8-13.
Using the **Auxiliary View** dialog box to create a view. Pick a point of reference for the auxiliary view.

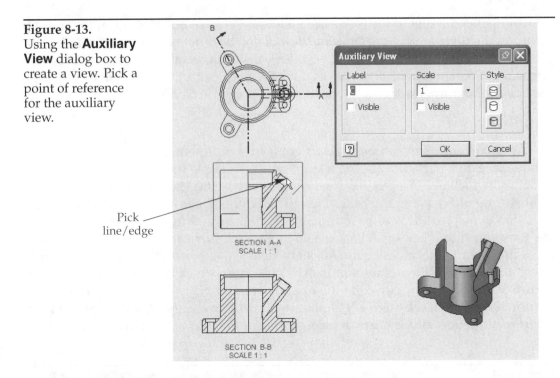

Pick line/edge

A full complete front view is needed to derive the auxiliary view we want. Right-click on Front View in the **Browser** and pick **Auxiliary** from the **Create View** cascading menu. Now pick the line/edge selected for the previous auxiliary. However, this time select this line/edge in the front view. Move your cursor to the upper right and place the view where you can for now. This is a true view of that angled feature, however, some changes are required.

Now delete the front view used to derive the auxiliary view. Right-click on Front View in the **Browser** and select **Delete** in the pop-up menu. This accesses the **Delete View** dialog box. See **Figure 8-14.** Do not pick **OK** at this point because both the front view and the auxiliary view will be deleted. Pick the **>>** button to expand the dialog box. Highlight the dependent view C and left-click on Yes in the **Delete** column changing it to No. Now pick **OK**. Move this view up next to SECTION A-A.

Generally, the entire part is *not* shown in auxiliary views. They usually only show the feature in question. This is covered in the section, *Editing the Drawing Views*, in this chapter.

Detail Views

A *detail view* focuses on a small feature of the part that cannot be shown or dimensioned properly in a drawing view. The detail view is typically drawn at a large scale—think of it as zooming in on the feature. Right-click on SECTION A-A and choose **Detail** in the **Create View...** cascading menu. This drawing view will be the

Figure 8-14.
The **Delete View** dialog box is used to select which views and their dependent views will be deleted.

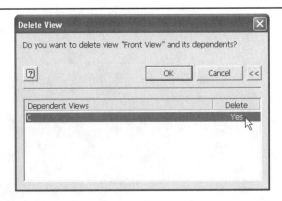

source for the detail view. This accesses the **Detail View** dialog box. Changes to the scale and the view label are made using this dialog box. In addition, you are prompted at the bottom of the screen to Select the start point of the fence. The start point, Point 1, is really the center of the detail circle. See **Figure 8-15.** Pick the Point 2 to define the detail circle (or fence). You are then prompted to Select a location for the view. Move your cursor to the upper right and pick a point to place the view. Please notice the label and scale of the detail view.

In the parent view, click on the circle. Use one of the green points and drag the circle out to include more of the part. Notice that the detail view changes to reflect the change in the circle. Save the file as Example-08-01.idw.

Broken Views

Create a new drawing using the Standard.idw template and change the sheet size to D. Right-click on Sheet:1 in the **Browser** and choose **Edit Sheet...** in the pop-up menu. In the **Edit Sheet** dialog box, change the size to D. Now rename the sheet to D-Size. Right-click on D-Size in the **Browser** and pick **Base View...** in the pop-up menu. See **Figure 8-16.** The file Example-08-02.ipt will be used as the basis for this IDW. Search for this file and make sure it is in the **File:** text box. Select Front in the **Orientation** list so a front view of the part is created. This view is the base view as well. Place the view by picking a point on the right side of the sheet and exit the dialog box.

Figure 8-15.
Pick two points that will be used to define the detail view.

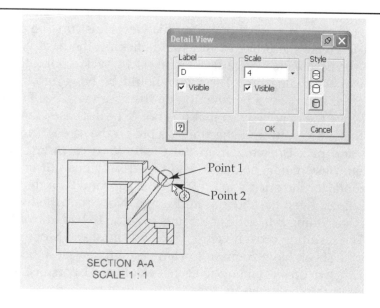

Figure 8-16.
A broken view is developed from a base view.

Figure 8-17.
The view runs off
the sheet.

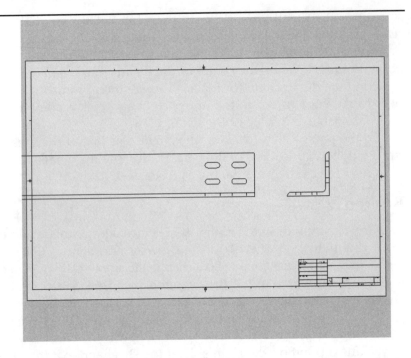

The broken view will be projected to the left of the base view. The **Broken View** tool does not create views by itself—it works on existing views only. In order to create a broken view, a normal orthographic view must first be created. This view will be used to create the broken view. Right-click on the base view and choose **Projected** from the **Create View** cascading menu. Then, pick a point to the left of the base view and exit the dialog box. The view should be hanging off the sheet as you would expect, or it may even be covering up the base view. See **Figure 8-17.**

Move the new view so that it is not over the base view. Then, pan and zoom so that the entire view and sheet are visible on the screen. Now, right-click on the new view and pick **Broken...** from the **Create View** cascading menu. This accesses the **Broken View** dialog box. See **Figure 8-18.** Choose **Structural Style** for the style, **Horizontal** for the orientation, 1.00 for the gap, and 2 for the number of symbols in the break lines. Keep the slider at mid-range. Now, pick the two points shown in Figure 8-18. Everything *between* these points is going to be removed from the view.

The resulting broken view will be off the sheet. Move the view to the right so it is on the sheet. Now zoom in to show the broken view. See **Figure 8-19.** Notice that there are two break symbols in each break line. Also, notice there is no specific broken view entry in the **Browser**.

If the style of the break needs to be changed after the view is created, the following steps have to be taken.

1. Hover the cursor over the break until it is highlighted and a green dot appears.
2. Right-click and select **Edit Break...** from the pop-up menu. This accesses the **Broken View** dialog box.
3. You can change style and display settings, but not orientation. Select the rectangular style and change the gap to 2.00". The rectangular style is used for nonstructural shapes.

When any view is projected from a view that was used to create a broken view, the new view is also broken. This is true whether the new view was created using the projected, auxillary, section, or detail tool.

To create an unbroken view of the same part, a new base view has to be added to the drawing. Right-click the drawing and select **Base View...** in the pop-up menu. Add another front view of the same part, this time place the base view off the drawing.

Figure 8-18.

Zoom out to see all of the view. The **Broken View** dialog box is used to define the type of broken view. Two points are used to define the break lines.

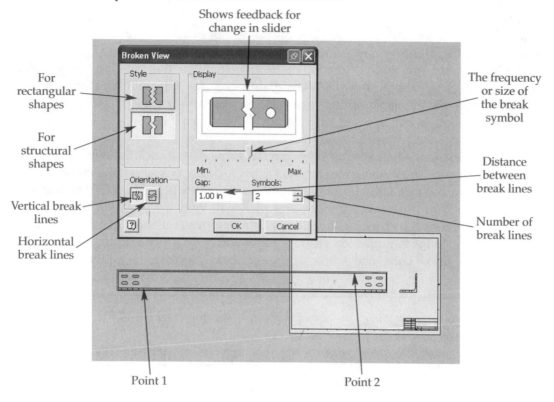

Figure 8-19.

The resulting broken view.

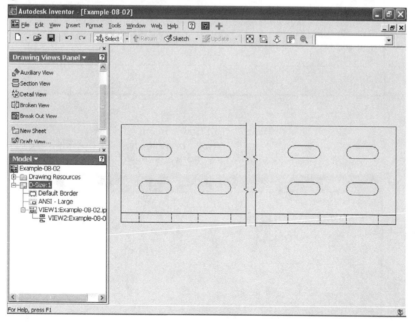

You now know how to create drawing views, but it is important to know how to edit them as well. Open the drawing Example-08-01.idw that you created earlier. This drawing should consist of a base view and a number of drawing views. Make sure you are not highlighting any geometry and right-click on a drawing view. Choose **Edit View...** from the pop-up menu, or double-click on the view in the graphics window. This accesses the **Drawing View** dialog box. See **Figure 8-20**. Depending on the view type, the available options are enabled or grayed out.

This dialog box can be used to "clean up" drawing views, so that they are more presentable and comply with standards.

1. Right-click on the half-section view, SECTION A-A, and pick **Edit View...** from the pop-up menu. Then choose the **Options** tab in the dialog box, check the **Tangent Edges** check box, and pick **OK**. The mounting foot should look better now. See **Figure 8-21**.
2. Right-click on the auxiliary view, and pick **Edit View...** from the pop-up menu. Change the style by picking the **Hidden Line Removed** button. On the **Options** tab, check the **Tangent Edges** check box. Enter VIEW C-C in the **Label** text box and check the **Show Label** check box. The view should look much better.

Figure 8-20.
A—The **Drawing View** dialog box is used to edit the view. B—The **Option** tab contains many options that affect the display of the view.

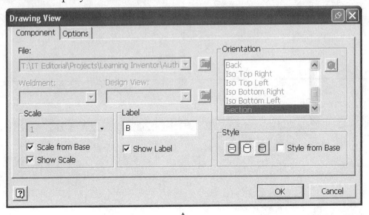

A

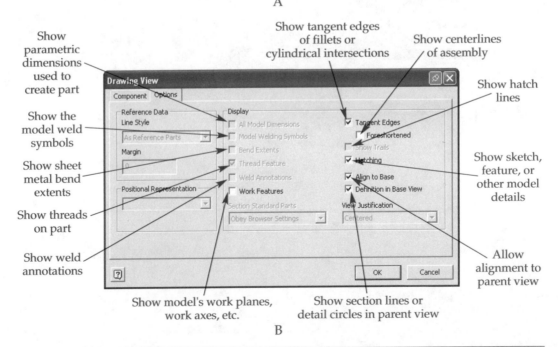

B

Figure 8-21.
A—The mounting foot with **Tangent Edges** check box unchecked, or off. B—The mounting foot with **Tangent Edges** check box checked, or on.

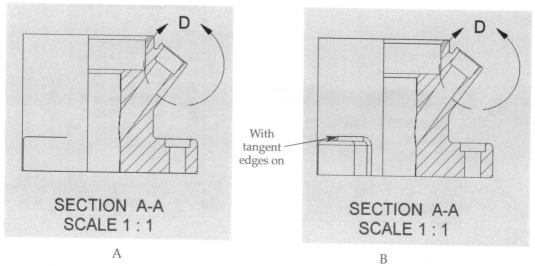

3. The hatching, or section lining, in the detail view can be confusing. Right-click on the detail view, and pick **Edit View...** from the pop-up menu. On the **Options** tab uncheck the **Hatching** check box, and pick **OK** to exit. The detail view should be easier to read.

Repositioning Annotation

In the top view, the section view labels A and B need to be dragged into better positions beside their cutting line arrows. Just click and drag them into position. You can do this to any piece of annotation in Inventor. It will retain its new position relative to the view.

Editing a Drawing View's Lines

There are other ways to edit drawing views, in addition to changing the properties of a given view. The lines or arcs that make up the graphics of a drawing view can be changed. The visibility, color, or linetype can be changed. To change color or linetype, select the line(s), right-click, and choose **Properties** from the pop-up menu. This accesses the **Edge Properties** dialog box. See **Figure 8-22A**.

In the dialog box, you will notice By Layer is the default for **Line Type** and **Line Weight.** The line type and weight can be changed using these drop-down menus. To change color select the color. To select multiple lines or arcs, hold down the [Shift] key and then select them.

To change the visibility of a particular view's lines or arcs, select the line or arc and right-click. Then choose **Visibility** in the pop-up menu, **Figure 8-23A**. You can also use window selection method to select the lines of arcs. It can be a bit tedious to get all of them. Do this for the auxiliary view until you get the finished view shown in **Figure 8-23B**.

PROFESSIONAL TIP

As an alternative, you could create a section view with a cutting line that does not pass through the part, but instead simply looks at it. This is similar to the old drafting technique of creating a view. The term *view* as used here is not to be confused with *drawing view* as we have been using it throughout this chapter.

Figure 8-22.
A—Changing the line style of the view. B— Changing the color of the view.

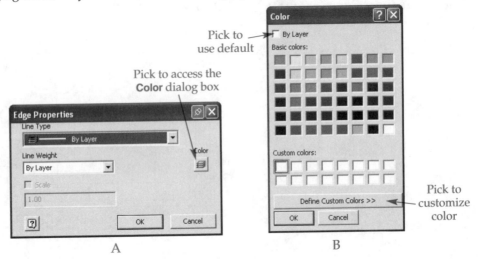

Pick to use default

Pick to access the **Color** dialog box

Pick to customize color

A

B

Figure 8-23.
A—Changing the visibility of a line or arc. B—Final detail view with all other lines invisible.

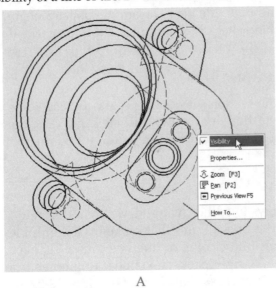

A

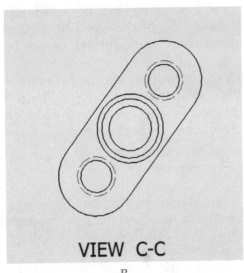

VIEW C-C

B

Figure 8-24.
Changing the properties of a label.

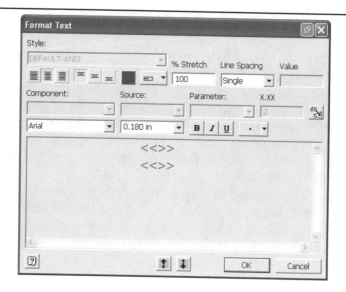

A drawing view's label can also be moved. In the graphics window, pick and drag the label, "VIEW C-C", up so it is closer to the view's visible edges. Furthermore, the label can be edited. Right-click on the label and choose **Edit View Label...** in the pop-up menu. This accesses the **Format Text** dialog box where the label's properties are controlled. See **Figure 8-24.** The double chevrons represent the default text. While you cannot delete the default text, you can edit it as to font type and size, etc. Also, you can add more text by simply typing after the double chevrons. Save your drawing under a new name.

PROFESSIONAL TIP

If the drawing view uses hidden lines and their scale just does not seem right, perhaps too small, then pick **Standards...** from the **Format** pull-down menu. Change the scale in the **Line Weight** area of the **Common** tab. Pick **Apply** to affect the drawing's hidden lines. This will affect the current drawing only. More on drafting standards in *Chapter 9.*

Changing the Model

Inventor is capable of bidirectional associativity. This means the *drawing* will follow any changes made to the *part*, or the *part* will follow any changes made to the *drawing*. If it is not already open, open Example-08-01.ipt. Change the extrusion height of Extrusion1 to 2.5 and update the part. Without saving the part, switch Windows to the Example-08-01.idw drawing you have created. There will be a pause and your drawing should have changed to reflect the changes made to the part. To better see this happen, change your windows setup to the **Arrange All** option. Now undo and redo the changes just made.

This illustrates the concept of bidirectional associativity. Conversely, you could change a model dimension in the drawing and thereby affect changes to the part. The drawing does not have any model dimensions showing, however. In *Chapter 9* you will have a chance to learn about dimensions and other types of annotations. Undo your changes to the part and switch back to the drawing to ensure that the undone changes have taken effect.

Changing the View Orientation

Sometimes the part is not oriented properly to Inventor's default coordinate system. This happens quite frequently as the developers of Inventor seem to disagree with the drafting industry as to what exactly is the top or the front of a part. If you notice, Inventor's default IPT template has the Z-axis pointing out of the screen toward the lower-left corner. Beginners will unknowingly begin to create their part based on this erroneous orientation. This orientation is typically used for engineering analysis in the 3D special effects industry, but is not used in the design or drafting industry.

It is recommended that you create your own template that uses the proper orientation of the coordinate system. Inventor's definition of a *front view* is the drafting industry's definition of a top view, or plan view of the XY plane. Instead, the drafting convention is for the Z-axis to point up indicating depth. If viewed along the Z-axis looking down at the part, this is referred to as the top view. All other viewpoints are derived from this top view. Other times, the person who created the part just did not follow convention and the part, may not be aligned along any of the coordinate system's axes. At times it will be necessary to change the view's orientation using the **Drawing View** dialog box.

PRACTICE 8-1

❑ Open Practice-08-01.ipt and examine the part and its orientation to the coordinate system. The part is oriented properly to the Z-axis, but it could be rotated 90° clockwise about the Z to bring the X-axis aligned with the part's long side. It is difficult to rotate a part once it is created. The only option is to change the view's orientation on view creation.

❑ Begin a new IDW based on the English template and on C-size sheet.

❑ First, produce the traditional top view of the part in the upper left-hand corner of the sheet. Select the **Base View** tool and try out all of the orthographic orientations. You will find that none of them fit our requirements for the top view. Of course, the isometric orientations do not either.

❑ While the **Drawing View** dialog box is open, select the **Change View Orientation** button. This opens a new interface called **Custom View**. From the toolbar along the top of the screen pick **Rotate** and then press the *[Spacebar]* to go into **Common View** *mode*. Pick the green arrow that is looking at the "front" of the Part. See **Figure 8-25.** You now have the correct orientation. Hit the [Esc] key to exit the **Rotate** tool. Take a moment to examine the rest of the tools available in this interface. You have used most of them before. The **Rotate at Angle** icon brings up the **Incremental View Rotate** tool, which is unique to this environment.

❑ Once you have the correct orientation, exit the interface by picking the **Exit Custom View**, which is the green check mark in the upper-left corner.

❑ In the **Drawing View** dialog box, pick a point in the lower left-hand corner of the drawing sheet. This is what is traditionally known as the front view. With the correct base view, the front view, it is a simple matter of creating the projected views.

❑ Right-click on the base view, and pick **Projected** from the **Create View** cascading menu. Pick three points— above for the "top" view, a point to the right for the "side" view, and a point to the upper-right for the "isometric" view. Refer to **Figure 8-26** for the correct layout.

❑ Right-click and choose **Create** to make all of the views.

❑ Double-click on the isometric view and choose the **Shaded Style**.

Figure 8-25.
A—Using the
Custom View to get
a proper front view.
B—The proper front
view.

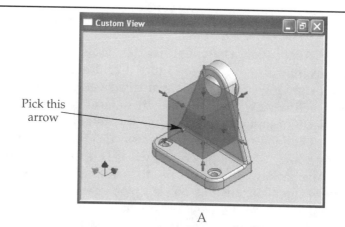

Pick this
arrow

A

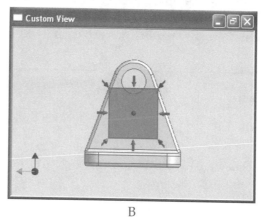

B

Figure 8-26.
Front view, top
view, side view, and
isometric view of
the part

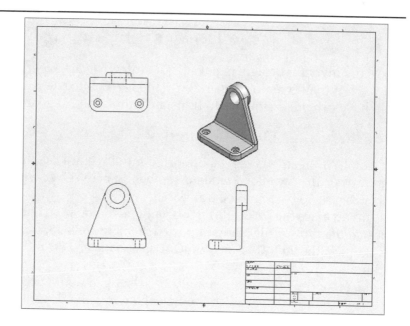

**PROFESSIONAL
TIP**

The **Custom View** interface is very helpful to reorient the part.
However, sometimes all that is needed to do is to change the
current view of the part in Inventor. Open the part, use the
Rotate tool to achieve the proper orientation, begin a new
drawing (IDW), create a base view, and choose the **Current**
orientation from the list.

Inventor's Options for Drawings

Selecting **Application Options...** in the **Tools** pull-down menu accesses the settings for Inventor drawings. See **Figure 8-27.** Remember, these are global settings that affect all drawings. **Section Standard Parts** controls the sectioning of standard parts in the drawing view of assemblies. By default, the **Obey Browser Settings** option is selected. Section standard parts is off by default in the drawing browser. The setting can be changed to **Always** or **Never.** If the **Precise View Generation** check box is checked, as it is by default, precise views are created as opposed to faster approximate views. Unchecking it is done only to speed up the process of creating drawing views. **Retrieve model dimensions on view placement** will cause any applicable model dimensions to appear on the view on creation. If unchecked, it will be necessary to select **Model Dimensions** on the **Options** tab in the **Drawing View** dialog box. If **Display line weights** is checked, lineweights defined in the current drafting standard will be used. If not checked, all visible lines are displayed using the same lineweight.

Figure 8-27.
The **Drawing** tab is used to change the global settings that affect all drawings.

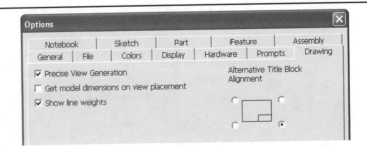

File Transfer with AutoCAD

Many Inventor designers have prior AutoCAD 2D experience. With this in mind and the fact that Inventor does not use the DWG file extension, the Autodesk developers have provided a means to translate AutoCAD 2D geometry.

Import AutoCAD 2D Geometry

AutoCAD geometry can be imported into Inventor via the **Open** dialog box. Pick **Open...** from the **Inventor Standard** toolbar. Specify DWG files as the file type and select Example-08-03.dwg to be translated.

This accesses the **DWG File Import Options** dialog box. The first page is where you indicated the type of file being imported. See **Figure 8-28.** It is a 2D AutoCAD geometry, so pick the **AutoCAD or AutoCAD Mechanical File** radio button. Pick **Next** to continue.

The second page is used mainly to select geometry. See **Figure 8-29.** Geometry can be selected by the layer. In the **Selection** area the default selection is **All.** If **All** is unchecked, then geometry can be selected in the preview image. Use the defaults and pick **Next** to continue.

The final page determines the output based on the selections chosen. See **Figure 8-30.** Will it be used to create a new Inventor Drawing (.IDW)? Perhaps the geometry will be used in the resources of an IDW—the title block and the border. This is how you could bring in your company's custom title blocks and borders. Leave it set to **New Drawing.** The **Promote Dimensions To Sketch** check box is used to place the AutoCAD dimensions as sketch dimensions on a sheet in a drawing file. These dimensions are associated with the sketch and change if you move or edit sketch geometry. By default, dimensions are placed as drawing dimensions. Drawing dimensions do not

Figure 8-28.
AutoCAD and Mechanical Desktop files can be imported into Inventor.

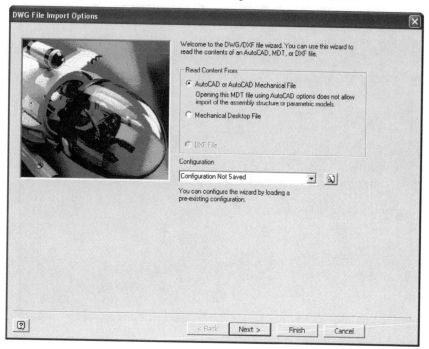

Figure 8-29.
The geometry, the whole file or parts of the file, can be selected using this page.

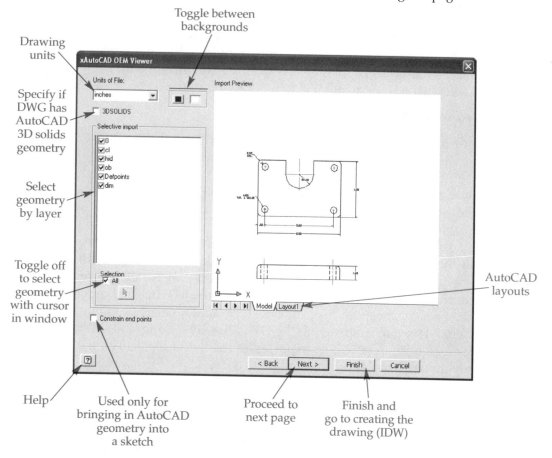

Figure 8-30.
This page is used to define what AutoCAD geometry should be used.

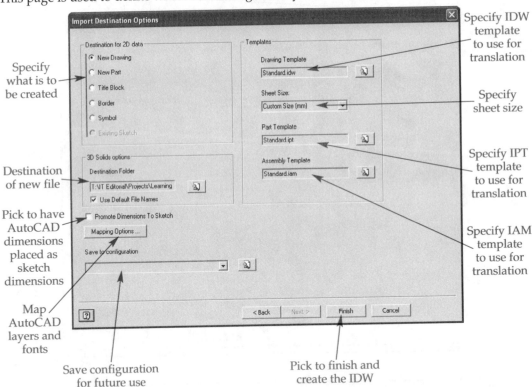

Specify what is to be created

Destination of new file

Pick to have AutoCAD dimensions placed as sketch dimensions

Map AutoCAD layers and fonts

Save configuration for future use

Specify IDW template to use for translation

Specify sheet size

Specify IPT template to use for translation

Specify IAM template to use for translation

Pick to finish and create the IDW

change or control features or part size. The **Mapping Options** button is used to specify how the AutoCAD layers and fonts should be mapped to Inventor equivalents. There is no need to change the mapping options. The **Templates** area is where Inventor templates can be selected. The defaults are fine for most work. Pick **Finish** to create the IDW file. Notice the **Browser** is populated with an ImportDraftView with the names of the AutoCAD layers listed underneath. Save the drawing under a new name and exit it.

Export Inventor 2D IDW Geometry

Inventor provides for the export of IDW geometry via the **Save Copy As** dialog box. Your finished Example-08-01.idw should be open. If not, please open it now. Pick **Save Copy As** from the **File** pull-down menu.

In the **Save As Type** drop-down list pick DWG files (*.dwg). Now pick the **Options...** button to access the wizard. This wizard outlines the process of specifying the different options available to translate the Inventor drawing to AutoCAD format. On the first page choose AutoCAD 2004 Drawing in the **File Version** area. See **Figure 8-31A.** Check the Pack and Go check box if you wish to save as a Zip file. Pick **Next** to continue.

You have the option of translating the IDW using an AutoCAD template file (.DWT) to control the outputted appearance. See **Figure 8-31B.** The **Source Data** area indicates the sheet or sheets to be exported. The **Model Geometry Only** check box determines what data that is written to the DWG file. By default, the check box is not selected. This means all data will be exported to the DWG file. Check the check box to export only model, base, and projected view data.

The **Export Options** area allows you to scale the data and write the data to Model Space or Layout. The **Mapping options** button accesses the options that control the mapping of Inventor styles to AutoCAD layers and linetypes to AutoCAD equivalents. For this example there is nothing to change in the **Mapping options** dialog box, so pick **Cancel** if it was accessed. The **AutoCAD Template File** textbox allows you to specify a DWG or DWT file used in creating the AutoCAD file. Layer properties of the template files are maintained when mapping options for layers may differ.

Figure 8-31.
A—This page defines the file version for the exported material. B—The output file appearance is controlled with this page.

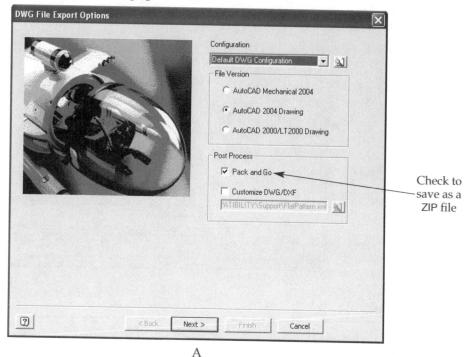

Check to save as a ZIP file

A

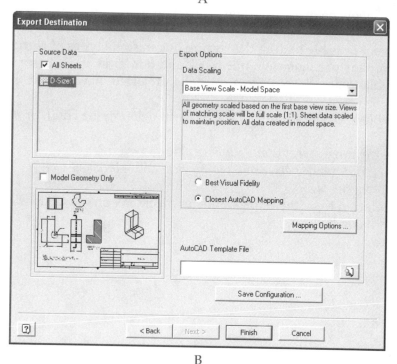

B

The **Save Configuration...** button allows you to save the settings in the wizard dialog boxes to a configuration file. This configuration file can then be used to set options when exporting other Autodesk Inventor drawings.

When done, pick **Finish** and give the file a descriptive name, such as Example-08-01-Done.dwg. The **Figure 8-32** shows the results of translating our Example-08-01.idw to AutoCAD 2004 format.

Figure 8-32.
The resulting
AutoCAD 2005
drawing file shown
open in AutoCAD
2005.

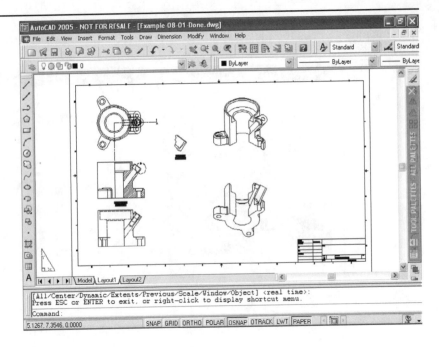

Chapter Test

Answer the following questions on a separate sheet of paper.

1. Is it possible to have more than one sheet per Inventor drawing file?
2. How can a view be made invisible so it is still there and influencing derived views?
3. Why is it *not* possible to drag the cutting line in some types of section views?
4. Can a company use their own title block and border in an Inventor file?
5. Is it possible to detach an orthographically projected view from its parent view? How?
6. Is it possible to delete a parent view without deleting the child, or derived, view?
7. What are the three styles that a view can be shown in?
8. What is the difference between a *broken view* and a *breakout view*?
9. What is a *tangent edge*?
10. What is referred to by the **Show Contents** check box on the **Options** tab of the **Edit View** dialog box?

Chapter Exercises

Exercise 8-1. *Beam.* Start a new IDW file in English units and use a C-size sheet. Using Exercise-08-01.ipt, create the drawing views of the beam as shown below. Inventor should automatically select an appropriate scale for the views. If not, use a scale of .12. The base view is the traditional front view from which all others are derived.

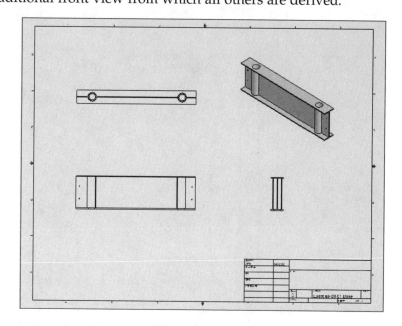

Exercise 8-2. *Bevel Gear.* Start a new IDW file in English units and use a C-size sheet. Using Exercise-08-02.ipt, create the drawing views of the bevel gear as shown below. Use a scale of 1 for the views. This is an exercise in creating a section view. Make sure to set the style of the two isometric views to shaded.

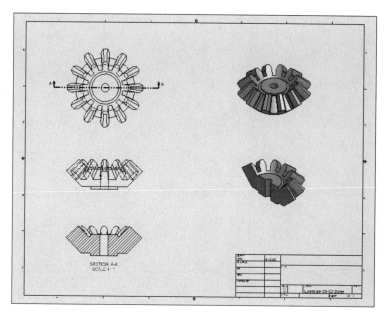

Exercise 8-3. *Support Guide.* Start a new IDW file in English units and change the sheet size to E. Using Exercise-08-03.ipt, create the drawing views shown below. Use a scale of .5 for the views. The isometric view is at a smaller scale than the others and is shaded. This is an exercise in creating views at scale for larger parts.

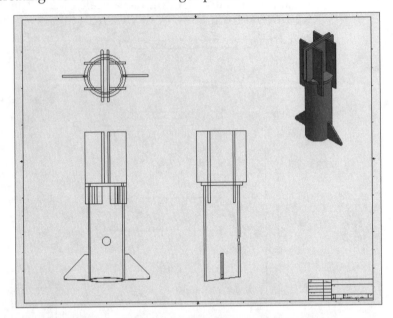

Exercise 8-4. *Tubular Brace.* Start a new IDW file in English units and use a C-size sheet. Using Exercise-08-04.ipt, create the drawing views shown below. Use a scale of .25 (Quarter) for the front view, which you will create first as the base view. From this view, generate the two section views and others. This is an exercise in visualizing section views.

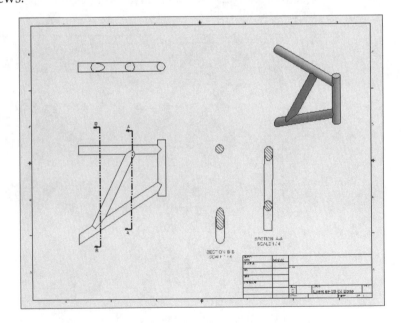

Exercise 8-5. *HVAC Transition.* Start a new IDW file in English units and use a C-size sheet. Using Exercise-08-05.ipt, create the drawing views shown below. Inventor should automatically scale down the base view as it is created. If not, the scale shown here is .14. The top and the isometric view were derived from the front view. Use the front of the part for the initial base view orientation.

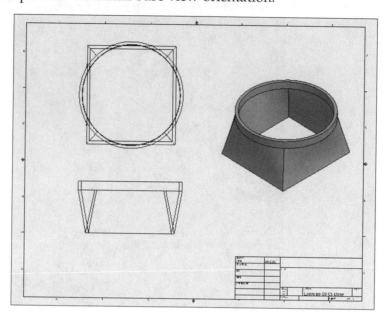

Exercise 8-6. *Tubular Support.* Start a new IDW file in English units and use a C-size sheet. Using Exercise-08-06.ipt, create the drawing views as shown. Use a scale of .5. Use the top of the part for the initial orientation. This is an exercise in creating a broken view.

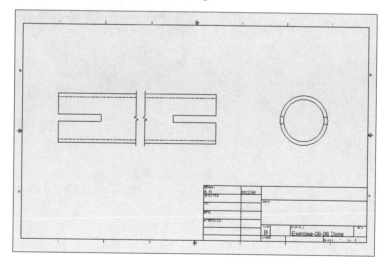

A number of parts are used to create an assembly like this one.

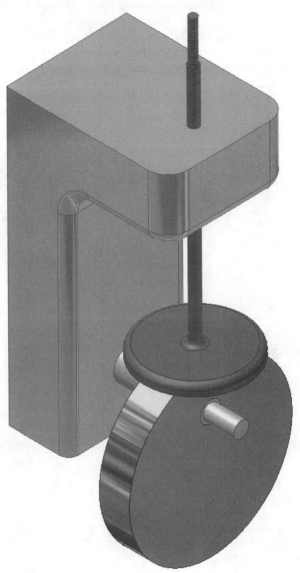

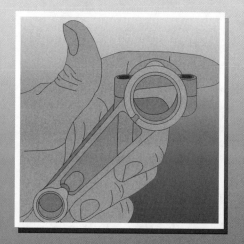

CHAPTER 9

Dimensioning and Annotating Drawings

Objectives

After completing this chapter, you will be able to:

* Apply dimensions to describe drawing views.
* Understand the different types of annotation and how to apply them.
* Specify or edit drafting standards.

User's Files

*The following is a list of files that you will need to work through this chapter. These files can be found on the **User's Files** CD included with this text.*

Examples	Exercises	
Example-09-01.ipt	Exercise-09-01.idw	Exercise-09-06.idw
Example-09-01.idw	Exercise-09-02.idw	
Example-09-01W.iam	Exercise-09-03.idw	
Example-09-02.iam	Exercise-09-04.idw	
Example-09-02W.iam	Exercise-09-05.idw	

Preparing to Annotate a Drawing Layout

Open Example-09-01.ipt and examine the part. This part file is identical to one from *Chapter 8*. Work will continue on the drawing views created in that chapter. Notice the XYZ coordinate system and the part's orientation to that coordinate system. Also, open Example-09-01.idw. This will be the main drawing that will be worked on in this chapter.

Adding Annotation to a Drawing

While studying the following sections, keep in mind that the **Drawing Annotation Panel** is actually "hidden" under the **Drawing Views Panel**, which is the default visible panel when you start a new drawing (IDW). To switch between panels, pick the **Drawing Views Panel** title bar and pick **Drawing Annotation Panel** from the drop-down menu, or right-click on the **Drawing Views Panel** and choose **Drawing Annotation Panel**.

This section covers what drafting standards are and how they are implemented into Inventor. Understand that *all* of the annotations used in this chapter are governed by a drafting standard. To access the drafting standards, pick **Styles Editor** from the **Format** pull-down menu. This accesses the **Styles and Standards Editor** dialog box, which is used to make changes to a drafting standard or create a new one based on an existing standard. See **Figure 9-1A.**

Be careful, if you make changes and elect to save your edits, the changes will overwrite the installed standard (drafting style). You may not want to do this, but a new standard can be created based on the current one. Picking the **New...** button accesses the **New Style Name** dialog box. See **Figure 9-1B.** This dialog box is used to create a new standard.

Many engineering firms in the United States, which use ANSI (American National Standards Institute) drafting standards, also work on international projects. These projects may require ISO standards to be in effect for all work. For this chapter, ANSI will be used for the most part. However, a custom standard based on ANSI will also be used.

Figure 9-1.
A—The **Styles and Standards Editor dialog** box is used to choose a drafting standard and create a new one based on an existing standard. B—**The New Style Name** dialog box is used to name the new style.

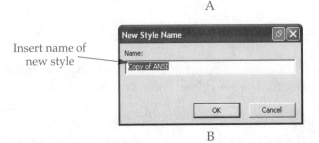

A

B

Figure 9-2.
The **Document Standards** dialog box shows the active standard.

Make sure ANSI is selected

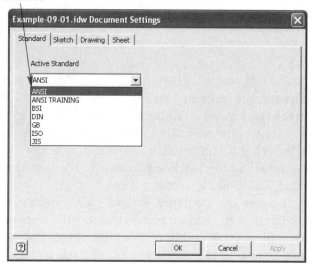

Layers

Inventor provides layers for organizing your annotation. Choose **Active Standard** from the **Format** pull-down menu and make sure ANSI is selected in the dialog box. See **Figure 9-2.** Then, proceed to using the **Styles Editor** to change the color property of the Center Mark(ANSI) layer in preparation for the next section. Make sure that Active Standard is showing in the drop-down list in the upper right, otherwise the listing will show all the layers for all the drafting styles. Notice that there are many different types of drawing objects that are controlled from this dialog box. You will be using this dialog box throughout this chapter. Take a moment to examine the listing before proceeding.

Expand the plus sign next to Layers and click on Center Mark(ANSI). On the right-hand side notice each of the properties of the layers are available to be edited: On/Off, Color, Linetype, Lineweight, check box for Scale by Lineweight. Click on the color swatch (it should be black) and select a teal color. Do this again using the same color for the Centerlines(ANSI) layer. Save your changes and exit the dialog box.

NOTE

The **Edit Layers** button on the **Standard** toolbar provides a shortcut to the **Layer Styles** dialog box.

Adding Centerlines

Every circular feature in a drawing view usually has a centerline. The only exceptions are fillets or small design radii. On the **Drawing Annotation Panel** there are four centerline tools. See **Figure 9-3.**

- **Center Mark** is usually used for the plan view of a circle or arc and requires *one* pick point to place.
- **Centerline** is used to draw or define a centerline. This requires *two or more* pick points to define, followed by choosing **Create** from the right-click pop-up menu.
- **Centerline Bisector** is used for side views showing the circle/arc as a cylinder. This requires just *two* pick points.
- **Centered Pattern** is for circular arrays/patterns, such as bolt circles. This requires *as many* pick points as it takes to define, followed by choosing **Create** from the right-click pop-up menu.

Figure 9-3.
The tools used to
create centerlines
and center marks.

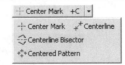

Adding Centerlines to the Top View

Zoom in on the top view. Pick the **Center Mark** tool and select the outermost circle of the part in the top view. You should see a big center mark appear. Then select the *outside arc* of one of the mounting feet and notice a partial center mark is created. Now on one of the other mounting feet, select the circle that represents the hole. Notice that a full center mark is created. See **Figure 9-4**. These center marks for the mounting feet are not right. Use **Undo** to remove them.

This time use the **Centered Pattern** tool. Pick the tool and select the outermost circle of the part in the top view. Next, select the outside arc of the 3 o'clock position mounting foot. Going counterclockwise, select the outside arc of the 10 o'clock mounting foot and then the outside arc of the 7 o'clock mounting foot. There should be a large arc attached to your cursor. Right-click, and choose **Create**. A partial centerline circle is created. Press [Esc] to exit the **Centered Pattern** tool.

To complete the centerline circle, pick the patterned centerline just created and drag the lower end point to meet the start point (3 o'clock). This is known as a *bolt circle*. Drag the endpoints of the center mark cross, so that the centerlines end outside of the bolt circle. See **Figure 9-5**.

PROFESSIONAL TIP

If the values used to create center marks need to be edited, use the **Styles and Standards Editor** dialog box. Expand Center Mark and double-click on Center Mark(ANSI) to change the numeric values.

Adding Centerlines to the Section Views

The section views are good examples of circular features shown in side view or in elevation. It depicts the circular feature as a cylinder. Typically, the **Centerline** or **Centerline Bisector** tool is used for this type of view. Please take note that both section views have an object line running down the center of the part in each view. This is because of the type of section views that they are. See **Figure 9-6**.

Figure 9-4.
The full center mark
in the top view.

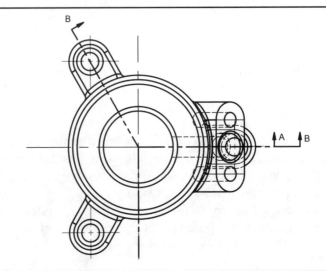

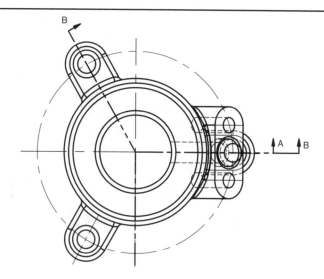

Figure 9-5.
The completed centerline circle.

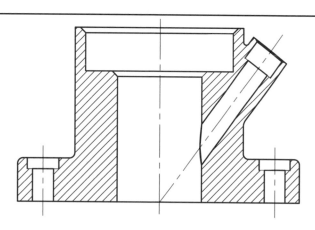

Figure 9-6.
Both section views have an object line running down the center of the part.

For the large centerline running down the center of the part, you could use either tool. Use the **Centerline** tool and select first the bottom center point and then the top center point of the large cylindrical feature that makes up the main body of the part. Right-click, choose **Create**, and then press [Esc] to clear the tool. Notice there is a centerline running through the part. Click and drag the endpoints of the centerline to extend it away from the part. Use the same method for the two small bolt holes in Section B-B. To finish up, use the **Centerline Bisector** tool to apply a centerline to the angled hole. Pick the tool and select first one of the angled sides of the angled hole, then select its mirrored edge. A centerline appears. Click and drag it into shape. These centerlines will be used to dimension the angle.

Adding Centerlines to the Auxiliary View

Pan and zoom so that the auxiliary view is centered in the graphics window. Centerlines will be placed on this view. Because of the projection of this view, the circular edges are shown as elliptical edges. However, Inventor interprets the elliptical edges as circular, so they can be used as a basis for centerlines.

Access the **Center Mark** tool and select the outer elliptical edge that represents the body of the part. Continue selecting the circles that represent the boltholes of the mounting feet. See **Figure 9-7.** These can really help a view make more sense.

Use the same tool and select the three circles in the face of the angled feature. Select them one at a time, thus creating three center marks that are not aligned with the axis of the face. **Undo** three times.

Figure 9-7.
Centerlines shown
in the auxiliary
view.

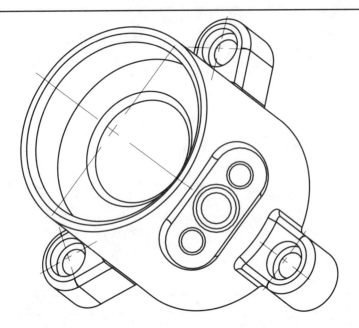

This time use the **Centerline** tool. Pick the tool and select the leftmost dashed circle that represents the tapped threads of this tapped hole. Continue on and select the largest circle that shows the central counterbored hole, then finish selecting the rightmost tapped hole. Right-click, choose **Create**, and the center marks should be created and be aligned with the face's axis. See **Figure 9-8**.

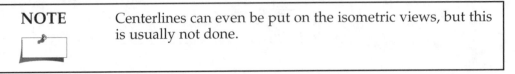

NOTE   Centerlines can even be put on the isometric views, but this is usually not done.

Left-click on the center mark that you placed earlier on the base view. Notice that there are two drop-down lists now active on the right-hand side of the **Inventor Standard** toolbar. The first one on the left is for the layer of the annotation object. It should read By Standard (Center Mark(ANSI)) because the layer for this center mark is determined by the drafting standard in effect when it was created. You can change the annotation to another layer by using this drop-down list. You can also turn the Center Mark(ANSI) layer off.

The drop-down list on the right is for the standard that the annotation is to use. It should read By Standard(Center Mark(ANSI)). That indicates what drafting standard is defining the center mark. Scroll down the list and examine the rest of the list. These other available drafting standards could be used as the basis for this center mark. Try picking each one of them in turn and notice changes happening to the linetype of the center mark.

Figure 9-8.
A— Center marks
not aligned with the
face's axis.
B—Center marks
aligned with the
face's axis.

A   B

Dimensioning in Inventor

There are two types of dimensions used in Inventor. *Parametric dimensions* drive the part's design and can be shown in a drawing view. For example, change the numeric value of a parametric dimension and the shape of the part will change. These dimensions are not placed by the user, but are retrieved by Inventor from the sketches controlling the part's features upon view creation. This can be toggled on/off from Inventor's application options (typically not), or right-click on any drawing view and choose **Get Model Dimensions** from the **Get Model Annotations** cascading menu.

Reference Dimensions are *driven* by the design of the part. For example, if the part is changed, the dimensions update to use the new value. Inventor refers to these as *general dimensions* on the tool panel.

Dimension Styles

Inventor uses a dimension style to control the appearance of the dimensions in your drawing. To control and make changes to the dimension styles, pick **Styles Editor...** in the **Format** pull-down menu. This accessed the **Styles and Standards Editor** dialog box. See **Figure 9-9.**

Change the drop-down list in the upper-right corner to read Local Styles. Notice that there are a number of predefined dimension styles. Normally, changes should not need to be made. Try to use ANSI and its dimension styles as much as possible without edits. There are other choices, such as ISO or JIS, if ANSI is not your company or national drafting standard. A company may find it necessary to make their own dimension style based on the national drafting standard for their current project. Keep in mind that as changes are made, the drawing's dimensions update to reflect the state of the settings.

If you want dimensions to appear in a different color, changing the color property of the Dimension layer will not work. This is because the drafting standard controls the color of the lines and arrows that make up a dimension. The text drafting standard controls the dimension text. You will need to use the **Styles and Standards Editor** dialog box to change the color of the dimension lines and the text. Make sure that Active Standard is showing drop-down list in the upper-right corner of the **Styles and Standards Editor** dialog box. Expand the Dimension listing and click on DEFAULT-ANSI. There are a number of tabs offering a wide array of settings. Pick the **Display** tab and then pick the color swatch. From **Color** dialog box, pick a navy blue. Next, pick the **Text** tab and notice that there is no color swatch here. However, you can see that text is controlled by the DEFAULT-ANSI text style. Save your changes. Now, expand the Text listing and pick DEFAULT-ANSI. Notice on the right there is a color swatch. Use it to select the same navy blue as before. Save these changes and exit. Now, you are ready to add dimensions.

Figure 9-9.
The **Styles and Standards Editor** dialog box.

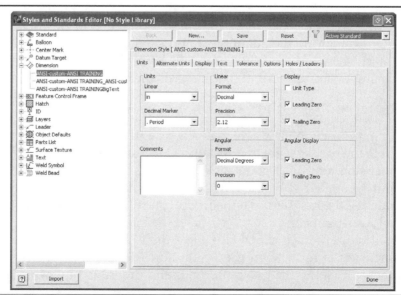

Figure 9-10.
The dimensioned views.

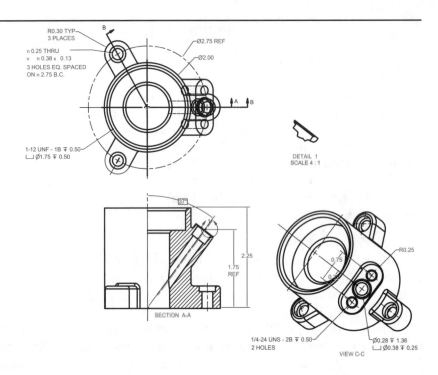

Dimensioning the Top View

As dimensions are added to this part, do not hesitate to move the drawing views to get more space. **Figure 9-10** shows how the views should look once they are dimensioned. Using the **General Dimension** tool, select the large bolt circle centerline and place the dimension text to the upper right. Do the same with the large circle for the body of the part. Continue and dimension the outside arc of the lower-left mounting foot. When done with these dimensions, make sure to press [Esc]. Dimensions can be moved quite easily by picking them and then pressing and dragging one of the green endpoints. Try this with the dimensions that you have placed.

You also have the ability to quickly change the standard that a particular dimension uses to control its appearance. Select a linear dimension and choose a different dimension style from the far right drop-down list on the **Inventor Standard** toolbar. Try one of the international dimension styles. This will change the inch dimension text to its metric equivalent. There are also dual dimension styles. Try one of these out.

Now, right-click the last dimension you placed and choose **Text** from the pop-up menu. This accesses the **Format Text** dialog box. See **Figure 9-11.** You can add text to the default dimension. Place the cursor at the end of the <<>> in the text box. These chevrons indicate the default text and cannot be overwritten. Type TYP, press [Enter] to start a new line, and then type 3 PLACES. Pick **OK** to exit the dialog box to see the changes. Do the same for the bolt circle diameter to add REF to the end of the default text.

Dimensioning the Section View A-A

In the section view, which serves as a front view, a few dimensions are needed to define the part's height and the angled feature. For the height dimension, use **General Dimension** tool. In Section A-A, select the lower-right corner of the part and then select the upper-right corner. Drag the 2.00 dimension to the right as shown in Figure 9-10. To define the height of the theoretical starting center point of the angled hole, select the lower-right corner of the part and the intersection of the angled centerline and the profile edge of the inclined face. Drag the 1.75 dimension to the right of the view. This dimension is for reference only, so add REF to the dimension text. For the angle dimension, select the two centerlines and drag the dimension above the view.

Figure 9-11.
Adding text to the default dimension.

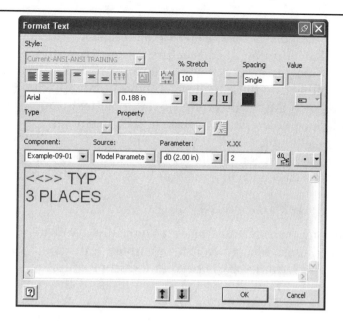

Dimensioning the Auxiliary View

The auxiliary view was created so the inclined face could be shown true size and thus dimensioned. Use the **General Dimension** tool and pick the endpoints of the center-lines to place the dimensions. Add the radius dimension and the dimensioning is done.

Adding Hole or Thread Notes

A *hole note* is typically used to describe the properties of a given hole. For example, how deep is the hole drilled, is it countersunk or counterbored, or is it tapped or not. The part in Example-09-01.idw uses no less than seven of Inventor's parametric hole features. Three of these are tapped with internal threads. Hole notes are needed to help describe these features and complete the drawing views.

With the top view centered in the screen, pick the **Hole Note** tool and select the hole on the upper-left mounting foot. Drag the hole note to the left of the view. When finished, press [Esc] to clear, right-click on the hole note text, and choose **Text** from the pop-up menu. Edit the text to add 3 HOLES EQ. SPACED ON Ø2.75 B.C. as shown in **Figure 9-12.** The hole note should appear in the color that was specified for text.

The hole note leader does not extend properly. It does not touch the actual edge of the selected hole. This is because the hole is buried in a parametric pattern/array and is beneath the feature level the **Hole Note** tool can see. If you want to prove this, select the 3 o'clock mounting hole on the right side of the view and place the hole note.

Figure 9-12.
Adding special symbols to the dimension.

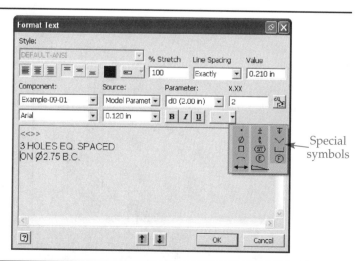

Special symbols

Continue using the **Hole Note** tool to place another hole note describing the large central bore hole in the part and the three small holes shown in the auxiliary view. Notice that since the thread feature is not part of the large bore hole, the hole note does not include tapped thread information. You will have to add the thread information either by editing the text as above or later placing the text with the **Leader Text** tool. Also, on the auxiliary view, additional text is required for the hole note on the small tapped holes. Add 2 HOLES to this hole note.

Repositioning a hole note is quite easy. Click on the note once to activate the green grips at several points on it. Click on the green grip on the leader (not one at the arrow's end) and drag the hole note around the hole.

Surface Texture Symbols

Surface texture symbols are typically used to describe the quality of the machined finish on a given face or surface. A thorough study of surface texture technology and terminology is beyond the scope of this textbook. For example, the face of the angled feature and the inside faces of the central bore diameter are to be machined. See **Figure 9-13.**

Zoom in on Section B-B. Pick the **Surface Texture Symbol** tool and pick a point on the line that represents the angled face. This will start the leader callout portion of the symbol. If you do not want a leader, pick a start point, right-click, and choose **Continue** to proceed directly to the **Surface Texture** dialog box. See **Figure 9-14.** Otherwise, pick a second point, then right-click and choose **Continue**. In the dialog box, leave all the defaults and enter 32 for the **A'** value and pick **OK** to exit.

To continue with the central bore diameter again select the tool, pick one point, drag out the leader, and pick a second point. Then right-click and pick **Continue**. Enter 63 for the **A'** value and pick the **OK** button. To get the multiple leaders on this surface texture symbol make, press [Esc]. Now select the symbol and right-click. Choose **Add Vertex/Leader** from the pop-up menu and pick two points to define each leader. Bring each leader back to the same common point as shown in Figure 9-13.

Feature Control Frames

A *feature control frame* uses geometric dimensioning and tolerancing (GD&T) symbols to define the geometric tolerancing characteristics of a part. See the appendix for GD&T symbols. The part must fall within the tolerances specified in the feature control frames or be rejected. With Example-09-01.idw still open, zoom in on and center Section B-B. **Figure 9-15** shows the section with the feature control frames applied.

Locate and select the **Feature Control Frame** tool, pick the **Feature Control Frame** button on the **Drawing Annotation Panel**, or simply press the [F] key. The process of adding a feature control frame is similar to the one used with all the other symbol

Figure 9-13.
Section B-B with surface texture symbols.

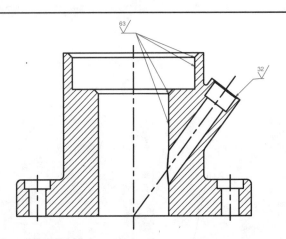

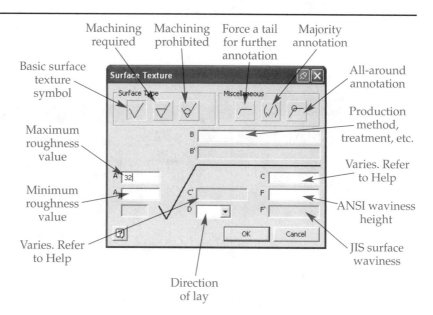

Figure 9-14.
The **Surface Texture** dialog box.

Machining required

Machining prohibited

Force a tail for further annotation

Majority annotation

Basic surface texture symbol

All-around annotation

Production method, treatment, etc.

Maximum roughness value

Varies. Refer to Help

Minimum roughness value

ANSI waviness height

Varies. Refer to Help

JIS surface waviness

Direction of lay

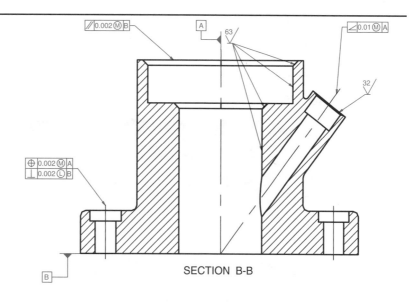

Figure 9-15.
Section B-B with surface texture symbols.

SECTION B-B

annotation tools. Select the tool, pick a start point for the leader arrow, pick a second point to specify length and direction of the leader, right-click and choose **Continue**, and specify values in a dialog box.

To apply a feature control frame to the centerline of the angled feature, select the tool and pick the end of the centerline. Move the cursor to the right and pick a second point to specify length and direction of the leader. Now, right-click and choose **Continue** in the pop-up menu. This accesses the **Feature Control Frame** dialog box. See **Figure 9-16.** Pick the **Sym** button to access the symbols menu. Select the angle symbol for this feature. In the Tolerance 1 box, type .01 and insert the maximum material condition symbol. Leave Tolerance 2 blank and type A for Datum 1. Press **OK** and the feature control frame is complete.

The angle of this centerline (and the feature that uses it) is now specified to be controlled in reference to Datum A. However, Datum A has yet to be defined. First, in order for this feature control frame to be valid, the angle dimension in Section A-A has to be changed to reflect that it is now a basic dimension. Right-click on the dimension and choose **Tolerance** from the pop-up menu. This accesses the **Dimension Tolerance** dialog box. See **Figure 9-17.** Select **Basic** and pick the **OK** button to exit the dialog box. This specifies that the dimension is a nominal value and the real value is

Figure 9-16.
A—The **Feature Control Frame** dialog box. B—The symbols menu.

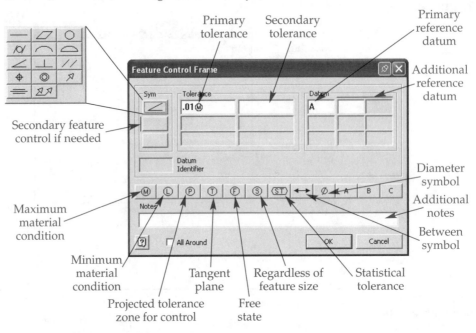

Figure 9-17.
The **Dimension Tolerance** dialog box.

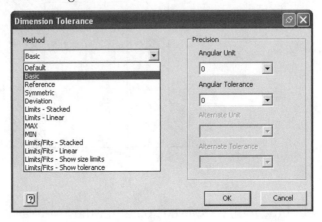

controlled by the feature control frame and the value of 0.01 at maximum material condition. This means the dimension could be as much as 37.01° or as little as 36.99°. This is a simplified explanation, but will suffice for the scope of this section.

Now, the datums need to be specified. Typically there are three datums that are perpendicular to one another. The **Datum Identifier Symbol** tool will be used to insert two datums, A and B. Pick the **Datum Identifier Symbol** button on the **Drawing Annotation Panel** and pick the end of the centerline of the main bore diameter. Move the cursor to the left and pick another point. Move the cursor up and pick a third point. This accesses the **Format Text** dialog. There should already be an A in the text box, so pick **OK** to exit the dialog box. Continue on and specify Datum B by picking the bottom surface of the part. Follow the same step for placing the datum, but change the letter to B in the text box.

Now that the two datums are in place, the remaining feature control frames can be added. To specify the top surface of the part is to be parallel to Datum B, select the **Feature Control Frame** tool. Now, pick the top surface of the part. Move cursor to the left and pick a second point to specify length and direction of the leader. Now,

Figure 9-18.
Using the **Feature Control Frame** dialog box to make the top surface of the part parallel to Datum B.

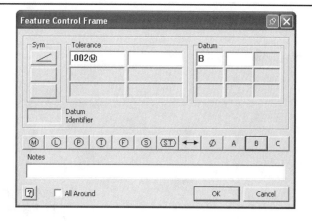

right-click and choose **Continue** in the pop-up menu. In the **Feature Control Frame** dialog box, specify the **Parallelism** symbol and enter the values as shown in **Figure 9-18.** Pick the **OK** button to exit the dialog box.

The **Feature Control Frame** tool should be active, so continue on to specifying the centerline of the mounting hole to be positioned in relation to Datum A and perpendicular to Datum B. Pick the end of the mounting hole centerline. Move the cursor to the left and pick a second point to specify length and direction of the leader. Now, right-click and choose **Continue** in the pop-up menu. In the **Feature Control Frame** dialog box, specify the **Position** symbol for the primary feature control and **Perpendicularity** symbol for the secondary feature control. Enter the remaining values as shown in **Figure 9-19.** Pick the **OK** button to exit the dialog box. Press [Esc] to exit the **Feature Control Frame** tool.

Take a good look at the resulting feature control frame. See **Figure 9-20.** Notice that the alignment is not correct. To fix the alignment, select **Standards...** from the **Format** pull-down menu. Access the **Control Frame** tab and pick the **End Padding** button to turn it on. See **Figure 9-21.** Pick the **OK** button to exit the dialog box and the feature control frame should be okay now. See **Figure 9-22.**

The **Datum Target** tool is used to insert datum target symbols. They provide a means to specify a point or region on the part to be considered as a measuring reference point or plane for gauge-based tolerance checking. Remember, GD&T is used primarily for large production runs of parts to help uniformity of shape across perhaps millions of parts. If a part does not pass the tolerance check, it is rejected. These datum target symbols specify the area on the part that is to be used for this all-important tolerence checking. These areas are typically ones that will not be radically affected by the various manufacturing methods used to produce the part. Therefore, these areas provide for a credible foundation for measuring gauges to be placed against.

Figure 9-19.
Using the **Feature Control Frame** dialog box to specify the centerline of the mounting hole to be positioned in relation to Datum A and perpendicular to Datum B.

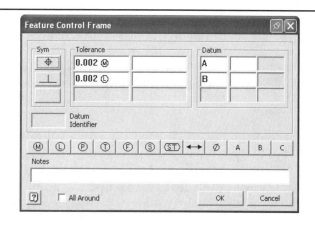

Figure 9-20.
The resulting
feature control
frame is misaligned.

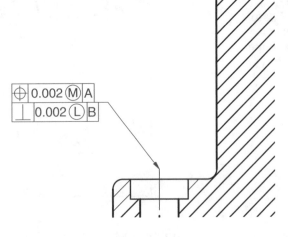

Figure 9-21.
The **End Padding**
button in the
Control Frame tab
is used to correct
alignment.

Figure 9-22.
The feature control
frame is now
aligned.

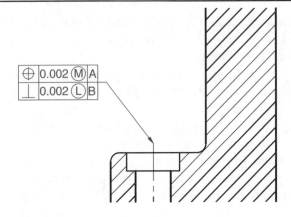

Adding Text

Everyone working with prints eventually learns that what is written on a drawing is as important than what is drawn. The **Text** tool on the **Drawing Annotation Panel** is used to place text.

CAUTION

Text used in Inventor is based on a text style. Any changes made to the text style will be reflected in the text on the drawing. To change the text style, pick the **Styles Editor...** in the **Format** pull-down menu. This accesses the **Styles and Standards Editor** dialog box. See **Figure 9-23**.

Keep in mind, text can be used for whatever situation calls for it—sketches in part modeling, text for the emboss operation, text used on drawing borders and title blocks, etc. With this example, text will be used to finish the GD&T symbol callout for mounting holes. Select the **Text** tool and pick a point below the dual feature control frame for the mounting hole. This accesses the **Format Text** dialog box. In the text box,

Figure 9-23.
The **Text Styles** dialog box is used to make changes to text styles.

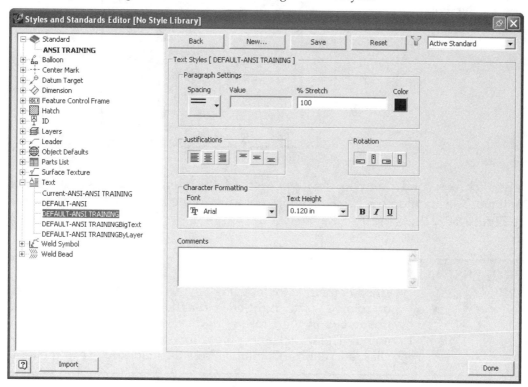

type TYP. 3 Holes. Use the tool again in practice and place several more text objects on the drawing.

The **Leader Text** tool is the same as adding text. However, a leader is placed before the **Format Text** dialog box and text is entered.

> **NOTE**
>
> The **Format Text** dialog box is only for the text object being created. Text styles govern the global appearance of text in your drawing.

Symbols

Symbols are user-defined, saved symbols that can be inserted into the drawing at any time. You can create symbols by expanding **Drawing Resources** in the **Browser**. Now, right-click on **Sketch Symbols** and pick **Define New Symbol** in the pop-up menu. This will put Inventor into Sketch mode and accesses the **Drawing Sketch Panel**. Use the tools to create a symbol, right-click on the sketch screen, and choose **Save Sketched Symbol**. Give the new symbol a meaningful and recognizable name. The new symbol can be found under **Drawing Resources** in the **Browser**. To exit Sketch mode, pick the **Return** button on the **Inventor Standard** toolbar.

To insert a symbol, select the **Symbols** tool in the **Drawing Annotation Panel**. This accesses the **Symbols** dialog box. See **Figure 9-24.** Here you can select which symbol to insert, scale the symbol, and rotate the symbol. The symbol can be placed with a leader if the **Leader** check box is checked. In Example-09-01.idw there is a symbol named Mark Part. It is meant to simulate a custom piece of annotation that an internal corporate drafting standard requires to be placed on piece part drawings. Pick Mark Part, do not scale or rotate, and check the **Leader** check box. Pick the **OK** button and pick a point on the vertical cylindrical edge of the upper-right isometric drawing

Figure 9-24.
Symbols are added
using the **Symbols**
dialog box.

view. Pick a second point to the upper-left of the view, right-click, and choose **Continue**. You should now have a symbol pointing to the isometric view of the part. See **Figure 9-25.** Press the [Esc] key to exit the tool.

PROFESSIONAL TIP

Another method for inserting a symbol without a leader, scaled, or rotated, is to right-click on the symbol under **Drawing Resources** in the **Browser** and pick **Insert** from the pop-up menu. Then, pick the points for symbol placement, right-click, and choose **Continue**. Press the [Esc] key to exit the tool.

Revision Table and Accompanying Revision Tag

The *revision table* is used to document the revisions that have been made to a drawing. It typically is located in the upper-right-hand corner of the drawing border, but it could be located in any of the corners.

The vertical height of the part needs to be revised. If the 2.00 dimension on the height was a parametric dimension, then the change could be made there. However, it is driven so the change has to be made to the extrusion distance. Open the part Example-09-01.ipt and change the distance of Extrusion1 to 2.25, update the part, and save. Open Example-09-01.idw if it is not already open. Notice the drawing views have updated. Some of the annotations may now be on top of the geometry in the views. If this is the case, simply move them as needed.

Figure 9-25.
The symbol
pointing to the
isometric view of
the part.

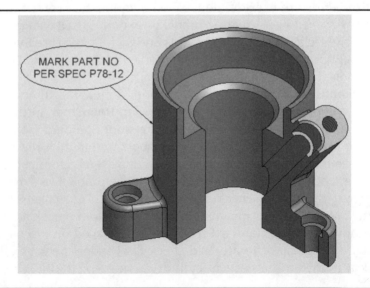

MARK PART NO
PER SPEC P78-12

Figure 9-26.
The edited revision table.

| | REVISION HISTORY | | | |
ZONE	REV	DESCRIPTION	DATE	APPROVED
1	1	Changed Dimension 2.00 to 2.25	10/27/XX	AFD

Pick the **Revision Table** tool from the **Drawing Annotation Panel**. Place the table in the upper-right corner of the drawing border. It will "snap" into place. It is automatically filled out with default information. To edit the properties, right-click on it, pick **Edit** in the pop-up menu, and make changes in the **Edit Revision Table** dialog box. The text in the revision table can be edited by right-clicking on that particular text. Change the description as shown in **Figure 9-26** and put your initials in the APPROVED box.

Before the revision is finished, a revision tag must be placed next to the new 2.25 dimension. This provides a visual locator to the change listed in the revision table. Pick the **Revision Tag** tool from the **Drawing Annotation Panel**. Pick only one point next to the 2.25 dimension, right-click, and choose **Continue**. Press [Esc] to exit the tool. We do not need it to draw a leader for us. See **Figure 9-27.** That is it for this example. Save the work with new name.

Placing Weld Symbols

Typically the edges for a joint are prepared to receive the weld. Weld symbols are used to describe the type and size of the weld to be made at a joint. Open Example-09-02W.iam and examine the weldment assembly. Look at the weldment features on the **Browser**. This assembly and another assembly will be used to create a drawing layout and apply weld symbols. There is no association between these weldment features and those weld annotations that will be applied.

A finished drawing showing weldment features and annotations is shown in **Figure 9-28**. To create this drawing, begin a new drawing using a C sheet size. Create a base view in the upper-left corner of the sheet from Example-09-02.iam at a scale of .5. Make sure the orientation is Top View. Create another view projected down from that base view. From this last view, create another view projected to the right. You should now have three orthographic views all based on Example-09-02.iam. Notice that there are no weldment features at the joint between the tee and the reducer. To create the rendered isometric view use Example-09-02W.iam. You will see why we use two assemblies in a minute. The isometric view is made by creating a base view and selecting the current view of Example-09-02W.iam. Place it in the upper-right corner of the sheet.

Figure 9-27.
Placement of the revision tag.

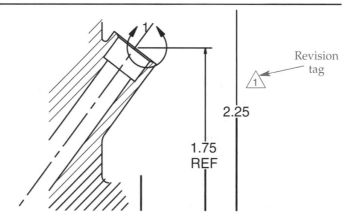

Figure 9-28.
Drawing showing weldment features and annotations.

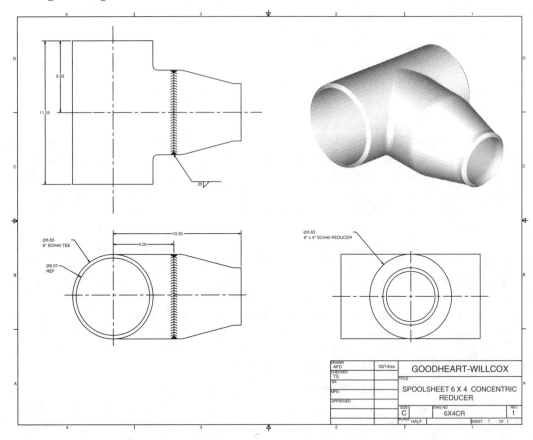

Applying the weld annotations

There are three annotations for weld symbology available to you in a drawing layout. The first one is the traditional weld reference symbol. The use of the weld reference symbol is similar to the other symbols. To add the necessary weld symbol to the welded seam in example drawing, select **Welding Symbols** tool from the **Drawing Annotation Panel**. Then pick a beginning point on the drawing view, pick another point to determine leader line length and direction, and right-click and pick **Continue**. This accesses the **Weld Symbol** dialog box. See **Figure 9-29**. It is presumed that the reader understands weld symbol nomenclature and terminology. The particular symbols on the **Arrow Side** and the **Other Side** tabs will not be covered.

Pick the **Arrow Side** tab and use the values shown in **Figure 9-30**. As values are entered or weld symbols specified in the dialog box, notice the visual feedback in the graphic window. Pick **OK** to place the weld symbol.

The other two annotations that are used to depict a welded seam are produced using the **End Treatment** and **Caterpillar** tools. The **End Treatment** tool is used to place the filled-in triangles that have been used for decades. Because of the two views, four will be placed. Select the **End Treatment** tool from the **Drawing Annotation Panel**. This accesses the **End Treatment** dialog box. See **Figure 9-31**. Pick the **V-type** button and then pick the end point of the seam, which is the intersecting point where the two pipes come together. Pick another point on the seam to determine direction, and right-click and pick **Apply**. Repeat this for the three more points on the two views. The numeric values on the **Options** tab are used to control the size of the end treatment symbol. The default values work for this drawing, so there is no need to change them.

The **Caterpillar** tool is used to create a drafted symbol for the weld. Because of the two views, two will be placed. Select the **Caterpillar** tool from the **Drawing Annotation**

Figure 9-29.
The **Weld Symbol** dialog box.

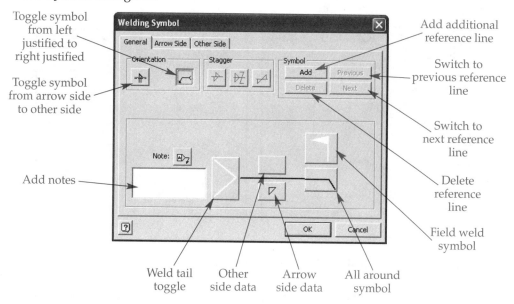

Toggle symbol from left justified to right justified

Toggle symbol from arrow side to other side

Add notes

Add additional reference line

Switch to previous reference line

Switch to next reference line

Delete reference line

Field weld symbol

Weld tail toggle

Other side data

Arrow side data

All around symbol

Figure 9-30.
The **Arrow Side** tab and the values to be used.

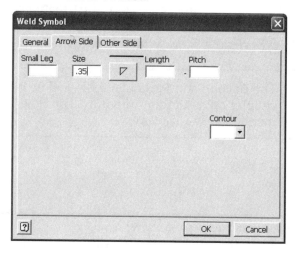

Figure 9-31.
The **End Treatment** dialog box.

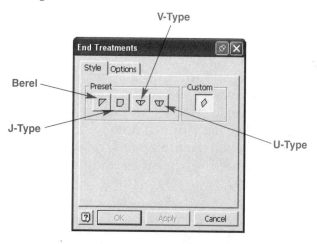

V-Type

Berel

J-Type

U-Type

Figure 9-32.
A—The **Weld Caterpillar** dialog box. B—The **Seam Visibility** check box and the values to be used for this example.

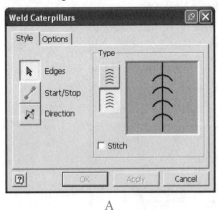
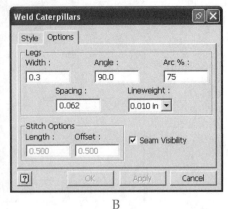

A   B

Panel. This accesses the **Weld Caterpillar** dialog box. See **Figure 9-32A**. On the **Options** tab, check the **Seam Visibility** check box and use the values shown in **Figure 9-32B**. Pick the line that represents the welded seam in the top view, and right-click and pick **Apply**. Do the same for the seam on the front view. Pick the **Cancel** button to close the dialog box and exit the tool.

If Example-09-02W.iam had been used for all the views, then we would never have had a visible seam to apply our caterpillar and end treatment to. Of course, that particular assembly was used for our isometric view, because it visually completes our weldment drawing. Save your work with a new name.

NOTE

The **Balloon** and **Parts List** tools will be dealt with in the chapter on assembly drawings. The **Hole Table** tool was mentioned in *Chapter 6*.

Chapter Test

Answer the following questions on a separate sheet of paper.

1. What is the difference between a parametric dimension and reference dimension?
2. What is a basic dimension?
3. How are properties, such as color or linetype, of drawing annotations controlled?
4. What purpose do centerlines serve in a drawing view and how many different tools does Inventor have to apply them?
5. What is a hole note and why does Inventor have a separate tool to create one? Why not just use the **Leader Text** tool?
6. What is a sketched symbol and when would you use one?
7. What is a surface texture symbol and why is it used?
8. What is the purpose of a weld symbol?
9. What is a drawing revision and why is it used?
10. Are the weld symbols (weld reference symbol, caterpillar and end treatment) associated with the referenced weldment assembly that was used to create the drawing? Why or why not?

Chapter Exercises

Exercise 9-1. *Support Beam.* Open Exercise-09-01.idw and refer to the drawing below as annotations are added to the drawing. Feel free to move the drawing views apart to make room for the dimensions. You may find it necessary to tweak the values for the linetype scale and the sizes used to control the center marks. Furthermore, none of the holes seen in the part were created with the **Hole** tool so adding hole notes to them will not work. This is a very common occurrence in the field. Use the standard dimensioning tool to annotate them. If you like, you can fill in the titleblock by adding text in the **iProperties** dialog box.

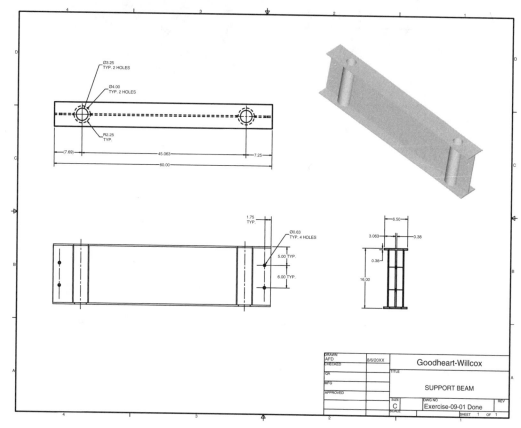

Exercise 9-2. *Mass-Produced Part.* Open Exercise-09-02.idw and refer to the drawing below as annotations are added to the drawing. This exercise concerns primarily with GD&T symbology and surface finish annotation. The application of basic dimensions is shown where a GD&T feature control frame demands it.

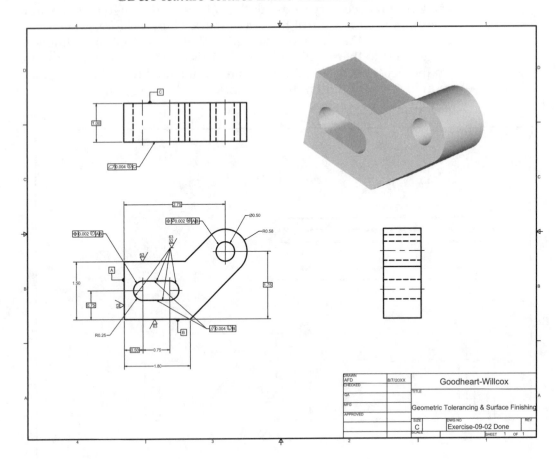

Exercise 9-3. *Precision Part.* Open Exercise-09-03.idw and refer to the drawing below and add the dimensions and other annotations.

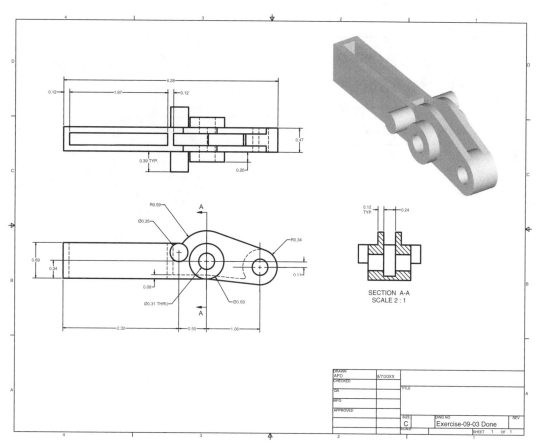

Exercise 9-4. *150 lb Flange.* Open Exercise-09-04.idw and refer to the drawing below as dimensions are added. Use a hole note and annotate the bolt circle.

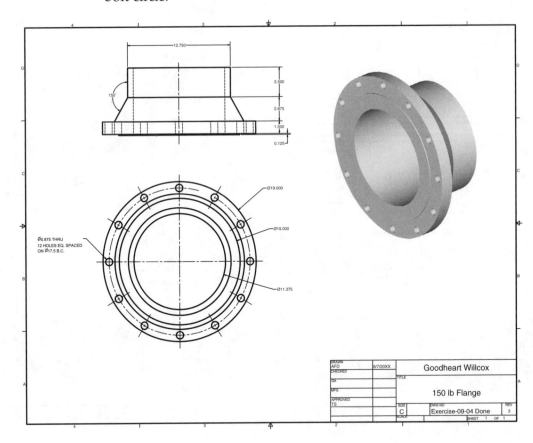

Exercise 9-5. *Brace Shaft with Broken View.* Open Exercise-09-05.idw and refer to the drawing below as dimensions are added to a drawing with a broken view. It is poor drafting practice to have the two diameter dimensions to hidden lines in the side view. What could be done to have the two inner diameters appear as visible lines in the view?

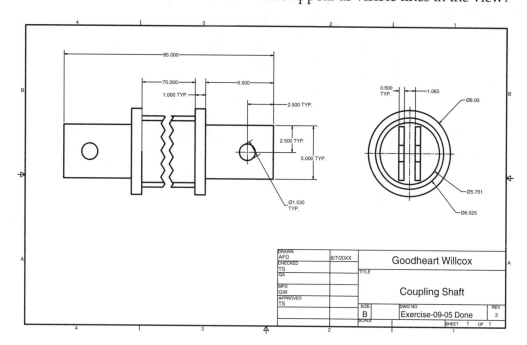

Exercise 9-6. *Hand Wheel.* Open Exercise-09-06.idw and refer to the drawing below as you add dimensions. You will need to edit the views to turn on tangential edges for a more pleasing aesthetic appearance to the views. The fillet radii cannot be dimensioned using Inventor dimensioning tool. Use the **Leader Text** tool instead.

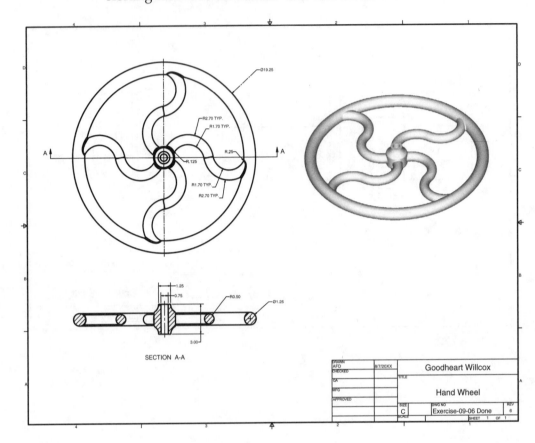

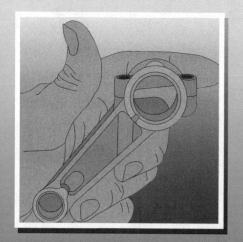

CHAPTER 10

Sweeps and Lofts

Objectives

After completing this chapter, you will be able to:

* Explain the difference between an extrusion, sweep, and loft.
* Explain the process for creating a sweep.
* Create 3D sketches.
* Create 2D and 3D sweeps.
* Define inline work features.
* Create inline work features.
* Create lofts.

User's Files

*The following is a list of files that you will need to work through this chapter. These files can be found on the **User's Files** CD included with this text.*

Examples	Practice	Exercises
Example-10-01.ipt	Practice-10-01.ipt	Exercise-10-01.ipt
Example-10-01b.ipt	Practice-10-02.ipt	Exercise-10-02.ipt
Example-10-02.ipt		Exercise-10-03.ipt
Example-10-03.ipt		Exercise-10-04.ipt
Example-10-04.ipt		Exercise-10-05.ipt
Example-10-05.ipt		
Example-10-06.ipt		
Example-10-07.ipt		

As you have seen, many complex parts can be created using extrusions and revolutions. However, Inventor has two other modeling techniques that are very powerful—sweeps and lofts. A *sweep* is a solid object created by extruding, or "sweeping," a profile along a path. With the **Extrude** tool, the path is an implied straight line the length of which is set by the extrusion terminator. A *loft* is similar to a sweep except that you can use multiple profiles to create a solid of varying cross sections. In lofts, the path is called a *rail*. There may be multiple rails, or none at all.

Creating Sweep Features

Sweep features are typically created for piping or tubing, but are not limited to these uses. You may be able to quickly create a feature as a sweep that would take much longer to create as an extrusion. For example, a picture frame with mitered corners is best modeled as a sweep. Sweeps have a variety of applications in modeling piping, tubing, and similar parts.

Creating a Sweep Path

The first step in creating a sweep is to create the path. The path is simply a sketch that is dimensioned and constrained as any other sketch. The path may be open or closed, depending on the shape of the final part. For example, if the part is piping, then the path will be open. If the part is a picture frame, then the path is closed. Open Example-10-01.ipt and examine the two sketches set up there. Notice that the path lies on a plane. Therefore, it is considered a *2D path*. Later in this chapter, you will learn about 3D paths and the ways of creating them. The path was created on the XZ plane, but any sketch plane can be used. This is an example of piping and, as such, it is best to plan the path so that it originates at a known point, such as the origin.

Creating a Cross Section

Like the path, the cross-sectional profile is a sketch that is dimensioned and constrained. However, the profile sketch must be separate from the path sketch. In order to use the **Sweep** tool, there must be two unconsumed sketches—the path and the profile. In this example, the profile was sketched on the XY plane. The sketch is of two concentric circles centered on the origin. Since the path is on the XZ plane and starts at the origin, this ensures that the cross section is at the start of the path and intersects the path. Inventor requires that the cross section *intersect* the path in some way. It is not necessary for the cross section to be at the start of the path. It may intersect the path somewhere along the path's length.

Creating the Sweep

With the path and the profile created, it is time to create the sweep using the **Sweep** tool. Pick the **Sweep** button in the **Part Features** panel. This accesses the **Sweep** dialog box. See **Figure 10-1.** With the **Profile** button in the **Shape** tab on, pick the profile in the graphics window. Select the area between the two concentric circles. Since the cross section is an ambiguous sketch, you must select the profile. If unambiguous, the profile is automatically selected. Next, pick the **Path** button to turn it on and select the path in the graphics window. The path is the S-shaped line.

In the **Output** area of the **Shape** tab, you can determine what type of part is constructed. Usually, you will be creating solids. Therefore, make sure the **Solid** button is on. However, if you want to create a surface, pick the **Surface** button.

Figure 10-1.
A—The **Shape** tab of the **Sweep** dialog box. B—The **More** tab of the **Sweep** dialog box.

Pick to select the profile
Pick to select the path
Select the operation
Set the output to solid or surface

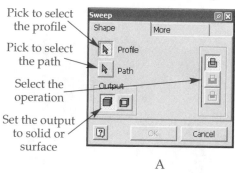

Enter the angle

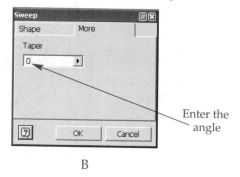

A   B

The three buttons on the right side of the **Shape** tab determine which type of operation is performed. The buttons are **Join**, **Cut**, and **Intersect**. These operations are the same as the corresponding operations in the **Extrude** tool. Since this sweep is the first feature of the part constructed, the only available option is **Join**.

The **More** tab in the **Sweep** dialog box is where you can set a taper angle. This allows you to gradually increase or decrease the size of the profile as it is extruded along the path. A negative angle decreases the size, and a positive angle increases the size.

You may have noticed when you picked the **More** tab that a preview of the profile appears along the path. Notice how the profile follows the path, adjusting its orientation so it is always perpendicular to the path. Now, pick the **OK** button in the dialog box to create the sweep. See **Figure 10-2**.

Sketch not perpendicular to the path

In the previous example, the cross section was sketched perpendicular to the path. To see the importance of this, open Example-10-01b.ipt. The path is the same as in the previous example. The profile is also the same, however, notice that the sketch is *not* perpendicular to the path. It is angled at 30° to the path. Using the same procedure described above, create a sweep of the profile. Notice that the resulting cross section of the part is *elliptical*, **Figure 10-3A.** If you rotate the view, you can clearly see the end of the pipe is elliptical, **Figure 10-3B.** This is because the circular profile is

Figure 10-2.
Two concentric circles were swept along a path to create this pipe run.

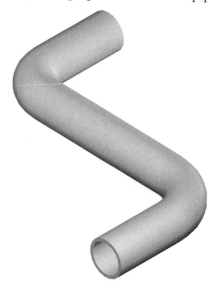

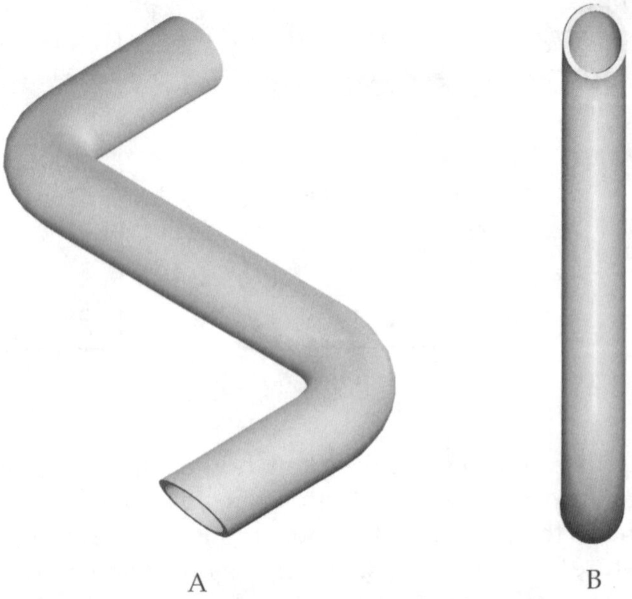

Figure 10-3.
A—If the circle profile is not perpendicular to the path, an elliptical cross section is created.
B—Rotating the view clearly shows the elliptical shape.

A

B

projected onto a plane that is perpendicular to the path. When a circle is viewed at an angle, it appears as an ellipse. Since the circular profile is at an angle to the path, the projected profile is an ellipse, which is then extruded along the path. It is important to remember this as you sketch paths and profiles.

Closed Paths

In the two previous examples, the paths were open. However, the path can be closed. Open Example-10-02.ipt. This file contains two unconsumed sketches—a profile and a path. Notice that the path is a rectangle, in other words, it is closed. Use the **Sweep** tool to create the part as a solid. The resulting part is picture frame molding. Pay particular attention to how the profile is extruded around the corners. Sharp corners are created. Even if a circle is swept on this path, sharp corners are created.

Practical Example of Sweeps

Look at **Figure 10-4.** This is a guard used in an industrial application to prevent debris from entering a pipe inlet. Now, open Example-10-03.ipt. The two bottom rings are already created in the file. Also, the file contains several sketches and work planes. One of the sketches—Shared Sketch for Rings—is used to control the attitude of the entire part in an assembly. This function is not discussed in this chapter. You will now model one of the spokes as a sweep, from which a circular pattern will be created to complete the part.

Creating the Sweep Path

Figure 10-5 shows the path needed to create one spoke. To create the path, refer to the figure and use the following procedure. The path is created as an arc.
1. In the **Browser**, double-click in the sketch Sketch for Path to make it the active sketch plane.
2. Using the **Project Geometry** tool, project the Z axis and the Work Plane for Cross Section onto the sketch plane. To project each, pick their name in the **Browser** with the **Project Geometry** tool active.
3. Project the top and bottom silhouette curves of the top ring. You may want to change the display to "sliced graphics" to better see the projected curves. Also, use the **Look At** tool to display a plan view of the sketch.

Figure 10-4.
This is a guard for a pipe inlet. A single spoke was created as a sweep, then a circular pattern was created.

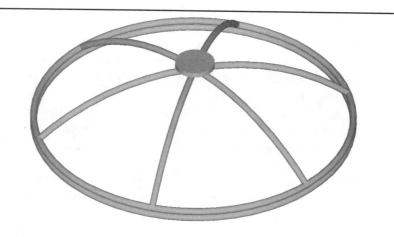

Figure 10-5.
Drawing the path for one spoke.

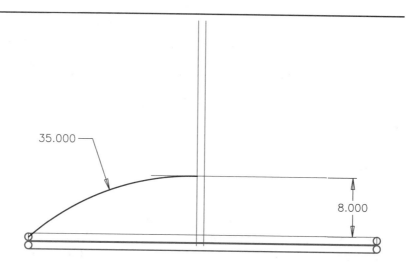

35.000

8.000

4. Connect the endpoints of the projected silhouette curves with construction lines. The midpoint of the left-hand vertical line will be the start point of the path.

5. Next, draw a construction line above and parallel to the projected silhouette lines. The length of the line and its horizontal location are unimportant. However, dimension the line so it is 8″ above the top of the top projected curve. Make note of this dimension name; you will use it later when creating the profile.

This construction line will be used to control the vertical location of the arc's endpoint. The projected Work Plane for Cross Section will be used to control horizontal location of the arc's endpoint. This work plane is where the sweep should end. Continue as follows.

6. Pick the **Three point arc** tool in the **2D Sketch Panel**. Draw an arc starting at the midpoint of the left-hand vertical line connecting the two projected silhouette curves. Place the other endpoint at the intersection of the horizontal construction line and the projected Work Plane for Cross Section line.

7. Dimension the arc to a radius of 35″.

8. The path for one spoke is now complete. "Finish" the sketch by right-clicking in the graphics window and picking **Finish Sketch** in the pop-up menu. Now display the isometric view.

Creating the Cross Section

Next, you need to create the profile of the spoke. The profile is simply a circle that is properly constrained. Refer to **Figure 10-6** and use the following procedure.

1. Make the Sketch for Cross Section the current sketch plane by double-clicking on it in the **Browser**.
2. Project the Work Plane for Path sketch onto the current sketch plane. Using the **Project Geometry** tool, pick on its name in the **Browser**. This is done so there is a centerline reference to which the profile can be attached. Also, project the silhouette curve for the top of the top ring.
3. Display a plan view of the sketch plane. The easiest way to do this is use the common view of the **Rotate** tool. Then, pick the arrow that points straight down onto the sketch plane.
4. Draw a circle. Apply a coincident constraint between its center and the projected work plane. Also, dimension its diameter to 1".
5. Now, you need to vertically locate the center of the circle. Apply a dimension between the circle's center and the projected silhouette curve of the top ring. Edit the dimension and enter the *name* of the dimension you recorded earlier.
6. "Finish" the sketch and display the isometric view.

Since the dimension controlling the vertical location of the path's endpoint is critical to the entire sweep, the vertical location dimension for the profile is tied to it. In this way, the path can be edited and the profile will automatically have the same vertical location. This is one method to "attach" the sketch plane profile to the path. For another method, refer to *Practice 10-1*.

Creating the Sweep

Now the path and profile are created. Pick the **Sweep** button in the **Part Features** panel. Make sure the **Profile** button is on in the **Shape** tab of the **Sweep** dialog box. Then, select the circle as the profile; zoom as needed. Next, pick the **Path** button in the dialog box and select the arc as the path. Finally, make sure the **Join** and **Solid** buttons are on in the dialog box and then pick the **OK** button to create the sweep. See **Figure 10-7.**

If you have trouble picking the path, it might be because you inadvertently drew the arc as a construction line. If this is the case, edit the sketch and change the arc to the Normal linetype. Then, "finish" the sketch and try the sweep again.

Figure 10-6.
Drawing the profile
for one spoke.

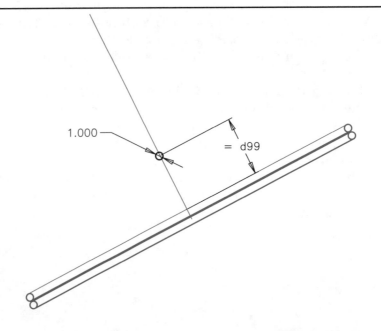

Figure 10-7.
The first spoke is
created.

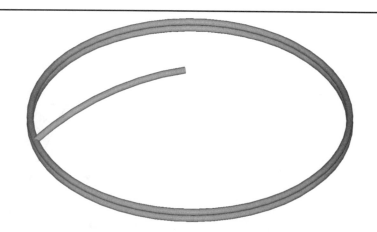

Completing the Part

There are no more sweeps required to finish the part. Pick the **Circular Pattern** button in the **Part Features** panel. Then, select the sweep as the feature. For the rotation axis, pick the **Rotation Axis** button in the **Circular Pattern** dialog box and pick the name Work Axis through Rings in the **Browser**. This is a work axis that was set up for creating the pattern. Finally, enter the number of features as 6 and the rotation as 360°; pick **OK** to create the pattern.

To create the disc on top of the part, create a new sketch using Vertical Work Plane as the sketch plane. This work plane was set up for creating the disc. Project the Work Axis through Rings and the silhouette curve of one of the spokes. Then, draw three lines representing half of the disc. Dimension them as shown in **Figure 10-8.** "Finish" the sketch. Finally, use the **Revolve** tool and revolve the lines around the Work Axis through Rings.

PRACTICE 10-1

❑ Open Practice-10-01.ipt. This file contains a sketch of the path for a paper clip. You must create the profile and sweep it.

❑ First, create a work plane on the endpoint of the path. Using the **Work Plane** tool, pick the endpoint of the path and then the path.

❑ Start a new sketch on the work plane.

❑ Project the endpoint of the path onto the sketch plane.

❑ Draw a circle with its center point on the projected path endpoint.

❑ Dimension the circle's diameter to .05".

❑ "Finish" the sketch.

❑ Using the **Sweep** tool, sweep the circle along the path.

Figure 10-8.
Creating the profile
for the disc, which
will be revolved.

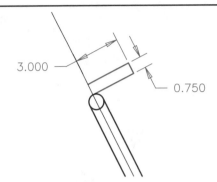

3.000

0.750

3D Sweeps

The sweeps that you have created so far followed planar paths. Although 2D sweeps have many applications, there are also many applications for sweeps whose paths do not lie on a plane. These applications include wiring, piping, tubing, and so on. A 3D sweep can be created in a part file (IPT) or assembly file (IAM).

In order to create a 3D sweep, the path must be created as a 3D sketch. So far, you have only created 2D sketches. A 3D sketch is created with lines and work features, such as work points and work planes. Creating a 3D sketch is the hard part. Once that is done, the **Sweep** tool takes care of the rest. The basic principle of creating a 3D sketch is to define work points and then connect the "dots" with a curve.

Creating a 3D Sketch

Open Example-10-04.ipt. Notice the work planes in the **Browser**. Each is named according to the offset values used to create it. The finished version of the example is shown in **Figure 10-9.** As you can see, these work planes will be instrumental in creating the 3D sweep.

To start a new 3D sketch, pick the arrow next to the **Sketch** button on the **Inventor Standard** toolbar. Then, pick **3D Sketch** from the drop-down list. See **Figure 10-10.** Notice that the **3D Sketch** panel is displayed in the **Panel Bar** where the **2D Sketch Panel** is normally displayed. Also, the sketch is listed in the **Browser** as 3D Sketch*x*. First, you need to create seven work points.

1. Select the **Work Point** button in the **3D Sketch** panel. Any three intersecting planes that do not share the same edge view intersect at a single point. Therefore, pick the following three work planes in any order to define the first work point: XZ plane, YZ plane, and Work Plane Start. Remember, if the plane is not visible in the graphics window, it can be selected in the **Browser**. A work point, a yellow diamond, appears at the intersection of the three work planes. See **Figure 10-11.**
2. Pick the **Work Point** button again. To create the second work point, pick the XZ plane, the YZ plane, and Work Plane 6in.
3. To create the third point, pick the **Work Point** button again and pick the YZ plane, Work Plane 6in, and Work Plane 42in.
4. Pick the **Work Point** button again. Then, pick Work Plane 6in, Work Plane 36in, and Work Plane 42in.
5. Pick the **Work Point** button again. Then, pick Work Plane 36in, Work Plane 42in, and Work Plane 54in.
6. Pick the **Work Point** button again. Then, pick the XZ plane, Work Plane 36in, and Work Plane 54in.
7. Pick the **Work Point** button again. Then, pick the XZ plane, Work Plane 30in, and Work Plane 54in.

Figure 10-9.
The completed pipe
run.

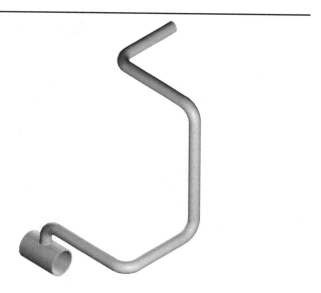

Figure 10-10.
Starting a new 3D
sketch.

Pick to start a
3D sketch

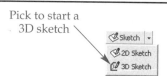

Figure 10-11.
Creating work
points. The points
are created in order
from A to G. Notice
the effect of the
Auto-Bend feature.

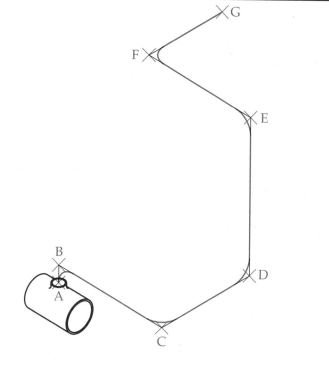

As you can see, a good command of visualization in 3D space is required to create a 3D sketch. Now, to finish the 3D path you need to draw a line connecting the work point. However, if you look at the final part in **Figure 10-9,** the corners are filleted. The Auto-Bend feature can be used to automatically fillet the corners as you draw a line in a 3D sketch. To set the Auto-Bend radius, select **Document Settings...** in the **Tools** pull-down menu. In the **Document Settings** dialog box, pick the **Sketch** tab, **Figure 10-12.** In the **3D Sketch** area of the tab, enter 6 in the **Auto-Bend Radius** text box. Since this is an English Standard (in).ipt file, you do not have to enter the unit abbreviation (in). Pick the **OK** button to apply the setting and close the dialog box.

Figure 10-12.
Setting the Auto-
Bend radius.

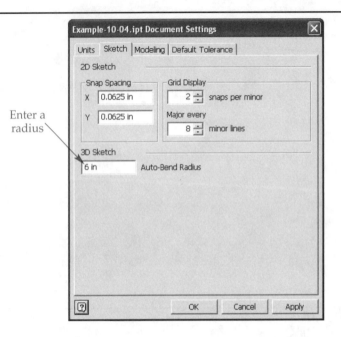

Enter a
radius

Now, you can draw the line. Pick the **Line** button in the **3D Sketch** panel. Right-click in the graphics window to display the pop-up menu. Notice that **Auto-Bend** is checked, which means it is active. See **Figure 10-13.** If you pick the menu item, it becomes inactive and the corners of the line will not be filleted. Leave **Auto-Bend** checked and press [Esc] once to cancel the pop-up menu. Then, in the graphics window, pick the work points one by one in the order in which you created them. As you connect the work points, notice that Inventor automatically applies a 6" radius fillet to each corner. After picking the seventh work point, press [Esc] to cancel the **Line** tool. Then, "finish" the sketch.

Finishing the 3D Sweep

The profile and path for a 3D sweep must be separate unconsumed sketches, just as with a 2D sweep. Pick the **2D Sketch** button from the **Sketch** drop-down list on the **Inventor Standard** toolbar. Then, pick Work Plane Start as the sketch plane. You can pick it in the graphics window or in the **Browser**. Zoom in on the pipe stub on top of the pipe tee. Using the **Project Geometry** tool, project the ID and OD circle of the pipe. Then, right-click in the graphics window and pick **Done** in the pop-up menu. Now, "finish" the sketch and zoom out.

Pick the **Sweep** button in the **Part Features** panel. With the **Profile** button active in the **Shape** tab of the **Sweep** dialog box, select the area between the two concentric circles as the profile. Then, pick the **Path** button in the dialog box and select the 3D line in the graphics window. Finally, make sure the **Join** and **Solid** buttons are on in the dialog box and pick the **OK** button to create the 3D sweep. The part now appears as shown in **Figure 10-9.**

Figure 10-13.
You can turn the
Auto-Bend feature on
and off in the pop-up menu.

Click

Figure 10-14.
The part tree for the completed pipe run.

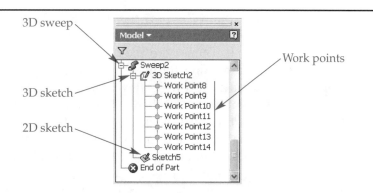

3D sweep

Work points

3D sketch

2D sketch

Expand the part tree for the sweep. Notice the structure. It contains two sketches, as a 2D sweep would. Also, notice that the work points are in the part tree for the 3D sketch. See **Figure 10-14.** All of these items can be renamed to be more meaningful in the design. For example, the default name Work Point*x* in a complex piping layout quickly becomes meaningless. Instead, name the work points by location, pipe type, or other meaningful naming convention.

Inline Work Features

In the previous example, all of the work planes were created before the 3D sketch was started. In this section, you will create the work planes as you are creating the 3D sketch. Work features created on the fly, or within an active tool, are called *inline work features.* As you will see, the method used in the previous example is easier than the one you will use in this section. However, both methods are valid for creating 3D sweeps in Inventor. Why bother learning the more difficult method? The method presented in this section is more widely used in creating assemblies. Also, it provides a more sophisticated process that is required in many real-world applications.

Open Example-10-05.ipt. This file is similar to the original file in the previous example. However, there is only one work plane—Work Plane Start. **Figure 10-15** shows the part tree for the completed 3D sweep in the **Browser**. Notice that some, but not all, of the work points have a work plane listed below them in the part tree.

Start a new 3D sketch by picking **3D Sketch** from the **Sketch** drop-down list on the **Inventor Standard** toolbar. Then, use the following procedure to create the work points for the 3D sketch of the path.

1. Pick the **Work Point** button in the **3D Sketch** panel. Then, select the XZ plane, YZ plane, and Work Plane Start. A work point appears at the intersection of the three planes. Note: This step is the same as in the previous example.

Figure 10-15.
The part tree for the completed pipe run created with inline work features. Notice that the work planes are below the work points in the part tree.

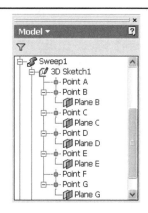

2. In the **Browser**, expand the part tree for the 3D sketch. Then, rename the new work point to Point A. This is done by picking the name and then picking it again; do not double-click. Then, type the new name in the text box that appears and press [Enter]. By renaming the work features as you create them, it is easier to keep track of items as you work through this procedure. Refer to **Figure 10-11.**

3. Pick the **Work Point** button in the **3D Sketch** panel. Then, pick the XZ plane and the YZ plane. For the third work plane, right-click in the graphics window and select **Create Plane** from the pop-up menu, **Figure 10-16.** Pick Work Plane Start in the graphics window and drag up. Release the mouse button and enter a value of 6 in the dialog box and press [Enter].

A new work plane is created offset 6" above Work Plane Start. At the same time, a work point appears at the intersection of the XZ plane, YZ plane, and the new work plane. Notice that the inline work plane was created last in Step 3. The new work plane is located under the new work point in the part tree. In the **Browser**, expand the part tree, rename the new work point to Point B, and rename the new work plane to Plane B. Also, turn on the visibility of Plane B. Continue as follows.

4. Make sure no work plane is selected. Select the **Work Point** button in the **3D Sketch** panel. Pick the YZ plane and Plane B. Then, right-click in the graphics window and select **Create Plane** from the pop-up menu. Pick the XZ plane, drag to the right, release the mouse button, and enter −42 in the dialog box. A work point appears at the intersection of the two existing planes and the new work plane.

5. In the **Browser**, rename the new work point to Point C and the new work plane to Plane C. Also, turn on the visibility of Plane C.

6. Make sure no work plane is selected. Pick the **Work Point** button in the **3D Sketch** panel. Pick Plane B and Plane C. Then, right-click and select **Create Plane** from the pop-up menu. Move the cursor over the YZ plane name in the **Browser** until the plane is outlined in red in the graphics window. Then, pick the YZ plane in the graphics window, drag to the upper right, release the mouse button, and enter 36 in the dialog box.

7. In the **Browser**, rename the new work point to Point D and the new work plane to Plane D. Also, turn on the visibility of Plane D.

8. Make sure no work plane is selected. Pick the **Work Point** button in the **3D Sketch** panel. Pick Plane C and Plane D. Then, right-click and select **Create Plane** from the pop-up menu. Pick Plane B, drag up, release the mouse button, and enter 54 in the dialog box.

9. In the **Browser**, rename the new work point to Point E and the new work plane to Plane E. Also, turn on the visibility of Plane E.

10. Make sure no work plane is selected. Pick the **Work Point** button in the **3D Sketch** panel. Pick the XZ plane, Plane D, and Plane E. An inline work plane is not needed.

11. In the **Browser**, rename the new work point to Point F.

12. Pick the **Work Point** button in the **3D Sketch** panel. Pick the XZ plane and Plane E. Then, right-click and select **Create Plane** from the pop-up menu. Pick Plane D, drag to the right, release the mouse button, and enter 30 in the dialog box.

Figure 10-16.
Creating an inline
work plane.

Pick to create an
inline work plane

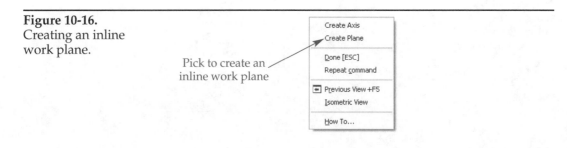

13. In the **Browser**, rename the new work point to Point G and the new work plane to Plane G. Also, turn on the visibility of Plane G.

Now, the sketch appears exactly as it did in the previous example after you created the seven work points. The rest of the procedure is the same as for the previous example. Set the Auto-Bend radius to 6 and draw a line between the work points. Then, "finish" the sketch, start a new 2D sketch on Work Plane Start, and project the ID and OD circle. "Finish" that sketch and sweep the circles along the path. The final part in this example is the same as in the previous example. However, the process is a lot more tedious. Remember, this more complex process is more useful when creating 3D sweeps in assemblies.

Editing a 3D Sweep

Editing a 3D sweep mainly involves editing the 3D sketches used for the sweep—the path and the profile. For example, you can change the offset values used to create the work planes. Since the work planes are used to create work points, and in turn the 3D path, this will alter the final part. You may also change the offset values to equations. In this manner, you can constrain the sweep to other features in the part. Additionally, if you right-click on the name of the sweep in the **Browser** and select **Edit Feature** from the pop-up menu, the **Sweep** dialog box is displayed. This allows you to select a different profile or path, change the output or creation method, or enter or change a taper angle.

Lofts

A *loft* is a solid or surface that is generated from a series of cross sections. The cross sections are created as sketches and do not need to be parallel to one another. A path is called a *rail.* The process of creating a loft is called *lofting.* Lofting is usually done to create organic shapes or parts with very complex, compound curves. Examples of where lofting is useful include a car body, ship hull, ergonomic mouse, or human hand. Other applications include modeling transitions between two known geometric shapes, such as a square-to-round transition found in HVAC ductwork.

Figure 10-17 shows another example that simply cannot be modeled as an extrusion, revolution, or sweep. You will create this hook, which is commonly found on any construction site crane. It serves as an excellent example of a loft. Open Example-10-06.ipt. This file contains three components already created—Shaft, Hilt, and Tip. It also contains 16 sketches on 16 separate work planes. These sketches will be used to create the loft. This example is ready to be lofted. Pick the **Loft** button in the **Part Features** panel. The **Loft** dialog box is opened. The next section discusses the features of this dialog box.

Curves Tab

The **Curves** tab is typically the only one in which you will make settings. See **Figure 10-18.** Notice the left-hand column is labeled **Sections**. The cross sections will be listed here as you pick them in the graphics window. Within the column is the label Click to add. Pick this label and you can then select the cross sections. The order in which the cross sections appear in this column determine how the loft is created. You can reorder the list by dragging a cross-section name and dropping it in a new location in the list.

Figure 10-17.
The curved portion
of this hook is
created as a loft.

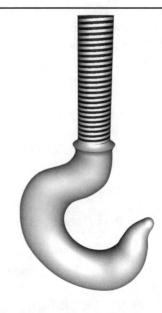

Figure 10-18.
The **Curves** tab of
the **Loft** dialog box.

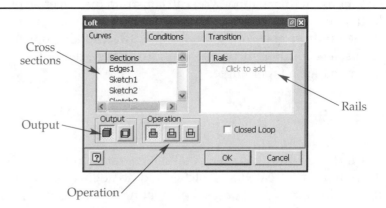

The right-hand column in the **Curves** tab is used to select rails. Rails can be used to further refine the shape of the loft. However, a rail is not required to create a loft. If you pick a series of cross sections, a rail is automatically generated.

Conditions Tab

The **Conditions** tab, **Figure 10-19,** is used if you need to change the boundary conditions at the end of cross-section sketches. This is done to affect the way the loft meets the existing part's edges. Any edges selected as cross sections appear in the list. To change the condition, highlight the edge in the list. Then, select the appropriate condition button—**Free Condition**, **Tangent Condition**, or **Angle Condition**. If **Angle** is selected, also set the angle. Finally, enter the desired weight. The weight determines how much this condition contributes to the final loft shape. A low value creates an abrupt change. A high value creates a gradual change.

Transition Tab

The **Transition** tab, **Figure 10-20,** is used if you need more control over the final loft shape. It may also be used if a set of cross sections is creating an undesirable loft even though the sketches are properly drawn. Using this tab, you can specify exactly the point on each sketch through which the rail (imaginary or real) should pass.

Figure 10-19.
The **Conditions** tab of the **Loft** dialog box.

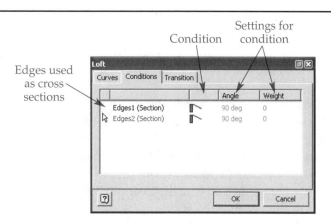

Figure 10-20.
The **Transitions** tab of the **Loft** dialog box.

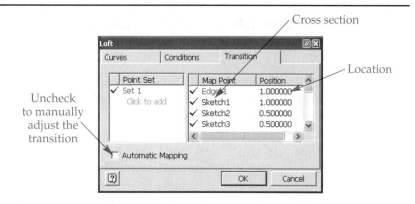

Finishing the Loft

In the **Curves** tab of the **Loft** dialog box, pick the Click to add label in the **Sections** column. You are ready to pick the cross sections in the graphic window. However, the first "cross section" to be picked in this example is not one of the sketches. In order to connect the loft to the Hilt feature, you must select the bottom edge of that feature. As you pick the 16 sketches in order, the area of each cross section is highlighted in cyan. Also, a green preview rail follows the outline of the resulting loft. The final "cross section" is the bottom edge of the Tip feature.

When all of the cross sections are listed in the **Sections** column, including the two edges, make sure the **Solid** and **Join** buttons are on. Then, pick the **OK** button to create the loft and close the dialog box. The loft is complete, but the part needs threads. Apply 3/4-10 UNC threads to the Shaft feature. The final part should look like Figure 10-17.

More on Lofts

Are lofts this easy? Not typically. There may be several hours of trial and error required to make all those sketches. If you look at the **Browser** in the previous example, you will notice a sketch called Shared Sketch for Layout. This sketch controls the work planes that in turn control the sketches for the loft. This is a clever way of controlling the sketches that make up a loft. Typically, there is no other effective method for constraining the loft sketches to the part.

The shared sketch in the previous example is based on the XZ plane. A 2D outline of the side view of the hook was drawn. At key points on the 2D outline, construction lines were drawn across the outline and trimmed to the outline. The construction lines were used as the basis for work planes perpendicular to the XZ plane. On the first work plane, a sketch was created and the construction line that controlled that work plane was projected onto the sketch. An ellipse was then drawn with its center at the midpoint of the projected construction line and its major axis endpoint at the endpoint of the same line. To create the shape of the cross section, the minor axis endpoint of the ellipse was dragged out. An elliptical dimension was applied with a value of .4″. This process was done on each work plane. The value of .4″ was used for the majority of the ellipses, except those near the Tip as the thickness narrows toward the Tip.

Why are so many sketches necessary? If you want the final loft shape to be smooth and conform to specifications, you need to limit the number of guesses Inventor is forced to make. The more cross sections you have, the more information Inventor has to perform the operation. As an experiment, undo the loft operation for the hook or delete it in the **Browser**. Then, repeat the process except only select a few of the sketches, every third or fourth. Compare the final loft shape to the previous one. The overall shapes may be similar, however, the second hook does not have nearly as smooth of a curve as the first one. Depending on which and how many sketches you select, you may even get an error because Inventor does not have enough information to make a smooth transition from one sketch to another.

Closed-Loop Loft

The hook in the previous example is an open-loop loft. The loft does not begin and end at the same point. A closed loop can also be used. Closed loops are typically used in sheet metal applications such as the airplane engine cowling shown in **Figure 10-21**. This is a very complex, closed-loop loft.

Open Example-10-07.ipt. There are 16 sketches on 16 different work planes. All the sketches are the same, just their orientation in 3D space is different. As in the previous example, the work planes and sketches are controlled by a shared sketch. This is also true for this example. Note: Each of the 16 sketches required several dimensions to locate and constrain the sketch. To improve visibility in the file, these dimensions have been deleted.

Pick the **Loft** button in the **Part Features** panel. In the **Curves** tab of the **Loft** dialog box, pick the Click to add label in the **Sections** column. Then, in the **Browser**, pick the first sketch (Sketch1). The sketches were created in order and, thus, appear in order in the **Browser**. Hold down the [Shift] key and pick the last sketch in the **Browser** (Sketch16). It may take a minute or so for Inventor to add the 16 sketches to the loft. Once all 16 sketches are listed in the **Sections** column of the dialog box, check the **Closed Loop** check box in the **Curves** tab. If you do not check this check box, the segment connecting the first and last cross sections will not be added. Instead, an open-loop loft is created. Finally, pick the **OK** button to create the loft. It may take Inventor some time to calculate the final part.

Figure 10-21.
This is a closed-loop loft.

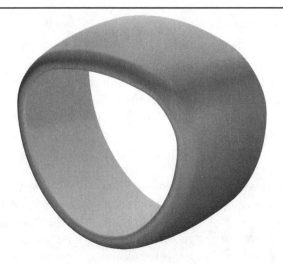

❏ Open Practice-10-02.ipt.
❏ Use the **Loft** tool to create a loft feature using all 15 of the sketches.
❏ The completed part is shown below.

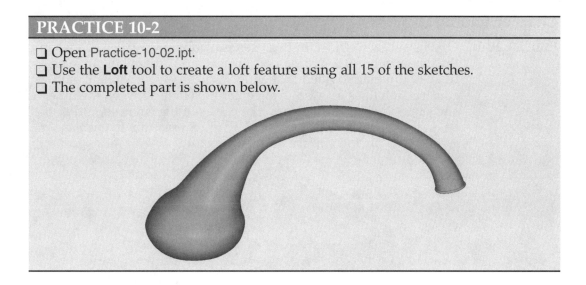

Editing a Loft Feature

You can edit a loft in basically the same way as for a sweep. Editing the work planes on which cross-section sketches are created will alter the loft's shape. You can also edit the individual sketches, which may have the most effect on the final shape and appearance. If you right-click on the loft name in the **Browser** and select **Edit Feature** from the pop-up menu, the **Loft** dialog box is displayed. You can alter the settings in this dialog box. For example, if you forgot to check the **Closed Loop** check box, you can edit the feature and check it. You can also change the operation, from **Join** to **Cut** for example. Editing the transitions by adjusting the point set is usually done after the cross sections are lofted. In this way, you can see the transitions that need adjustment.

Chapter Test

Answer the following questions on a separate sheet of paper.
1. What is the difference between an extrusion and a sweep?
2. What is the difference between a sweep and a loft?
3. What is a *rail?*
4. What is the difference between a 2D path and a 3D path?
5. What happens if you sweep a circle along a path if the circle is *not* perpendicular to the path?
6. If a circle is swept around a square corner, what shape does the corner have on the finished part?
7. Describe the function of the Auto-Bend feature.
8. How do you turn the Auto-Bend feature off?
9. What is the difference between a 2D sweep and a 3D sweep?
10. Define *inline work feature.*
11. What is the process of creating a loft called?
12. When are lofts typically created?
13. List the three tabs in the **Loft** dialog box.
14. Explain the difference between an open-loop loft and a closed-loop loft.
15. When are the transitions for a loft object typically edited? Why?

Chapter Exercises

Exercise 10-1. Open Exercise-10-01.ipt. In this exercise, you will create a groove in the top surface for a mating part. Create a .15″ diameter circle on the work plane provided in the file. Project geometry and create construction lines as needed to center the circle between the inner and outer surfaces of the part. The center of the circle should also be on the top surface of the part. Use the **Sweep** tool to cut the groove into the top surface, as shown below.

Exercise 10-2. Open Exercise-10-02.ipt. This file contains a sketch for the path of a hitch pin, shown below. Create a work plane on which you can create the sketch for the profile. The profile is a 1.4 mm circle and should start at the endpoint of the long straight section. A construction line is provided for help in creating the work plane.

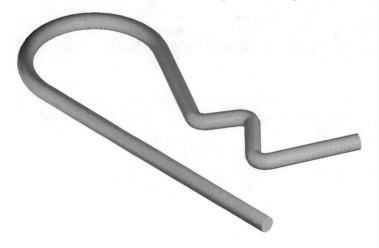

Exercise 10-3. Open Exercise-10-03.ipt. This file contains a center hub and outer ring for a valve handwheel. There is also a path for one of four spokes on the completed handwheel, as shown below. Create a work plane on which to sketch a profile. The profile is a 1″ diameter circle and should be located at the endpoint of the path that is inside the hub. Suppress the Hub feature as you work. Create one spoke as a sweep using the path and profile. Then, finish the part by a circular pattern of the spoke about the center of the hub.

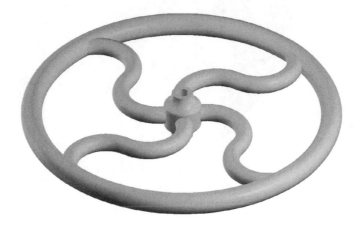

Exercise 10-4. Open Exercise-10-04.ipt. This is part of a cover for the engine on a gasoline lawn mower. You will be adding ventilation cutouts, as shown below. The file contains two unconsumed sketches.

A. Create a work plane perpendicular to the sketch Path for Surface Sweep at its lower-left endpoint. Then, start a new sketch on that work plane. Draw a vertical line up from the endpoint of the path. Its length is not important. "Finish" the sketch.

B. Sweep the line along the path. However, pick the **Surface** button in the **Sweep** dialog box before selecting the line profile and path.

C. Using the **Extrude** tool, extrude the closed profile in the sketch Sketch for Extrusion. Use the surface you just created as the termination. Also, in the **More** tab of the **Extrusion** dialog box, check the **Minimum Solution** check box. Finally, set the operation to **Cut** and pick the **OK** button to create the extrusion.

D. To complete the part, create a circular pattern of the extruded cutout. Use **Placement** values of 11 for the number and 5 for the degrees. Also, expand the dialog box and pick the **Incremental** radio button in the **Positioning Method** area. Then, pick the **OK** button to create the pattern.

Exercise 10-5. Open Exercise-10-05.ipt and examine the existing sketches. The final lofted shape will be a *Moebius strip*—a unique body that has only two sides, instead of the four that a 3D rectangular ribbon would have. A rubber band has four sides. Use the **Loft** tool to select all of the sketches to create a loft. You may find it easier to select them in the **Browser**.

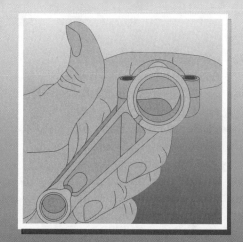

CHAPTER 11

Building Assemblies with Constraints

Objectives

After completing this chapter, you will be able to:

* Create and use projects.
* Create an assembly from existing parts.
* Apply mating constraints using various options.
* Apply insert constraints using various options.
* Apply tangent constraints using various options.
* Edit constraints placed in an assembly.
* Edit parts in place within an assembly.
* Place standard fasteners into an assembly.

User's Files

*The following is a list of files that you will need to work through this chapter. These files can be found on the **User's Files** CD included with this text.*

Examples

2nd Base.ipt	Bat Cover.ipt	Spring_Button.ipt
Base-Plate.ipt	Box.ipt	Support.ipt
Bearing-Block.ipt	Bottom.ipt	Thumb Screw - ANSI
Bushing.ipt	Bushing.ipt	B18.17 - 14 - 28 X 1
CR-Locating Pin.ipt	Clamp-L.ipt	14_1.ipt
Molding.ipt	Clamp-R.ipt	Thumb Screw - ANSI
	Coin.ipt	B18.17 - NO. 8 - 40 X
Exercises	Lid.ipt	38_1.ipt
Axle.ipt	Pin.ipt	Top.ipt
Ball-Socket.ipt	Rod.ipt	Wheel.ipt
Base-Plate.ipt		

Building an Assembly File

In this chapter, you will learn how to build an assembly file. An *assembly* shows how parts fit together to create the final product. For example, the assembly file shown in **Figure 11-1** contains five different parts—Base-Plate, Bearing-Block, Bushing, CR-Locating Pin, and a standard cap screw selected from Inventor's fastener library. Some of these parts, such as the cap screw, are placed in the assembly file multiple times. When a part is placed in an assembly file, it is called an *instance* of the part, which is similar to an OLE link to the original part file. All parts are positioned in an assembly with respect to each other using assembly constraints. Also, in this chapter you will create an Inventor project. A *project* is a system for keeping track of the locations of all the files in an assembly.

Figure 11-1.
This is an exploded view of the assembly you will create in this chapter.

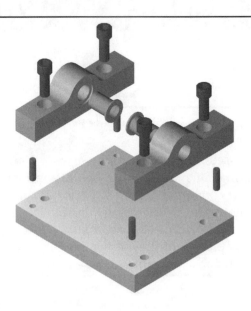

Creating a Project with Folders

The assembly that you will build in this chapter, like most assemblies, requires a variety of files and file types:

- The assembly file (IAM), showing all the parts put together.
- The part files (IPT) of the modeled components.
- A project file (IPJ), which keeps track of all the part (IPT) and assembly (IAM) files.

To keep these files organized and to get access to the correct folders, Inventor has a process called *projects* that locates and creates the required folders and builds the IPJ file. The reason for this seemingly complex process is that an Inventor assembly project may contain a large number of files, stored on many computers in a network, that are worked on by many people. This is called a *shared environment.* For this chapter, you are going to set up an isolated environment. All of the files will be contained in folders that are for your use only. Others will not share these folders or their contents.

At least two projects are created when Inventor is installed, one called Default and the other tutorial_files. To this point, you have worked with files located in the Default project. However, you can create new projects. Start by creating a new folder on the hard drive. This can be anywhere using any name, but for simplicity, create the folder off of the root and use your name as the folder name. In that folder, create a subfolder called Example-11-01. For example, c:*username*\\Example-11-01.

Figure 11-2.
The projects available are listed at the top of the **Open** dialog box when you pick **Projects** in the left-hand column.

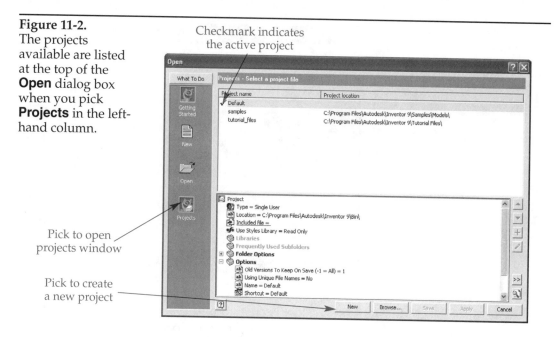

Checkmark indicates the active project

Pick to open projects window

Pick to create a new project

Next, start Inventor. To create a new project, you cannot have any Inventor files open. Then, select **Projects...** in the **File** pull down. Your screen should be similar to **Figure 11-2** with the Default project checked. Pick the **New** button at the bottom of the dialog box, *not* the **New** icon on the left of the dialog box. Then, in the first panel of the wizard, select the **New Single User Project** radio button in the **Inventor project wizard**. See **Figure 11-3**. Then, pick the **Next** button.

In the next panel of the wizard, fill in the following information as shown in **Figure 11-4.**

- In the **Name** text box, type Example-11-01.
- In the **Location** text box, type the path to the subfolder you created. You can pick the browse button (**...**) to locate the subfolder instead of typing the path.

Notice the **Project File to be created** field, which is shown in gray. Inventor will create a project file (IPJ) with the same name as entered in the **Name** text box. Pick the **Finish** button to complete the project wizard.

This will bring you back to the **Open** dialog box with the projects displayed. Double-click on the name Example-11-01 in the **Project name** column to make it the active project. See **Figure 11-5**. A check mark appears next to the name of the active project. Then, pick the **Cancel** button to close the **Open** dialog box. Note: This will not cancel any of the work you have just done.

Figure 11-3.
Creating a new project.

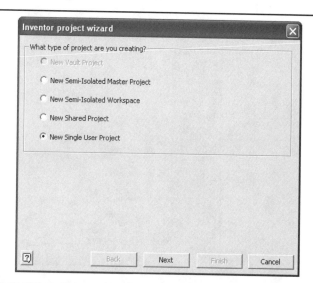

Figure 11-4.
Specifying the
project name and
the location of files.

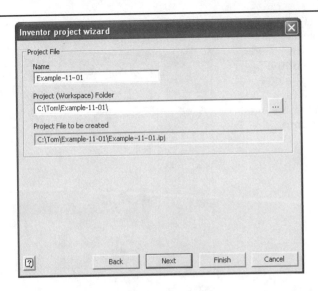

Now, the project file Example-11-01.ipj is created and placed in the Example-11-01 folder. The IPJ file is a data file that points Inventor to the locations of all the files. Do *not* double-click on the IPJ file as it will launch another session of Inventor. Using Windows Explorer, copy the five files in the Chapter11\Examples\Example-11-01 folder installed on your hard drive into the new Example-11-01 folder.

In Inventor, pick the down arrow next to the **New** button on the **Inventor Standard** toolbar. Select **Assembly** from the drop-down list. See **Figure 11-6.** This starts a new assembly file. Save the file as Example-11-01.iam. The location of the file is automatically the Example-11-01 folder. Right-click on Example-11-01 in the **Browser** and select **iProperties** from the pop-up menu. In the **Properties** dialog box, pick the **Project** tab. You can input additional information about this project similar to that shown in **Figure 11-7.** When done, pick the **OK** button to close the **Properties** dialog box.

Building the Assembly

Now, you are ready to build the assembly by placing the parts and adding assembly constraints. Pick the **Place Component** button in the **Assembly Panel** or press the [P] key. The **Open** dialog box appears listing the part and assembly files in the folder defined in the project. You can navigate to other folders if needed. The first

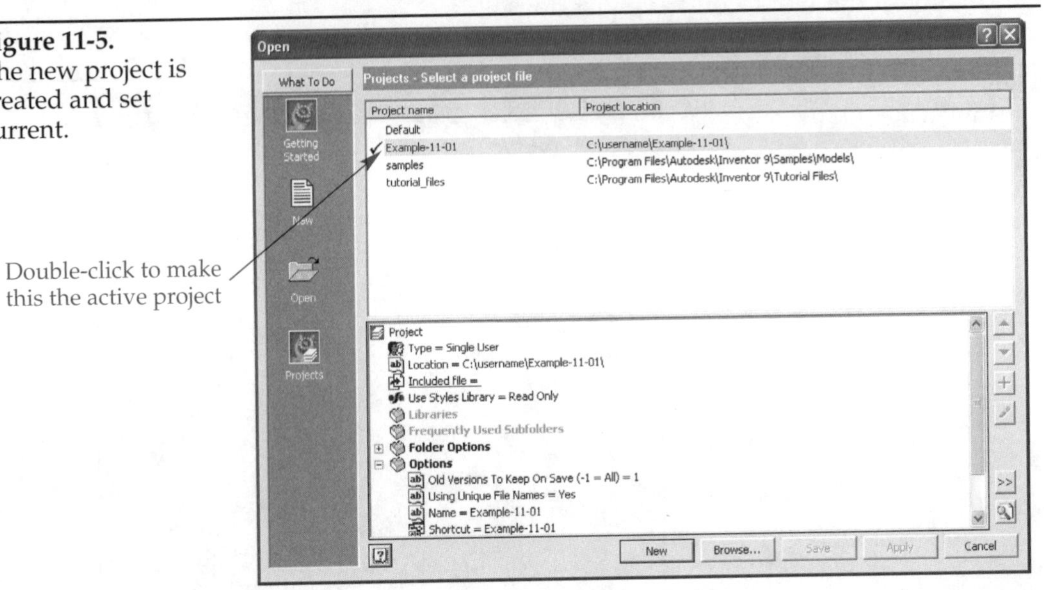

Figure 11-5.
The new project is
created and set
current.

Double-click to make
this the active project

Figure 11-6.
The **New** drop-down list contains several options; select **Assembly**.

Figure 11-7.
Additional information about a project can be included using the **Properties** dialog box.

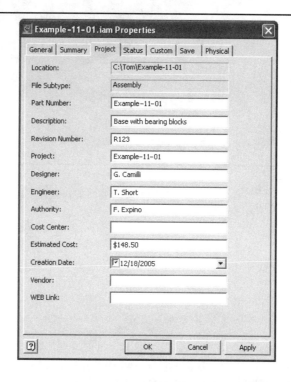

part you will place is the file Base-Plate.ipt. Highlight its name in the dialog box and pick the **Open** button. The part is placed in the assembly file. Notice how a copy of the part appears attached to the cursor. This allows you to place multiple instances of the part. Since only one instance of the base is needed, press [Esc] or right-click and select **Done** to end the command.

Look at the **Browser** and note the pushpin icon in front of the name **BASE-PLATE:1**. This means that the part is grounded—locked to the assembly's coordinate system—and cannot be moved. The first part in an assembly is automatically grounded and the following parts are typically constrained to it. In order to move or rotate the **BASE-PLATE** it is necessary to remove the ground; right-click on the part in the **Browser** and remove the check from **Grounded** check box.

The isometric orientation of the part within the assembly is the same as its orientation in the part file, which in this case is vertical. However, for this assembly, it would be better if the base appears horizontal. See **Figure 11-8**. There are three methods to accomplish this. First, you can use the **Rotate** tool to change the viewpoint and then redefine the isometric view. A second option is to use the **Rotate Component** tool. This is used in a similar manner as the **Rotate** tool, but the part is rotated while the viewpoint remains stationary. Unfortunately, this method requires you to drag the part into position, which may be difficult or inaccurate. The third and most accurate method uses constraints to position the component relative to the assembly file's axes, planes, or center point.

Expand the assembly file's **Origin** folder in the **Browser**. Be sure that the folder belongs to the assembly file and not the component file. See **Figure 11-9**. Make the X axis, XZ plane, and Center Point visible. To see the axes and center point, pick the **Move Component** tool and drag the component out of the way. Pick the **Place Constraint**

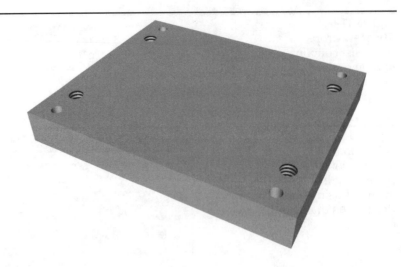

Figure 11-8.
The base is placed
in the assembly and
rotated to produce
an appropriate
view.

Figure 11-9.
The expanded file's
expanded Origin
folder.

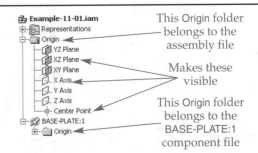

tool, pick the edge of the block at the bottom of the front face, then pick the axis—pick **Apply**. Now pick the bottom left corner of the block; you may have to hover over the corner and use the **Select Other** tool to pick the corner. After the corner is picked, pick the Center Point—again, pick **Apply**. Finally, pick the large front face of the component, and pick the XY plane. If only four holes are visible, pick the button that is not selected in the **Solution** area of the **Place Constraint** dialog box. When all eight holes are visible, pick **OK**.

CAUTION

When selecting different types of geometry while applying mating constraints, be aware that you can select only faces while the **Flush** button is on. If you need to select edges or points, press the **Mate** button in the **Solutions** area.

The component's position is constrained by the Center Point and its orientation is constrained by the X axis and the XZ plane. Since this component is the main part of the assembly, right-click the component's name in the **Browser** and check the **Grounded** check box. Of course, there is some redundancy because the component is grounded and completely constrained; therefore, you may remove the constraints if you choose.

Constraining Centerlines and Degrees of Freedom

The next parts in the logical assembly order are the four locating pins. The geometry in the part file complies with standards and has a flat on one side. With the **Place Component** tool, place the part CR-Locating Pin.ipt. Notice how the first instance is not automatically placed. Only the first part placed into the assembly is automatically placed. Place four instances of the part into the assembly, one near each corner of the base. The exact location is not important because later, constraints will be used to precisely locate the pins.

When an ungrounded part is placed in an assembly, it has six degrees of freedom. It can rotate about each of X, Y, and Z axes. It can also move along each of these axes. Putting it another way, it would take three coordinates (X, Y, and Z) and three angles to specify exactly where the part is in space and its orientation. As *constraints* are applied, degrees of freedom are removed. For example, a mating constraint applied face-to-face removes one linear and two angular degrees of freedom. A symbol that shows the degrees of freedom can be displayed by selecting **Degrees of Freedom** from the **View** pull-down menu. See **Figure 11-10**. Notice how the base does not have a degree of freedom symbol displayed—the part is grounded. To turn off the display of the symbols, select **Degrees of Freedom** from the **View** pull-down menu again.

The first pin is located by mating the centerlines of the pin and the hole. Then, the top of the pin will be mated to the top face of the base with an offset. Zoom in on one corner of the base so the hole and pin are visible. Change the display to wireframe. Pick the **Constraint...** button in the **Assembly Panel**. The **Place Constraint** dialog box is displayed, **Figure 11-11**. In the **Type** area of the **Assembly** tab, pick the **Mate** button if it is not already on (depressed). Move the cursor over the pin in the graphics window. There are 12 possible selections that will be displayed as you move the cursor around over the pin—three face arrows on the flat faces, six center points of arcs, two line edges of the flat, and one centerline. You may need to rotate the view or use the **Select Other** tool to see all of the selections. Select the centerline of the pin and then move the cursor over the hole and select its centerline. A preview of the pin aligned with, and placed inside of, the hole is displayed. Pick the **OK** button in the dialog box to place the constraint. See **Figure 11-12**.

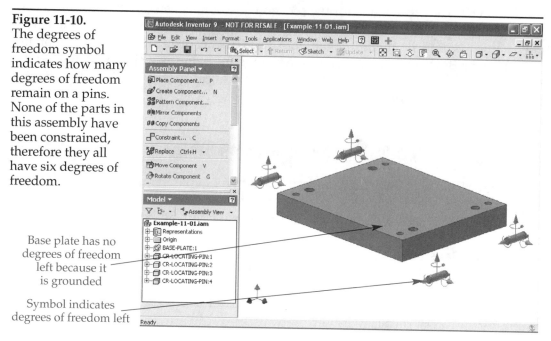

Figure 11-10.
The degrees of freedom symbol indicates how many degrees of freedom remain on a pins. None of the parts in this assembly have been constrained, therefore they all have six degrees of freedom.

Base plate has no degrees of freedom left because it is grounded

Symbol indicates degrees of freedom left

Figure 11-11.
Placing a mating constraint.

Mate button

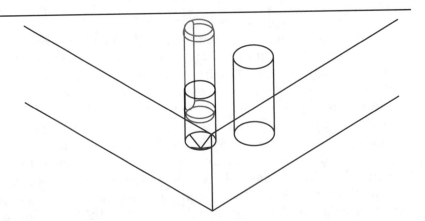

Figure 11-12.
The centerline of the pin has been aligned with the centerline of the hole using a mating constraint. While the pin extends into the hole, this depth is not constrained.

Four of the degrees of freedom are removed by the application of this constraint. The degrees of freedom left are a movement along and rotation about the axis of the pin. If you pick the pin in the graphics window, you can drag it up and down along the centerline of the hole. However, you cannot move the pin around the face of the base.

Now, apply another mating constraint between the top of the pin and the top face of the base. However, since the blind hole is .5″ deep, enter −.5 in the **Offset:** text box. This will make the pin extend .5″ into the hole. See **Figure 11-13.** There is still one rotational degree of freedom left. Removing it is not necessary as the orientation of the flat in the hole is not critical.

Now, expand the branch for the pin in the **Browser**. The name of the pins were amended with sequential numbers as you inserted them. The constraints on the pin are displayed. The offset value for the second constraint is listed with that constraint. The value can be edited by double-clicking on the constraint name and entering a new value. Simply highlighting the constraint name displays an edit box at the bottom of the **Browser** that you can use to change the value as well. If you expand the branch for the base, the constraints are also there because they are between the pin and the base. See **Figure 11-14.**

Using the Insert Constraint

The next pin will be assembled using a different constraint. The previous two-step procedure is so common in creating assemblies that Inventor combined them into one constraint called *insert*. Zoom in on the next pin and hole. Then, continue as follows.

1. Pick the **Constraint...** button in the **Assembly Panel**.
2. Pick the **Insert** button in the **Type** area of the **Assembly** tab. See **Figure 11-15.**

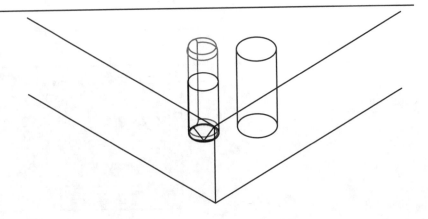

Figure 11-13.
Another mating constraint is applied between the end of the pin and the top face of the base. An offset was specified so the pin extends into the hole.

Figure 11-14.
The constraints on a part are displayed in the part tree in the **Browser**.

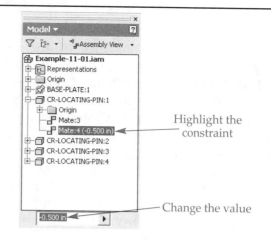

Highlight the constraint

Change the value

Figure 11-15.
Applying an insert constraint.

Insert button

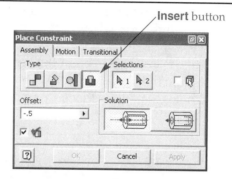

3. In the graphics window, pick the end face of the pin. The direction arrow should be parallel to the centerline. Then, pick the top circle of the hole.

The preview shows the pin aligned with the centerline of the hole. Also, notice that the pin is aligned so that the direction arrows point to each other. This is because the **Opposed** button in the **Solution** area of the dialog box is on. Pick the **Aligned** button in the **Solution** area. The parts are aligned so that the arrows point in the same direction. Pick the **Opposed** button and continue as follows.

4. Enter –.5 in the **Offset:** text box. Do not press [Enter] yet as doing so will apply the operation.

5. Notice in the wireframe view that arrows and an axis are shown in red. This indicates that the constraint is applied both to the faces (red arrows) and the centerlines (red centerline).

6. Pick the **OK** button to apply the constraint and close the dialog box.

If you look on the **Browser**, you can see this one constraint takes the place of two separate mating constraints. However, it gets even better! Zoom in on the third pin and hole. Open the **Place Constraint** dialog box, pick the **Insert** button, and enter –.5 in the **Offset:** text box. Now, pick and hold on the end of the pin. The direction arrow should be parallel to the centerline. Drag the pin over the hole. As you touch the edge of the hole, the preview shows the pin inserted in the hole. Release the mouse button. Then, pick the **Apply** button in the dialog box to apply the constraint. This completes the operation, but the **Place Constraint** dialog box remains open, which allows you to apply another constraint. Using this "drag and drop" technique, insert the fourth pin. Then, pick the **OK** button in the dialog box to apply the constraint and close the dialog box.

Placing the Bearing Blocks

Now, place one instance of the part Bearing-Block.ipt into the assembly. Remember, its exact location is not important at this point. You may want to display a shaded view. With the **Rotate Component** tool, revolve the bearing block until the bottom face is visible. See **Figure 11-16.** This tool only rotates the selected part.

First, use a mating constraint to place the bottom of the bearing block on the top of the base. The mating constraint makes two planar faces parallel to each other. The offset distance in this case is zero since the two faces touch. Open the **Place Constraint** dialog box. In the **Type** area, pick the **Mate** button if it is not already on. Pick the top face of the base and then the bottom face of the bearing block. For each selection, the red direction arrow should be perpendicular to the face and pointing away from the part. In the **Solution** area of the dialog box, make sure the **Mate** button is on. This orients the part so that the direction arrows point at each other. If you pick the **Flush** button, the direction arrows will point in the same direction. Pick the **OK** button in the dialog box to place the constraint and close the dialog box. The bottom of the bearing block is mated flush to the top of the face, but not oriented correctly on the locating pins. See **Figure 11-17.**

The next logical constraint is to align the bearing block with the pins. However, for this example, first align the bearing block with the edge of the base so that it is square with the threaded holes. Open the **Place Constraint** dialog box. In the **Type** area of the

Figure 11-16.
After the bearing block is placed in the assembly, rotate it until the bottom face is visible.

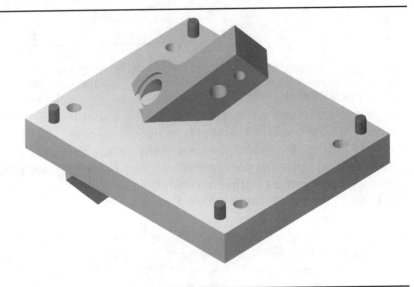

Figure 11-17.
The bottom face of the bearing block is constrained to the top face of the base. The pick points will be used to square the bearing block with the holes.

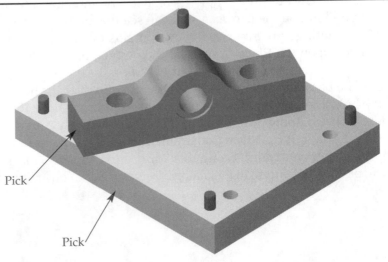

Assembly tab, pick the **Mate** button. Then, pick the two faces as shown in **Figure 11-17.** Notice how the preview shows the bearing block moved off the base. Now, pick the **Flush** button in the **Solution** area of the dialog box. The preview shows the bearing block on the base and square with, but not centered on, the threaded holes. Pick the **OK** button in the dialog box to apply the constraint and close the dialog box.

The blind holes of the block need to mate up with the pins, therefore this is the area where the next constraints will be placed. The block could be aligned by constraining one of its sides to the side of the base plate or constraining the counterbored holes to the threaded holes. But, these are not critical features that control the alignment on the real parts, so they should not be used to control the alignment on the model. Since this component is not grounded, you can use the **Rotate Component** tool to turn the block and see the holes in the bottom. This does *not* remove the constraints you have already placed. Then, continue as follows.

1. Open the **Place Constraints** dialog box. In the **Type** area of the **Assembly** tab, pick the **Mate** button.
2. Pick the centerline of a hole. Then, pick the centerline of the corresponding pin.
3. Pick the **OK** button. (If you pick **Apply,** you cannot use the **Rotate Component** tool.)
4. Rotate the block again and repeat step 2 for the other pin and hole.
5. Pick **OK** to apply the constraint and close the dialog box. See **Figure 11-18.**

You will use the **Rotate Component** and the **Place Constraints** tools almost every time a component is inserted into an assembly. If you rotate a component and cancel the **Place Constraints** tool, you can put the component back to its correct position by picking the **Update** button on the **Inventor Standard** toolbar.

Placing the Bushings

The **Insert** constraint will be used to constrain a bushing in the hole in the bearing block. Place one instance of the part Bushing.ipt into the assembly. Then, continue as follows.

1. Open the **Place Constraint** dialog box. In the **Type** area of the **Assembly** tab, pick the **Insert** button. Also, pick the **Opposed** button in the **Solution** area.
2. Pick and hold on the outer edge of the back of the flange as shown in **Figure 11-19.** Drag the bushing into the bearing block until the preview shows the bushing seated. The outer face of the bushing should be flush with the outer face of the bearing block.
3. Pick the **OK** button in the dialog box to apply the constraint and close the dialog box.

Figure 11-18.
The bearing block is properly constrained in the assembly. Notice how it is square to the base and aligned over the pins.

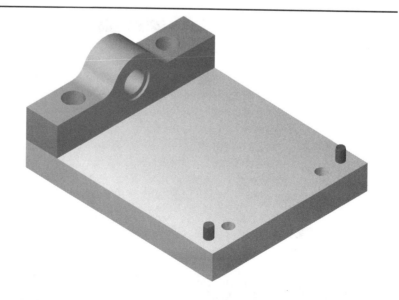

Figure 11-19.
Pick the outer edge
of the back of the
flange when
constraining the
bearing.

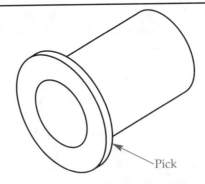

Pick

Determining the Cap Screw Specifications

Before putting in the cap screws that hold down the bearing block, the thread specifications need to be checked on the threaded holes in the base. There are two choices for accessing the base plate—open the part file or edit it in place within the assembly. To edit it in place, right-click on Base-Plate:1 in the **Browser**. The :1 part of the name indicates that this is the first instance placed, which in this case is the only instance. You can also right-click on the part in the graphics window. Then, select **Edit** in the pop-up menu. See **Figure 11-20A.**

All the other parts are displayed semitransparent in the graphics window and are grayed-out in the **Browser**. See **Figure 11-20B.** The part tree for the base plate is also displayed in the **Browser**. Right-click on Hole 3/8 Bolts in the **Browser** and pick **Edit Feature** in the pop-up menu. The **Holes** dialog box is displayed. The **Size** tab displays the pitch as 3/8-24 UNF with 24 threads per inch, which is correct for this example. Record this information; you will need it to select the correct cap screw. In the **Type** tab, verify that the termination is set to **Through All**. Pick the **OK** button to close the dialog box.

Now, you need to determine the proper length for the cap screw. Right-click on Extrusion1 in the **Browser** and select **Show Dimensions** in the pop-up menu. The extruded height is .75 inches; record this information. Now, right-click in the graphics window, right-click again, and select **Finish Edit** from the pop-up menu. This returns you to the assembly.

Figure 11-20.
By selecting **Edit**
from the pop-up
menu, you can edit
a part in place
within the assembly.

Select to edit
in place

Copy	Ctrl+C
Delete	
Edit	
Open	
Promote	Shift+Tab
Demote	Tab
Infer iMates...	
Replace Component	Ctrl+H
Create Note	
Component Selection	▶

A

Example-11-01.iam
Constraints
Representations
Origin
BASE-PLATE:1
　Origin
　Extrusion1
　Hole Dowl Pins
　Hole 3/8 Bolts
　End of Part
CR-LOCATING-PIN:1
CR-LOCATING-PIN:2
CR-LOCATING-PIN:3
CR-LOCATING-PIN:4
BEARING-BLOCK:1

B

Using **Edit
Feature,** you
can view the
specifications
for this
tapped hole

Using the edit in place method, edit Bearing-Block:1. Show the dimensions on Extrusion1. The distance from the bottom of the bearing block to the top of the holes for the cap screws is .875". Record this information. Edit the feature Hole for Cap Screws. The depth of the counterbore is .375". Record this information as well. Close the dialog box and return to the assembly.

Looking at the information you recorded, the cap screw is a 3/8-24 UNF thread. Also, the length must be longer than .5" (.875 – counterbore depth of .375) but no longer than 1.25" (.75 + .875 – .375). Generally, a thread engagement length equal to at least one diameter is sufficient. In this case, the diameter is 3/8 (.375), so the length should be between .875" (minimum of .5 + one diameter of .375) and 1.25" (maximum length). For this example, you will use 1" long cap screws. They provide more thread engagement while not exceeding the maximum length.

PROFESSIONAL TIP

When editing in place, if you change the part(s) in some way, such as changing the thread size, the referenced files are updated when the assembly is saved.

Placing Standard Parts

Inventor has a library that contains catalogs of a wide variety of standard parts and components. *Standard parts* conform to industry standards for size and other specifications. Augmented by Internet resources, such as the Thomas Register, these can greatly reduce the time required to build parts for an assembly. To display the parts in the Inventor library, pick **Model** at the top **Browser** to display the drop-down list. Then, select **Library** from the list. The Standard Parts catalog name appears in the **Browser**. This is the main catalog, with many subcatalogs. It may be easier to navigate through the catalog if you think of a folder structure on the hard drive. Double-click on the catalog to display the subcatalogs.

The Standard Parts catalog has subcatalogs for Fasteners, Steel Shapes, and Shaft Parts. You may want to float the **Browser** and resize it to better view the options. Double-click on the Fasteners name. Yet more subcatalogs are displayed. These are types of fasteners contained within the fasteners catalog. Pick **Icon View** from the view drop-down list to display representations of the part types. See **Figure 11-21.** Double-click on the name Screws and Threaded Bolts to display more subcatalogs, then double-click on the name Socket Head Types to display all of the socket head cap screws contained in the library. See **Figure 11-22.** You can tell that you are looking at parts, and not subcatalogs, by the detailed description below each part.

To specify size and thread type, double-click on the part. For this example, double-click on the Hexagon Socket Head Cap Screw - UNF. A new screen appears that allows you to make settings for the part. See **Figure 11-23.**

In the **Thread description** drop-down list, select 3/8-24-UNF. In the **Nominal Length** drop-down list, select 1. Since this assembly is an English Standard (in).iam file, the values in this drop-down list are in inches.

Figure 11-21.
You can display the library catalogs as lists or as icons. Icons are shown here. Note: The **Browser** has been floated and resized.

Pick to display icons

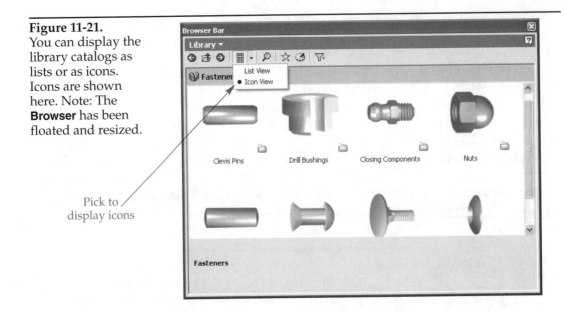

Figure 11-22.
Selecting a part from a library catalog.

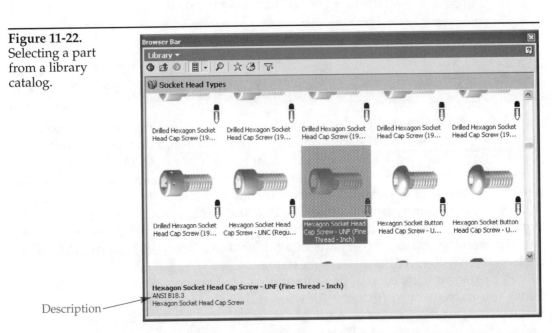

Description

Return the **Browser** to its docking position below the **Assembly Panel**. See **Figure 11-24**. Then, pick on the icon of the cap screw and drag it into the graphics window. When you release the mouse button when the cursor is in the graphics window, a preview of the cap screw appears next to the cursor. Now, you can place two instances of the cap screw. Pick once near each counterbored hole. Then, press [Esc] to end placing the cap screws. See **Figure 11-24**.

Apply an insert constraint between each fastener and its corresponding hole. When making selections, pick the bottom diameter of the head and the bottom diameter of the hole. The top of the cap screw should be flush with the top of the bearing block. See **Figure 11-25**. Save the assembly file.

In the **Browser**, change from the Library back to the Model. The two screws are listed in the part tree with very long names. You can change the name if you wish, but it is not necessary. Using Windows Explorer, look in the Example-11-01 folder. Notice that a file named ANSIB18.3-3_8-24-1 1_4.ipt has appeared. This is one cap screw. In the next chapter, you will create a special project and folder for standard parts that you use often or are unique to your company. For now, leave the cap screw file where it is.

Figure 11-23.
Entering specifications for the selected part.

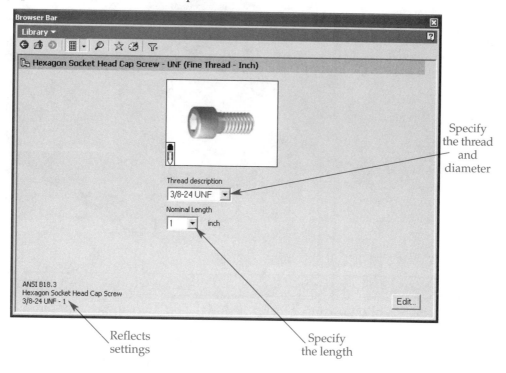

Specify the thread and diameter

Reflects settings

Specify the length

Figure 11-24.
Two instances of the standard part are placed into the assembly.

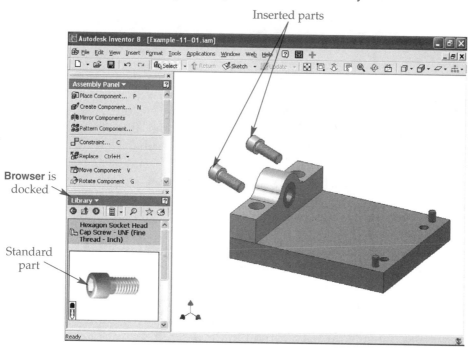

Inserted parts

Browser is docked

Standard part

Also, notice that there is a subfolder called OldVersions. This is where Inventor stores the previous versions of saved parts. There may also be an IDV file. This is a design view file, which will be discussed later in the book. As you can see, this simple assembly requires quite a few files.

Figure 11-25.
The two cap screws are properly located and constrained in the assembly.

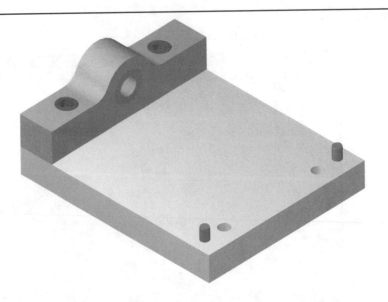

PRACTICE 11-1

❑ Place the second bearing block, bushing, and two more cap screws in the assembly.
❑ Constrain these parts to the assembly.
❑ Remember, the shoulder of the bushing should point to the interior of the part. The assembly is now complete.
❑ Save and close all files.

Constraining Edges of Parts

Create a new subfolder *username*\Example-11-02, where *username* is the folder you created earlier for the subfolder Example-11-01. Create a new project with the name Example-11-02. Specify the location as the subfolder you just created. Also, make it the current project. Next, using Windows Explorer, copy the files from the folder Chapter11\Examples\Example-11-02 that was installed on the hard drive, to the new Example-11-02 folder you just created.

Create a new assembly file and save it with the name Base with Molding in the Example-11-02 folder. Using the **Place Component...** tool, add one instance of the part 2nd Base.ipt. The part should be placed in the correct orientation, as shown in **Figure 11-26.**

To prevent selecting unwanted features during operations, suppress the hole features. To do this, right-click on the name 2nd Base:1 in the **Browser** and select **Edit** from the pop-up menu. Next, right-click on the name Hole Dowel Pins in the part tree for the base and select **Suppress Features** from the pop-up menu. The holes are no longer displayed on the part. Right-click in the graphics window and select **Finish Edit** from the pop-up menu. You are returned to the assembly.

Now, place one instance of the part Molding.ipt into the assembly. The part is not properly aligned with the base. However, as constraints are applied, the part will be oriented correctly. Apply a mating constraint between the flat bottom face of the molding and the front face of the base. You may need to rotate the view or the part to see the proper face on the molding. Apply another mating constraint between the right end of the molding and the right edge of the base. The faces should be flush. See **Figure 11-27.**

Figure 11-26.
The base is placed
into the assembly.

Figure 11-27.
The molding is
placed into the
assembly and
properly
constrained. Notice
the location of the
curved face in
relation to the top
face of the base.

The top edge of the molding's rectangular face is located at the top of the base. To locate the molding at the top edge of the base, apply a mating constraint between the top edge of both parts. When applying a mating constraint, it can be applied between faces, centerlines, points, and edges when the **Mate** button in the **Solution** area is on. Remember, when the **Flush** button is on, you can only select faces. Save the file and keep it open for the next section.

The Tangent Constraint

Now, suppose the upper curved face of the molding needs to be tangent to the top face of the base, instead of the edges being coincident. This can be achieved with the tangent constraint, which makes two curved faces or one curved and one planar face tangent to each other. Parts can be tangent to the inside or outside of each other. See **Figure 11-28.**

Figure 11-28.
When using the tangent constraint, you have two choices as to how the parts are aligned.
A—**Outside**. B—**Outside**. C—**Inside**. D—**Inside**.

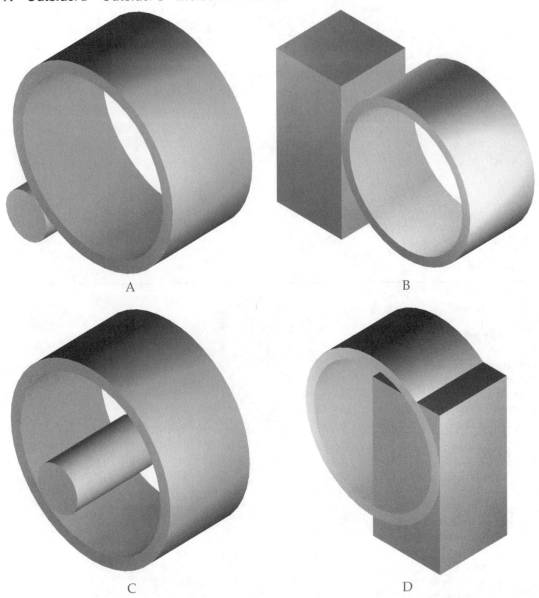

A

B

C

D

In the **Browser**, right-click on the last mating constraint applied and select **Delete** from the pop-up menu. If you do not delete this constraint, an error will be generated when the tangent constraint is applied. Open the **Place Constraint** dialog box and pick the **Tangent** button in the **Type** area. Then, pick the upper curved face and the top face of the base. In the dialog box, pick the **Inside** button in the **Solution** area. Do not enter an offset. Finally, pick the **OK** button to apply the constraint and close the dialog box. See **Figure 11-29.**

Figure 11-29.
The curved surface
is constrained
tangent to the top
face of the base.
Compare this to
Figure 11-27.

Chapter Test

Answer the following questions on a separate sheet of paper.

1. What is an Inventor project?
2. Define *shared environment.*
3. Name the two projects that are created when Inventor is installed.
4. Which type of Inventor files must be open to create a new project?
5. Define *assembly.*
6. What is an *instance?*
7. How does the **Rotate Component** tool differ from the **Rotate** tool?
8. When a part is placed in an assembly, how many degrees of freedom does it have?
9. What is a *grounded* part?
10. Briefly describe *editing in place.*
11. What is a *standard part?*
12. How are degrees of freedom removed from a part placed in an assembly?
13. How can you see the degrees of freedom on the parts in an assembly?
14. After a constraint has been applied to a part, what happens to the constraint when the **Rotate Component** tool is used?

Chapter Exercises

Exercise 11-1. Create a new project named Exercise-11-01 with the path set to Chapter11\Exercises\Exercise-11-01 that was installed on the hard drive. Start a new assembly file. Assemble and constrain the parts as shown below. The wheel is offset .125″ from the bushing. The exploded view is shown for reference only.

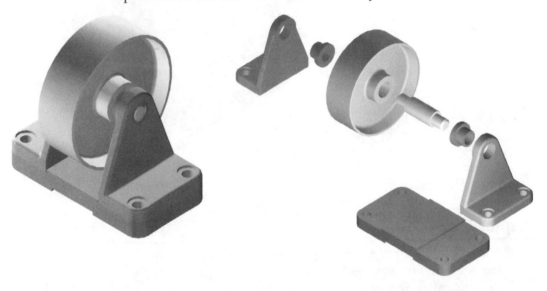

Exercise 11-2. Create a new project named Exercise-11-02 with the path set to Chapter11\Exercises\Exercise-11-02 that was created on the hard drive. Start a new assembly file. Assemble and constrain the sheet metal box as shown below. Put the coin anywhere inside the box.

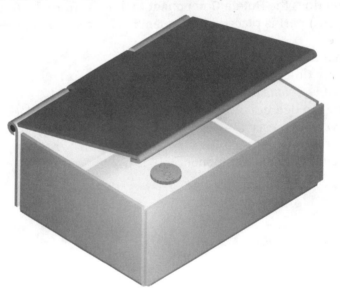

Exercise 11-3. Create a new project named Exercise-11-03 with the path set to Chapter11\Exercises\Exercise-11-03 that was installed on the hard drive. Start a new assembly file. Assemble and constrain the ball socket clamp as shown below. You can mate the work axis of the ball socket and the clamps to align them with the ball. Then, apply the tangent constraint. The exploded view is shown for reference only.

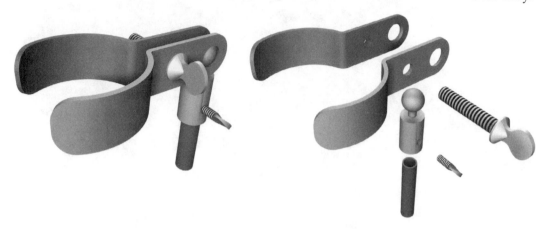

Exercise 11-4. Create a new project named Exercise-11-04 with the path set to Chapter11\Exercises\Exercise-11-04 that was installed on the hard drive. Start a new assembly file. Assemble and constrain the door-bell remote switch as shown below. Then, place two ISO 7050 F-H 2.2 × 9.5 mm flat head Phillips screws in the two countersunk holes in the bottom. Use the drag and drop method to insert the screws in the holes. The exploded view is shown for reference only.

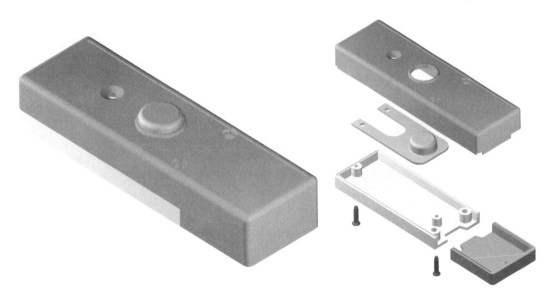

Assemblies are created using existing parts that are assembled using constraints.

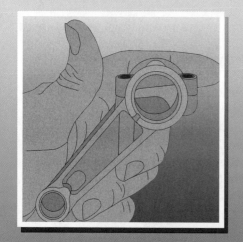

CHAPTER 12

Working with Assemblies

Objectives

After completing this chapter, you will be able to:

* Add paths to a project.
* Create a part from within the assembly.
* Apply the angle constraint.
* Drive constraints.
* Constrain work planes.
* Apply assembly constraints to sketches.
* Create adaptive parts.
* Set the visibility of parts in an assembly.
* Create and use design views.

User's Files

*The following is a list of files that you will need to work through this chapter. These files can be found on the **User's Files** CD included with this text.*

Examples

Example-12-01A.ipj	Example-12-05-Assembly.iam	Slider Clamp.iam
Con_Rod.ipt	Fly_Wheel.ipt	**Practice**
EM1203_Control.ipt	Piston.ipt	Practice-12-01-Tank.iam
Example-12-03-Assembly.iam	Piston_Pin.ipt	
Example-12-04-Adaptivity.iam	Single Cylinder Engine.iam	
	Single Cylinder Engine2.iam	

Setting Additional Paths in Projects

In many projects, the part and assembly files may be in different folders or even on different computers. For example, the Slider Clamp assembly shown in **Figure 12-1** has an assembly file and the part files for the Slider, Input, and Coupler Link are located in the Chapter12\Examples\Example-12-01A folder. The Slider Clamp Base Assembly is a subassembly, indicated by the icon in front of the name in the **Browser**, and it is in the folder Chapter12\Examples\Example-12-01 folder. Notice that the grounded symbol was removed from the base assembly to make the subassembly symbol more visible.

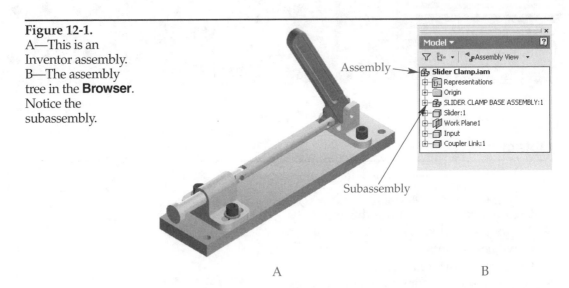

Figure 12-1.
A—This is an Inventor assembly.
B—The assembly tree in the **Browser**. Notice the subassembly.

A   B

There are two methods of managing Inventor projects. Remember from previous chapters, a project is a system that creates and manages a logical file structure that keeps track of the locations of all the files in an assembly so that a team of people can work together on a design. This structure is kept in a file called the project file, with an IPJ file extension. You have created and edited projects from within Inventor. Now you will edit a project directly from Windows Explorer using the **Inventor Project Editor**.

First, Inventor can be open but it is not necessary. However, all Inventor files must be closed. In Explorer, go to the folder Chapter12\Examples\Examples-12-01A and locate the project file Example-12-01A.ipj. Right-click on the file and select **Edit** from the pop-up menu, which brings up the **Inventor Project Editor**. See **Figure 12-2**.

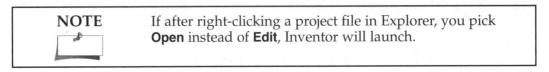

NOTE	If after right-clicking a project file in Explorer, you pick **Open** instead of **Edit**, Inventor will launch.

The available projects are listed in the upper section of the dialog box and the current project is indicated by a check mark. If Example12-01A is not the current project, double-click on it and make it so. The **Edit Project** area should look similar to **Figure 12-2**.

You will have to expand the dialog box and then the **Workspace** and **Workgroup Search Paths** to see them listed. Inventor will search for files in all of these specified folders when the assembly file is opened.

Figure 12-2.
The **Inventor Project Editor** can be used to edit certain aspects of a project from within Windows Explorer.

Current project

Search paths

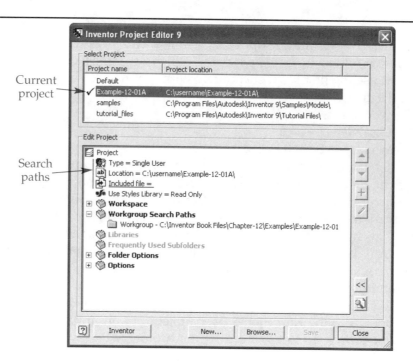

To add a new search path, highlight the name of the category, such as Workgroup Search Paths. Then, either pick the plus button on the right side of the dialog box or right-click and select **Add Path** from the pop-up menu. A new path appears as two text boxes, **Figure 12-3.** The first text box is used to name the path, such as Workgroup-Mech Engineering. The second text box is the actual path. Either type the path or pick the button on the right end of the text box to locate the folder in an Explorer-type browser.

Figure 12-3.
Adding a new path to the project.

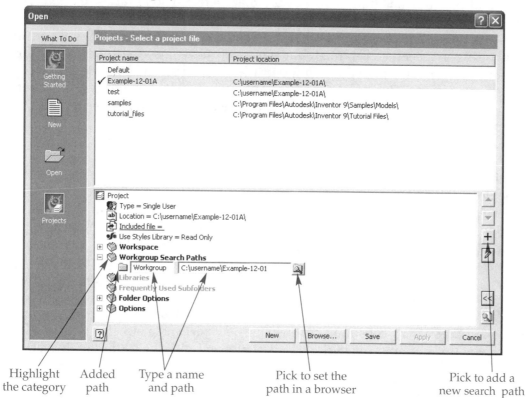

Highlight the category

Added path

Type a name and path

Pick to set the path in a browser

Pick to add a new search path

When using projects, file names are important. It is best to use unique names for all files. For example, suppose you have two files named Base.ipt located in two different folders in the search paths. When Inventor looks for the file named Base.ipt, it will use the first file it encounters with that name. The search order is:

1. Libraries.
2. Workspace.
3. Local Search Paths.
4. Workgroup Search Paths.

By using unique names, you can ensure that Inventor will always select the correct file. Keep your folder structures and file names as simple and logical as possible. However, use long, descriptive part and assembly names when needed. If you rename a file or move it out of the search path, Inventor will not be able to locate it. If Inventor cannot find a file, a dialog box is displayed asking you to locate the file.

You cannot be sure where your files will be loaded in every case. For example, they could be on a server, so you will be asked to create a new project for each example, exercise, and practice.

When you create a new project, you will have a choice of five options. For this textbook, use the **New Single User Project** option. Input the project name and then find or browse to locate the drive and folder where your files for the example are located. This will be the path for the "Project (Workspace) Folder."

Click the **Next** button and make sure that there are no libraries selected. Then click **Finish** and **Apply** to make this the current project. After setting up a project in this manner, Inventor will search for files in this folder and all its subfolders. You must be careful to assign unique file names for new parts and assemblies in a project. Duplicating file names may result in an error.

Creating Parts in the Assembly View

Make sure no files are open. Create a new *single user project* and name it Example-12-01.ipj. Make the *project (workspace) folder* the Chapter12\Examples\Example-12-01 folder and click the **Next** button. Make sure that there are no libraries selected. Then pick **Finish** and **Apply** to make this the current project.

Open the assembly Slider Clamp.iam in the Linkage subfolder. The links in this mechanism have been constrained, but pins need to be inserted through the three joints. So far, you have placed existing part files into an assembly. Now, you will be creating new parts directly in the assembly view. Start by picking the **Create Component** button in the **Assembly Panel**. The **Create In-Place Component** dialog box is displayed, **Figure 12-4.**

For this example, enter the name Short_Pin in the **New File Name** text box. The resulting IPT file will have this name. The **File Type** drop-down list is used to determine if the new component is a part or an assembly. Since the pin is a part, select **Part** in the drop-down list.

The path where the new component will be saved is listed in the **New File Location** text box. By default, this is the folder for the current project. You can type a new path in the text box, or you can browse to a different folder if you pick the **Browser...** button next to the **File Type** drop-down list. Using the **Browser...** button lets you look inside the folder to check current file names, helping to avoid duplicating a file name. For this example, leave the location.

When checked, the **Constrain sketch plane to select face or plan** check box automatically places a mate constraint between the new part and a face that you are asked to pick. For this example, uncheck this check box and pick the **OK** button to start the part.

Figure 12-4.
Creating a new
component from
within an assembly.

File name
Path
Uncheck

Now, you must select a sketch plane. If you pick in a blank area of the graphics window, the sketch plane is plan to the view. The resulting sketch is parallel to the screen. If you pick a face on one of the existing parts, the sketch plane is placed on that face. Pick the top of the base as the sketch plane. Then, create a sketch as follows.

1. Draw a circle at the origin.
2. Dimension the circle to a diameter of .25".
3. "Finish" the sketch.
4. Extrude the circle a distance of .5".
5. From the **Format** pull-down menu, select **Active Standard**.
6. In the dialog box, change the **Active Material Style** to **Bronze, Soft Tin** and pick **OK**.
7. In the **Browser**, double-click on the icon next to the main assembly name Slider Clamp.iam to return to the assembly.

A new part is created and added to the assembly. The pin is on the screen, but it is not yet saved. Save the assembly file and when prompted, pick **OK** to save its dependents. Now, you need to constrain the pin to the slider. Turn off the visibility of the coupler link to make it easier to see and select the slider. Then, continue as follows.

1. Place an insert constraint between the pin and the hole in the slider. Enter an offset of –.125". See **Figure 12-5.**
2. Using the **Place Component** tool, place an instance of Short_Pin into the assembly. In order to do this, you must have saved the assembly and its dependents.
3. Apply an insert constraint between the second pin and the hole in the input link. Use an offset of –.125".
4. Turn on the visibility of the coupler link.

Using this same procedure, create another part called Long_Pin. Make its diameter .25" and its length .75". Apply an insert constraint between the pin and the hole between the base assembly and the input. The end of the pin should be flush with the outer face. See **Figure 12-6.** Save the assembly file and its dependents.

Figure 12-5.
A—Applying an
insert constraint
between the pin and
the hole. B—The pin
is inserted into the
hole.

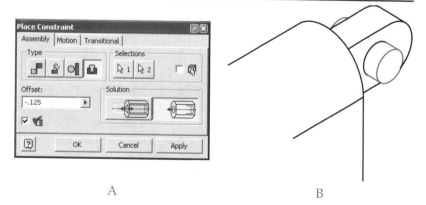

A   B

Figure 12-6.
The assembly is complete.

Angle Constraint

An angle constraint works on part edges and faces, work planes, and work axes. It sets an angle between any two of these features on two different parts. Right now, the input link of the Slider Clamp assembly is free to rotate about its pivot on the base. Pick the input and drag it around. Notice the constraints, but also notice that you can move the link to positions that are physically impossible in the real mechanism.

1. Pick the **Constraint...** button in the **Assembly Panel** to open the **Place Constraint** dialog box.
2. Pick the **Angle** button in the dialog box.
3. Using the two buttons in the **Selections** area of the **Assembly** tab, pick the narrow bottom face of the input and the top face of the base assembly.
4. In the dialog box, set the **Angle** to 120°. The angle is defined as the angle between the two arrows if you put them tail-to-tail.
5. The buttons in the **Solution** area change the arrow directions into or out of the part. Pick the button that results in the assembly configuration. See Figure 12-1A.
6. Pick the **OK** button to apply the constraint and close the dialog box.

Driving Constraints

To *drive* a constraint is to animate the constrained parts through a sequence of motion allowed by the constraint. By driving the angle constraint you just applied, you can animate the motion of the mechanism. In the **Browser**, expand the part tree for Input. Right-click on the angle constraint name and pick **Drive Constraint** from the pop-up menu. The **Drive Constraint** dialog box is displayed. Expand the dialog box by picking the **>>** button. See **Figure 12-7.**

1. In the **Start** text box, enter 120. This is the beginning angle value, or in this case, the maximum "closed" value for the link.
2. In the **End** text box, enter 30. This is the ending angle value. In this case, the link will rotate from 120° to 30°, or through 90° of motion.
3. The **Increment** area is used to set the angle for each step in the motion. Pick the **amount of value** radio button and enter 2 in the text box.

Figure 12-7.
Driving a
constraint.

Ending angle

Starting
angle

Controls

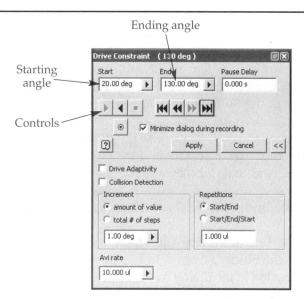

4. The **Repetitions** area is used to set how the parts cycle through the motion. Since the link oscillates (travels back and forth), pick the **Start/End/Start** radio button. Also, enter 5 in the text box so the cycle of motion is repeated five times.
5. The parts are currently at the end of the motion. Therefore, pick the **Reverse** button in the controls to animate the motion. If the parts were at the beginning of the motion, you would pick the **Forward** button.

Any constraint that has an offset, such as mating, flush, angle, tangent or insert, can be driven using this same procedure. This allows you to create all kinds of rotating and sliding animation. However, you can only drive one constraint at a time.

PROFESSIONAL TIP
By picking the **Record** button in the **Drive Constraint** dialog box, the motion can be saved to an AVI video file that can be played in Windows Media Player. In this way, you can share the animated part with others who do not have access to Inventor.

Constraining Work Planes and Axes

Constraints can be applied to work planes and axes as well as part faces. This can simplify the assembly process. In this example, you will be working with a simple model of a one-cylinder engine. All of the parts have work planes through their centers, which will be used to align the parts.

Make sure no files are open. Create a new *single user project* and name it Example-12-02.ipj. Make the *project (workspace) folder* the Chapter12\Examples\Example-12-02 folder and click the **Next** button. Make sure that there are no libraries selected. Then pick **Finish** and **Apply** to make this the current project.

In Inventor, open the file Single Cylinder Engine.iam from the linkage subfolder. This assembly currently includes the engine block and crank. The problem is how to align the center of the crank with the center of the bore without calculating offset values. Notice the two work planes. First, you will constrain the two planes to each other. Then, you will constrain the rotational centerline of the crankshaft with the work axis on the engine block. By doing this, the crankshaft is free to rotate and the resulting assembly is much more flexible. For example, if the block dimensions are changed, the crankshaft will remain centered within the bore.

1. Pick the **Constraint...** button in the **Assembly Panel**.
2. In the **Place Constraint** dialog box, pick the **Mate** button in the **Type** area and the **Flush** button in the **Solution** area. **Flush** is used because the flat on the crank is to be on the left side of the block.
3. Pick the work plane on the crank so the direction arrow points up. Pick the work plane on the block so the direction arrow points to the left.
4. Pick **Apply** in the dialog box to apply the constraint and leave the dialog box open.
5. With the **Mate** button in the **Type** area on, pick the **Mate** button in the **Solution** area.
6. Pick the centerline of the crankshaft and then the work axis through the bearing housings on the block.
7. Pick the **OK** button in the dialog box to apply the constraint and close the dialog box.

The crankshaft is centered under the bore. Also, the rotational centerline of the crankshaft is inline with the centerline of the bearings. See **Figure 12-8.** The crankshaft is free to rotate. Pick it and drag to make it rotate.

Now, the connecting rod needs to be placed into the assembly and constrained to the crankshaft. This will be easier to do if you turn off the visibility of the block. Place an instance of Con_Rod.ipt into the assembly. Constrain the large end of the connecting rod to the crankshaft using the process described above. The centerline of the large hole should be constrained to the centerline of the journal, not the centerline of the bearing housings. See **Figure 12-9.** Since the rod is symmetrical, it does not matter if you use a mating or flush solution on the work planes.

Using a similar process, place an instance of Piston_Pin.ipt, constrain its work plane to the work plane on the connecting rod, and then constrain its centerline to the centerline of the small hole in the connecting rod. Place an instance of Piston.ipt, constrain its work plane to the work plane on the connecting rod, and constrain the centerline of its hole to the centerline of the pin. See **Figure 12-9.** Your assembly may look slightly different, depending on where you picked planes and centerlines. Continue as follows.

Figure 12-8.
The crankshaft is properly located and constrained.

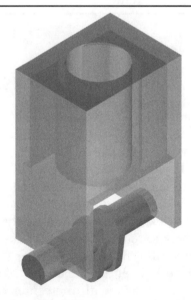

Figure 12-9.
The piston, piston
pin, and connecting
rod are added to the
assembly.

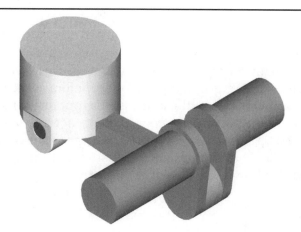

1. Turn on the visibility of the block.
2. Constrain the centerline of the piston to the centerline of the bore. If the piston is inside the block in your assembly, you may need to move the piston in order to acquire the centerline.
3. Place an instance of Fly_Wheel.ipt into the assembly.
4. Constrain the flat on the flywheel to the flat on the crank. Also, constrain the centerlines. Finally, constrain the end of the flywheel flush to the end of the crank.
5. Drag and rotate the flywheel. The piston goes up and down.

Now, automate the motion. Put an angle constraint between the work plane on the flywheel and the top of the block. Both direction arrows should point up; use the buttons in the **Solution** area of the **Place Constraint** dialog box as needed. Also, set the angle to 0°. Next, right-click on the constraint name in the **Browser** and select **Drive Constraint** from the pop-up menu. Set the values as shown in **Figure 12-10** and "run the engine."

Figure 12-10.
Driving the angle
constraint to rotate
the flywheel.

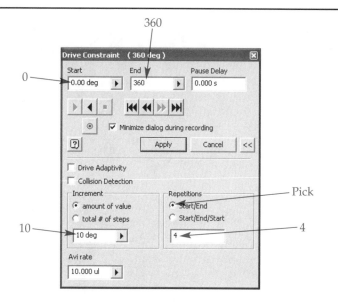

Assembly constraints can also be applied to circles, arcs, and lines in sketches. This is useful for conceptual designs where you are trying to develop the correct kinematical geometry, or geometry of motion, before constructing the detail parts. In this example, you will look at the kinematic layout of a scissors lift.

Make sure no files are open. Create a new *single user project* and name it Example-12-03.ipj. Make the *project (workspace) folder* the Chapter12\Examples\Example-12-03 folder and click the **Next** button. Make sure that there are no libraries selected. Then pick **Finish** and **Apply** to make this the current project.

Open the assembly Example-12-03-Assembly.iam in the linkage subfolder. This assembly contains eight parts: one base, one top, four instances of the link, one control, and one backboard.

Except for the backboard, each part is a simple sketch that has not been extruded; you can see the sketch dimensions. In the isometric view, **Figure 12-11,** you can see that only the backboard has depth. The circles on the ends of the links have mating constraints to the circles on the base and the top, and to each other at the center. The left edges of the top and the base have mating constraints so the parts stay lined up and parallel.

When the sketched circles are mated, they are not constrained in the Z direction. As such, in an isometric view, their motion can be confusing when you try to move them in just the XY plane. This is the reason for the backboard. A line in each of the parts is constrained to the backboard with a mating constraint. This keeps all the parts in the same plane and parallel to the backboard.

The base and the backboard are both grounded. The control link has an angle constraint with the top that keeps it horizontal. There is a mating constraint between the top and the base with an offset of 9", which has been suppressed. All of this means that you can drag the top up and down and the entire linkage will move. Be careful not to move the top up too far. The link motion may reverse. If this happens, undo the drag.

Figure 12-11.
The sketch for a scissors lift.

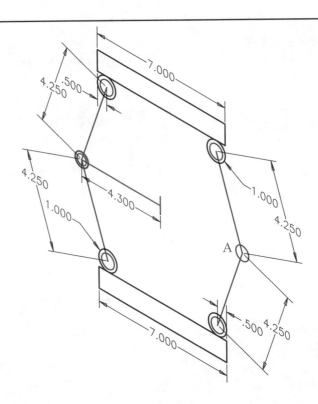

A second control link needs to be placed on the right side. It will be constrained to the existing control so they line up. Then, you will move the top down as if each of the controls is a hydraulic cylinder. Finally, you will measure the distance between the controls. The procedure is:

1. Place an instance of EM1203_Control.ipt into the assembly.
2. Using the **Look At** tool, display a plan view of the control you just placed.
3. With **Rotate Component** tool, rotate the inserted control approximately 180° about the screen's Z-axis.
4. Apply a mating constraint between the circle in the inserted part and the Circle A shown in Figure 12-11.
5. Apply a mating constraint between the long line in the inserted control and the face of the backboard.
6. Apply a mating constraint between the long line in the inserted control and the corresponding long line in the existing control. This mating constraint keeps the long lines in the two controls inline.
7. In the **Browser**, expand the assembly tree for the top. Right-click on the mating constraint with the 9" offset and select **Suppress** in the pop-up menu to unsuppress the constraint.
8. Edit the mating constraint on the top that currently has a 9" offset and change the offset to 3". See **Figure 12-12**.
9. Select **Measure Distance** from the **Tools** pull-down menu. Then, pick the short vertical line in each control. The distance between the two lines is displayed in the **Measure Distance** dialog box, **Figure 12-13**.

Figure 12-12.
The offset on the mating constraint between the top and the base is changed to 3.

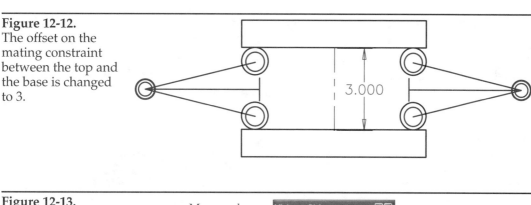

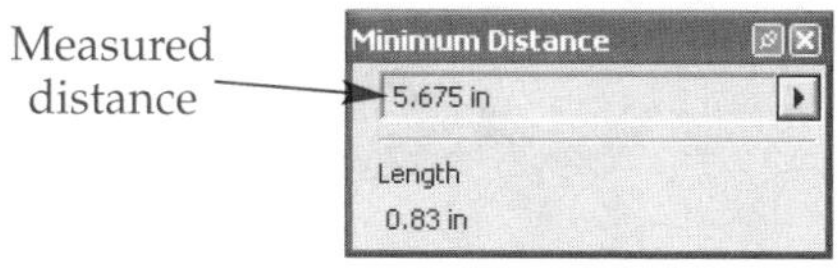

Figure 12-13.
Measuring the distance between the two controls.

Measured distance

Minimum Distance
5.675 in
Length
0.83 in

Adaptive Parts

An *adaptive part* has unconstrained features, the sizes and locations of which are controlled by another part. For example, you can locate features on a part based on the geometry of a second part. Then, if the second part's features are changed, those on the first part change as well. For instance, the diameter of a clearance hole can be based on the mating threaded hole diameter. The assembly process can be simplified by working in a 2D front view of the parts.

Adaptive Location

Make sure no files are open. Create a new *single user project* and name it Example-12-04.ipj. Make the *project (workspace) folder* the Chapter12\Examples\Example-12-04 folder and click the **Next** button. Make sure that there are no libraries selected. Then pick **Finish** and **Apply** to make this the current project.

In Inventor, open the file Example-12-04-Adaptivity.iam. This assembly is shown in **Figure 12-14** with the cap moved to expose the six threaded holes in the base. Matching clearance holes need to be placed in the cap. The holes in the base can be projected onto a sketch plane on the cap. Then, holes can be created in the cap, which will be adaptive. The procedure is:

1. Right-click on the cap name in the **Browser** and select **Edit** from the pop-up menu.
2. Display a wireframe view so you can see the holes in the base.
3. Start a new sketch with the top face of the cap as the sketch plane.
4. Project the circle tops of the six holes onto the sketch plane.
5. Using the **Point, Hole Centers** tool, place hole centers at the center points of all six circles
6. "Finish" the sketch.
7. Using the **Hole** tool in the **Part Features** panel, place .5" diameter holes at each of the points. Use the **Through All** setting.
8. Right-click in the graphics window and select **Finish Edit** from the pop-up menu.

A pair of circular arrows appears in the assembly tree in the **Browser** in front of the cap name, **Figure 12-15A**. These red and cyan arrows indicate that some feature of the part is adaptive. If you edit the cap, the adaptive symbol appears in front of the sketch for the holes feature and the sketch for the holes as well. See **Figure 12-15B**.

Now, right-click on the base name in the **Browser** and select **Edit** from the pop-up menu. Next, edit the sketch for Hole1 and change the 3" dimension to 1.5". "Finish" the sketch and "finish" the edit. The corresponding hole in the cap has moved to remain inline with the threaded hole in the base. See **Figure 12-16.**

Figure 12-14.
Adaptive clearance holes will be added to the cap. Note: The cap is shown out of place so the threaded holes in the base are visible.

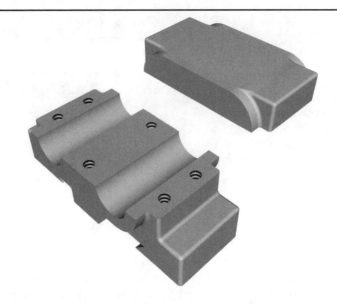

Figure 12-15.
A—The part has the adaptive symbol next to its name in the assembly tree. B—The feature of the part has the adaptive symbol next to its name in the part tree when the part is edited.

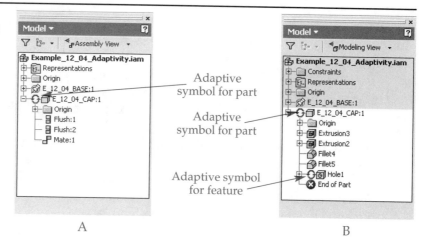

Figure 12-16.
When a threaded hole in the base is moved, the corresponding clearance hole in the cap moves as well.

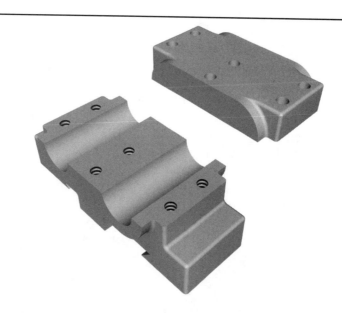

PROFESSIONAL TIP

Experienced users suggest using adaptivity sparingly. When you get the assembly the way you want it, remove adaptivity by right-clicking on the component and selecting **Adaptivity** in the pop-up menu; a check mark appears next to the menu item when adaptivity is on.

❏ Make sure no files are open. Create a new *single user project* and name it Practice-12-01.ipj. Make the *project (workspace) folder* the Chapter12\Practices\Practice-12-01 folder and click the **Next** button. Make sure that there are no libraries selected. Then pick **Finish** and **Apply** to make this the current project.

❏ In Inventor, open the file Practice-12-01-Tank.iam from the linkage subfolder.

❏ This assembly is a large tank with a welded pipe and flange. You will add a cover to the assembly. The cover contains holes that adapt to the locations of the holes and pin in the flange. The procedure is:

1. Zoom in on the flange. Using the **Create Component** tool, create a part called Practice-12-01-Cover.
2. Pick the front of the flange as the sketch plane.
3. Project the outside diameter of the flange onto the sketch plane.
4. "Finish" the sketch and extrude the circle 25 mm.
5. Start a new sketch on the front of the cover.
6. Project all 12 bolt holes from the flange, but not the smaller-diameter pin. Change to a wireframe display if needed.
7. Place a hole center at the center of each projected circle.
8. "Finish" the sketch. Using the **Hole** tool, place 25 mm diameter holes on each hole center using the **Through All** option.
9. Start a new sketch on the front of the cover. Project the 20 mm diameter locating-pin hole. Add a hole center to the center of the projected hole. "Finish" the sketch and place a 20 mm diameter hole on the hole center using the **Through All** option. "Finish" the edit.
10. Apply a mating constraint between the back face of the cover and the front face of the flange.
11. Apply an insert constraint between the locating-pin hole in the cover and the pin.

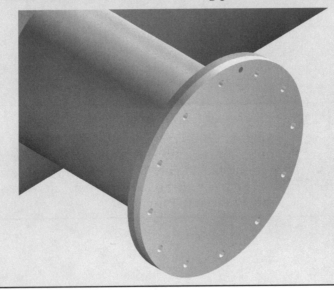

Adaptive Size

The second application of adaptivity ties the size of features in a part to the size of features in another part in the assembly. For example, the two plates shown in **Figure 12-17** are constrained with a 4″ offset. Each post is constrained with an insert constraint to each plate. The length of each post is adaptive. If the 4″ offset is changed to a different value, the posts will automatically change length.

Figure 12-17.
The four posts will be made adaptive so their length will change as the offset between the two plates is changed.

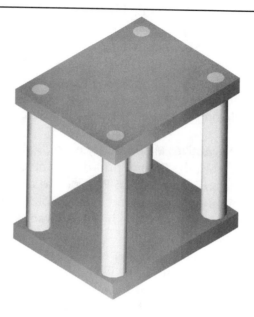

Make sure no files are open. Create a new *single user project* and name it Example-12-05.ipj. Make the *project (workspace) folder* the Chapter12\Examples\Example-12-05 folder and click the **Next** button. Make sure that there are no libraries selected. Then pick **Finish** and **Apply** to make this the current project.

In Inventor, open the file Example-12-05-Assembly.iam. This assembly contains the two plates shown in Figure 12-17. You will add the four posts and make their lengths adaptive.

1. Start a new English Standard (in).ipt file. (You do not need to close the assembly file.) Turn on the visibility of the XY plane and display the isometric view.
2. Draw a circle at the origin and dimension it to a diameter of .75".
3. "Finish" the sketch and extrude the circle 1" using the midplane option.
4. Right-click on the extrusion name in the **Browser**. Select **Adaptive** from the pop-up menu, **Figure 12-18.**
5. Start a new sketch on the XY plane.
6. Display sliced graphics.
7. Draw a circle at the center of the existing feature. Dimension the circle to a diameter of .5".
8. "Finish" the sketch.
9. Select **Parameters** from the **Tools** pull-down menu.

Figure 12-18.
Setting a feature to be adaptive.

Select to make the feature adaptive ➔

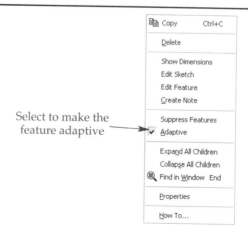

Notice in the **Parameters** dialog box that d1 is the extruded length of the first extrusion. The .5" diameter circle will be extruded so it protrudes out of each end of the .75" diameter shaft. The parameter d1 will be used in an equation for the extruded length so that the .5" diameter shaft is always longer than the .75" shaft. Continue as follows.

10. Extrude the circle using the midplane option. For the extents, select the **Distance** option and enter the equation d1 + 1.0 in the **Depth** text box. Pick **OK**. See **Figure 12-19.**

11. Change the **Active Material Style** to Brass, Soft Yellow by selecting **Active Standard...** from the **Format** pull-down menu.

12. Save the file as Example-12-05-Post.ipt and close the file.

13. In the assembly file, place one instance of the post at the lower-left corner.

14. Right-click on the post name in the **Browser** and select **Adaptive** from the pop-up menu.

Adaptivity was turned on in the part file for the .75" diameter shaft. However, the part also must be made adaptive in the assembly. Continue as follows.

15. Place an insert constraint between the .75" diameter shaft and the hole in the plate. Make sure you pick the large diameter. See **Figure 12-20.**

Figure 12-19.
The post is created as a part file.

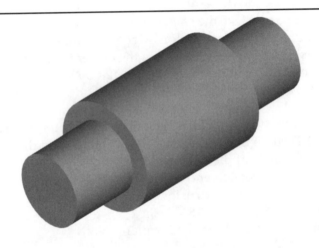

Figure 12-20.
The post part file is added to the assembly and inserted into a hole.

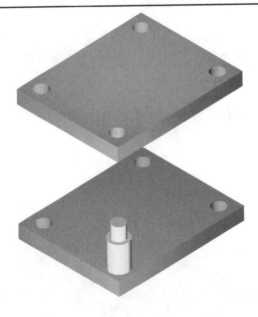

16. Place three more instances of the post into the assembly. Constrain each in the same way. These instances do not have to be set to be adaptive. In fact, **Adaptive** is grayed out in the pop-up menu for the three additional instances; they will match the behavior of the first instance.
17. Rotate the view to show the bottom of the assembly.
18. Place an insert constraint between the first post and the hole in the top plate, again selecting the large diameter.
19. All four posts change length to fit the space between the two plates. The assembly should now look like Figure 12-17.

To see the adaptability of the shafts, edit the mating constraint for the plates. Change the 4" offset to 5" and the posts adapt to the new value. You can also drive the offset in the mating constraint. In order to do this, you must check the **Drive Adaptivity** check box in the **Drive Constraint** dialog box. See **Figure 12-21.**

Figure 12-21.
Driving a constraint and the adaptability of parts.

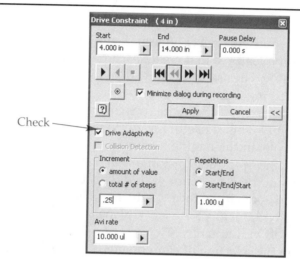

PROFESSIONAL TIP

Adaptive sizing needs to be used carefully; users have reported problems with large assemblies containing multiple adaptive parts.

Visibility and Design Views

Make sure no files are open. Create a new *single user project* and name it Example-12-06_Visibility.ipj. Make the *project (workspace) folder* the Chapter12\Examples\ Example-12-06 folder and click the **Next** button. Make sure that there are no libraries selected. Then pick **Finish** and **Apply** to make this the current project. In Inventor, open the file Single Cylinder Engine2.iam from the linkage subfolder.

Near the top of the **Browser** are three tools or menus. See **Figure 12-22.** These are:
- **Browser Filters**: Controls the display of items in the **Browser**. See **Figure 12-22A**. For example, if **Hide Work Features** has a check mark next to it, the **Origin** icon is not displayed in the **Browser**. Note that this does not affect the display of the model.
- **Design View Representations**: Pick **Other** to list all of the saved representations that you can restore. See **Figure 12-22B**.

Figure 12-22.
A—The **Browser Filters** drop-down list. B—The **Design View Representations** drop-down list.
C—The **Modeling and Position Views** drop-down list.

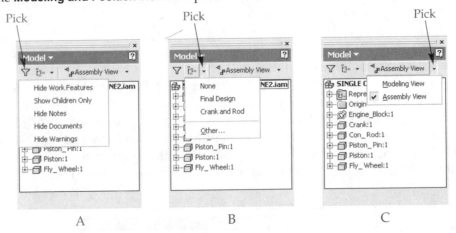

A   B   C

- **Modeling and Assembly Views**. See **Figure 12-22C**. The modeling view displays the features that create each part and any features that are added to the assembly. The assembly view displays the constraints that determine how the parts fit together.

The most important thing you can control in **Design View Representation** is component visibility. This is useful for working on components that are obscured by others. For example, in the engine assembly you might want to work on the piston pin and rod, but the block and piston are in the way. You can create a design view where the block and piston are not visible. This tool can also save settings of:

- **Work and sketch feature visibility.** This is an easy way to display the assembly without all of the work planes and sketches visible, without turning each one off individually.
- **Disabling the selection of components.** Disable the selection of components that you are not going to work on to make selection of relevant components easier.
- **Color and material style.** You can hide or override these for any component; for example, to emphasize a particular part.
- **Zoom and viewing angle.** The zoom factor and viewing angle are saved with the representation.
- **Views to document the assembly steps.** Save multiple design views as you build the assembly and you can "record" the process in views.

In the engine assembly, pick the **Design View Representation** button in the **Browser** and several choices are listed:

- **None.** This is the default and always available in every assembly. Selecting **None** displays the current state of the model.
- **Saved Representations.** Following **None** is a list of the representations that have been saved in this assembly file. In this case they are **Final Design** and **Crank and Rod**. Note that these are also listed under **Representations** in the **Browser**. See **Figure 12-23A**.
- **Other….** This opens the **Design View Representations** dialog box. See **Figure 12-23B**. To make a design view representation active, select its name and then pick the **Activate** button. New representations are created by typing the new name in the bottom text input box and picking **New**. There are two formats that new representations can be saved as:
  - **Public.** The information is saved in the design file. This format is associative and changes are displayed in the drawing files.
  - **Private.** The information is saved within a separate file with an IDV extension. This file is located in the folder shown in the path. Design view representation files, IDV's, are not associative to the drawing file.

Figure 12-23.
There are two places to select a design view representation. A— **Browser**. B— **Design View Representations** dialog box.

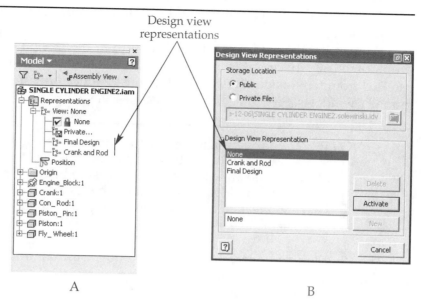

A

B

Figure 12-24.
Disabling a part so it cannot be selected.

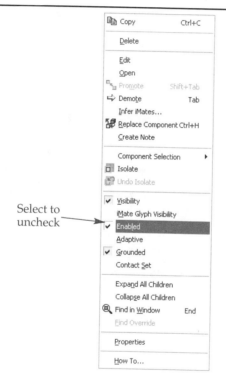

Select to uncheck →

Now, set up the assembly for another new view:

1. Turn off the visibility of the Engine_Block.
2. Right-click on Fly_Wheel in the **Browser** and deselect **Enabled** from the pop-up menu. See **Figure 12-24.** This disables the Fly_Wheel so it cannot be selected. A check mark appears next to **Enabled** when a part can be selected. Also, disable the Crank.
3. Turn off the work planes by selecting **Object Visibility** in the **View** pull-down menu and picking **Origin Planes** in the cascading menu to uncheck the item.
4. Pick the Con_Rod, either in the graphics window or the **Browser**, and change the color to Red (Bright).
5. Change the viewpoint and zoom in on the Con_Rod and Piston.
6. Open the **Design View Representations** dialog box.

7. Enter the name Piston and Rod and pick the **Save** button. Then, close the **Design Views** dialog box.

Now, in the **Browser**, select Final Design from the **Design View Representations** drop-down list. Notice how the parts appear in the graphics windows. Select Piston and Rod from the **Design Views** drop-down list. Notice how the parts appear in the graphics windows. Switch back to the Final Design view.

NOTE
> Any design view representations that are created in an assembly can be used in the drawing for that assembly. If you anticipate making an assembly drawing, you may want to build a representation that can be used in the assembly creation and as a view to be placed in the assembly drawing.

The **Modeling and Assembly Views** drop-down list controls whether or not the constraints are displayed. In the modeling view, the component features are displayed under the assembly in the assembly tree. All of the assembly constraints are located in a branch at the top of assembly tree, rather than under the corresponding part. See **Figure 12-25.** This view is very useful for editing a feature of a part. In the position view, which is the view you have used to this point, the constraints are displayed under the component itself and this is most useful for applying and editing assembly constraints.

Figure 12-25.
A—The modeling view. B—The position view.

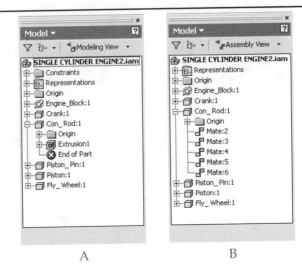

A   B

Chapter Test

Answer the following questions on a separate sheet of paper.

1. How do you open the **Inventor Project Editor**?
2. Which tool allows you to create new parts in the assembly view?
3. What does the angle constraint do?
4. What does it mean to *drive* a constraint?
5. How do you open the **Drive Constraint** dialog box?
6. Why would you apply a constraint between two work planes as opposed to the corresponding parts?
7. Give an example of where you would want to apply constraints to sketches.
8. What is an *adaptive part?*
9. What are the two types of adaptive parts?
10. Briefly describe how to save a new view in Inventor.

Chapter Exercises

Exercise 12-1. Make sure no files are open. Create a new *single user project* and name it Exercise-12-01.ipj. Make the *project (workspace) folder* the Chapter12\Exercises\Exercise-12-01 folder and click the **Next** button. Make sure that there are no libraries selected. Then pick **Finish** and **Apply** to make this the current project.

Start a new assembly file. Assemble and constrain the linkage as shown below. Apply an angle constraint between a face on the base and one on the Input Link. Drive the angle constraint between the base and the input link to automate the motion for the input between the stops.

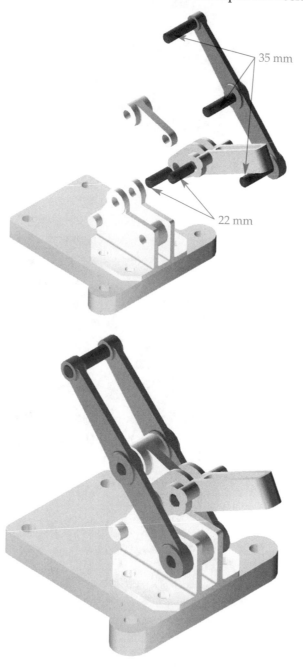

35 mm

22 mm

NOTE Need help? Run Exercise-12-01.avi to see how this assembly goes together.

Exercise 12-2. Make sure no files are open. Create a new *single user project* and name it Exercise-12-02.ipj. Make the *project (workspace) folder* the Chapter12\Exercises\Exercise-12-02 folder and click the **Next** button. Make sure that there are no libraries selected. Then pick **Finish** and **Apply** to make this the current project.

Start a new assembly file. Assemble the pivot as shown below. The shaft has a groove for the Woodruff key. Make this groove adaptive. Apply an angle constraint and drive it to rotate the shaft.

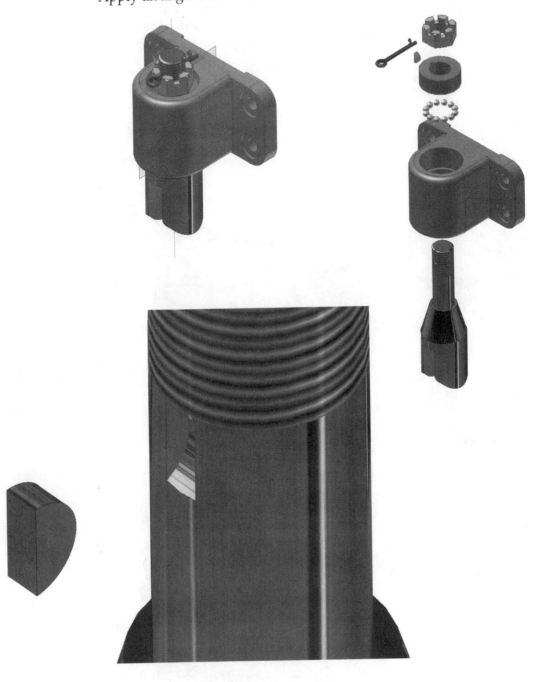

> **NOTE** Need help? Run Exercise-12-02.avi to see how this assembly goes together.

Exercise 12-3. Make sure no files are open. Create a new *single user project* and name it Exercise-12-03.ipj. Make the *project (workspace) folder* the Chapter12\Exercises\Exercise-12-03 folder and click the **Next** button. Make sure that there are no libraries selected. Then pick **Finish** and **Apply** to make this the current project.

Start a new assembly file. Assemble and constrain the holding fixture as shown below. Note that there are three subassemblies and in each one the parts are already constrained to each other.

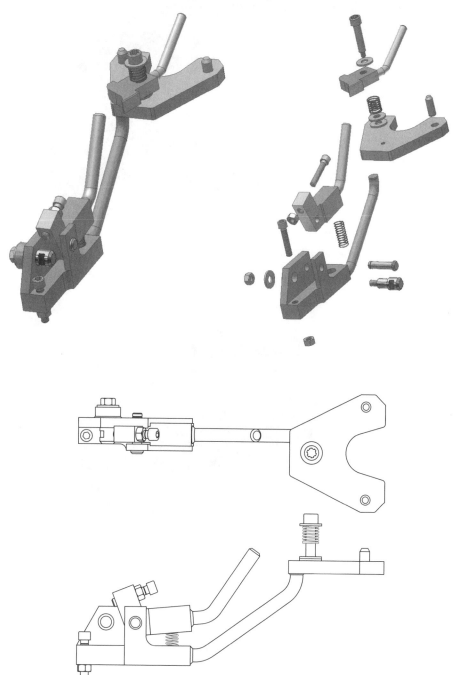

NOTE	Need help? Run Exercise-12-03.avi to see how this assembly goes together.

Exercise 12-4. Make sure no files are open. Create a new *single user project* and name it Exercise-12-04.ipj. Make the *project (workspace) folder* the Chapter12\Exercises\Exercise-12-04 folder and click the **Next** button. Make sure that there are no libraries selected. Then pick **Finish** and **Apply** to make this the current project.

Start a new assembly file. Assemble and constrain the base and cap. Create adaptive bolt and bushing holes in the cap as shown below. Fillet the cap as shown using the same radius as was used on the base. Place and constrain two bushings.

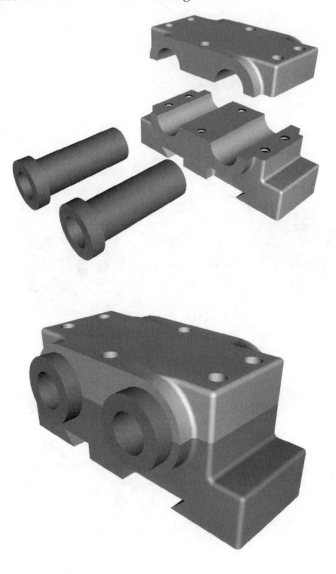

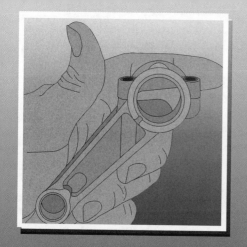

CHAPTER 13

Motion Constraints and Assemblies

Objectives

After completing this chapter, you will be able to:

✳ Understand and use the rotation motion constraint.
✳ Understand and use the rotation-translation motion constraint.
✳ Use the collision detector feature of the **Drive Constraint** dialog box.
✳ Understand and use the second solution of the rotation-translation motion constraint.
✳ Understand and use the transitional motion constraint.
✳ Apply assembly constraints to sketches.

User's Files

*The following is a list of files that you will need to work through this chapter. These files can be found on the **User's Files** CD included with this text.*

Examples

Example-13-01.iam
Example-13-02-Belt.iam
Example-13-03.iam
Example-13-04.iam
Example-13-05.iam
Example-13-06.iam

Practice

Practice-13-01.iam

Exercises

Exercise-13-01.iam
Exercise-13-02.iam
Exercise-13-03.iam

Exercise-13-04.iam
Exercise-13-05.iam
Exercise-13-06.iam
Exercise-13-07.iam

The Motion Constraints

Besides the assembly constraints for positioning components introduced in the last two chapters, there are three motion constraints that will relate the relative motion of one component to another. These are found in the **Place Constraint** dialog box. See **Figure 13-1**. They are as follows:

- **Rotation**. This is found on the **Motion** tab. Used for rotating part to rotating part, such as a pair of gears.
- **Rotation-Translation**. This is found on the **Motion** tab. Used for rotating part to translating part, such as a gear and rack.

Figure 13-1.
A—The **Place Constraint** dialog box showing the **Rotation–Translation** and **Rotation** constraint buttons. B—The **Transitional** constraint button.

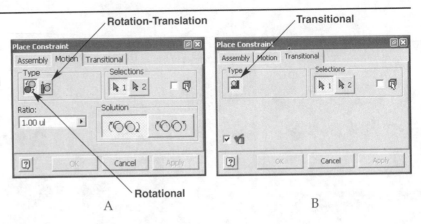

A

B

- **Transitional.** This is found on the **Transitional** tab. Used for surface contact, such as a cam and follower.

NOTE	The **Motion** tab has two choices for a solution and a box for a numerical value called the ratio. The **Transitional** tab has neither. Like the **Assembly** tab, these two tabs have a **Pick part first** check box.

The Rotation Constraint

Make sure no files are open. Create a new *single user project* and name it Example-13-01.ipj. Make the *project (workspace) folder* the Chapter 13\Examples\Example-13-01 folder and click the Next button. Make sure that there are no libraries selected. Then pick **Finish** and **Apply** to make this the current project.

Now, open the Example-13-01.iam file. This assembly has two friction wheels that turn together without slipping. This concept is the basis for gear design where the diameters of the wheels represent the *pitch diameters* of the gears. The small wheel has a 2″ diameter and has an angle constraint named DRIVE ME Angle. This can be seen if Small_Wheel is expanded in the **Browser**. See **Figure 13-2.** This constraint controls its angle to the base, so if the constraint is driven from 0° to 360° the wheel rotates once.

The diameter of the large wheel is 4″. It should rotate once for every two rotations of the small wheel. To create a constraint that will do this, pick the **Constraint** button in the **Assembly Panel**. This accesses the **Place Constraint** dialog box. See **Figure 13-3.** Pick the **Motion** tab where the **Rotation** is the default type. For the first selection, pick anywhere on the *curved* face of the small wheel. For the second selection, pick anywhere on the *curved* face of the large wheel. Make sure it is done in this order, because it makes a difference when the ratio is input. Since these wheels turn in opposite directions, pick the second solution labeled **Reverse**. Notice the rotational vectors on the face of the wheels do not necessarily show the correct directions. Also notice the ratio that was automatically calculated. If the flat faces of the wheels were picked, the rotational constraint would work but the ratio would not be 1:1. The ratio would have to be manually corrected.

Figure 13-2.
The angle constraint used to drive the wheel.

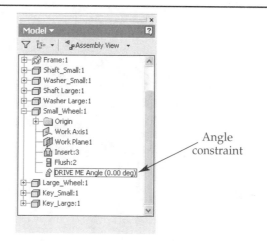

Angle constraint

Figure 13-3.
The **Motion** tab where the **Rotation** is the default type and the **Reverse** solution is selected.

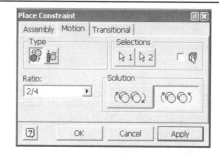

The value of the ratio may seem counterintuitive. You might think it is the angular velocity ratio of the first pick over the second—it is not. It is actually the ratio of the diameters, the first pick over the second. To emphasize that ratio, in the **Ratio** text box enter the fraction 2/4. Pick **OK** and the constraint will appear in the **Browser**. If the **Browser** is set for Modeling, it will appear under Constraints. If the **Browser** is set for Assembly, it will appear under Small_Wheel:1. Above the Rotation:1 constraint is DRIVE ME Angle. Use this constraint to drive the angle constraint on the small wheel from 0° to 360° with two repetitions. The small wheel will rotate twice and the large wheel will rotate once.

Both Wheels Turning in the Same Direction

Make sure no files are open. Create a new *single user project* and name it Example-13-02.ipj. Make the *project (workspace) folder* the Chapter 13\Examples\Example-13-02 folder and click the Next button. Make sure that there are no libraries selected. Then pick **Finish** and **Apply** to make this the current project.

Open the Example-13-02-Belt.iam file. Apply a rotation motion constraint using the Forward solution. For the first selection, pick anywhere on the face of the small pulley. For the second selection, pick anywhere on the face of the large pulley. Make sure it is done in this order, because it makes a difference when the ratio is input. Enter a ratio of 1/4. Pick Apply and Cancel and the constraint will appear in the **Browser**. See **Figure 13-4.** Now drive the angle constraint from 0° to 360° with four repetitions. The small pulley should rotate four times, while the large pulley rotates one time. Unfortunately, moving the belt is something to be figured out.

Figure 13-4.
The belt and pulley assembly.

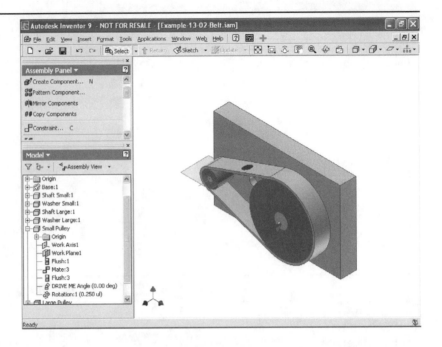

PRACTICE 13-1

❑ Make sure no files are open. Create a new *single user project* and name it Practice-13-01.ipj. Make the *project (workspace) folder* the Chapter 13\Practices\Practice-13-01 folder and click the Next button. Make sure that there are no libraries selected. Then pick **Finish** and **Apply** to make this the current project.

❑ Open the file Practice-13-01.iam. Two gears are going to be attached to the shafts and washers and nuts added. An angle constraint will be placed on the gear's work planes to line up the teeth and then this constraint is suppressed. Then a rotation motion constraint is added to make the gears move.

❑ The red pinion gear is already inserted on the shaft and has an angle constraint between the work planes so both components rotate together.

❑ Put an insert constraint between the washer and the gear.

❑ Put an insert constraint between the hex nut and the washer.

❑ The nut rotates with the gear, so put an angle constraint between one of the flat hex faces of the nut and the gear work plane. See the figure below.

Flat face of hex nut is first selection

Work plane is second selection

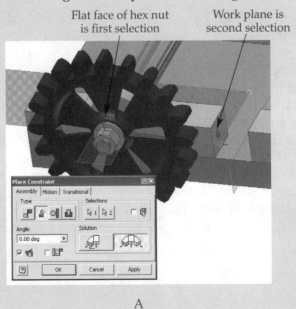

A

(Continued)

❏ Now place the Large_Gear, another nut, and a washer. Insert the gear on the shaft.

❏ Constrain the washer and nut as before.

❏ Line up the work planes on the shaft and gear with a mate constraint. This works because the planes are coplanar. A mate or angle constraint could also be used. The nut must be constrained with an angle.

❏ The visible work plane on the pinion is through the center of the tooth, and the visible work plane on the gear is through the center of the space. Put a mate constraint between these two planes and rename it SUPPRESS mate gear to gear. This is done because it will be suppressed when the gears are made to rotate. See the figure below.

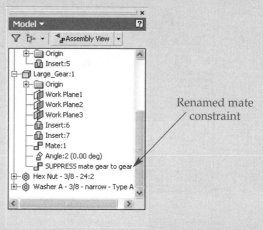

Renamed mate constraint

B

❏ Put a rotation motion constraint between the two gears, picking the red pinion first. Input a ratio of 20/40, the ratio of the number of teeth, and use the reverse solution. See the figure below.

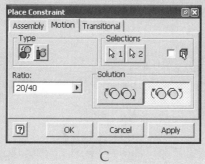

C

(Continued)

❑ Apply an angle constraint between the work plane on the pinion and the base. Pick any point on the base. See the figure below. Rename the constraint DRIVE ME Angle.

D

❑ If you try to drive the DRIVE ME Angle constraint on the pinion gear, you will get an error message. This is a result of the mate constraint called SUPPRESS Mate gear to gear locking the gears in place preventing them from rotating with respect to each other. Therefore, you must right-click on that mate constraint and pick **Suppress** in the pop-up menu. See the figure below. If this step is not done, you will receive an error message.

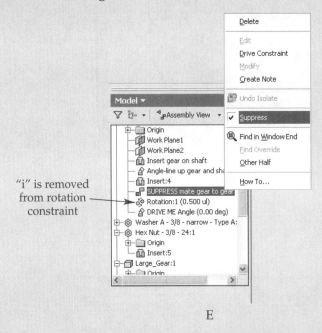

E

❑ Drive the angle constraint from 0° to 360° for 2 repetitions. The gear teeth will mesh, rotating the large gear once.

PROFESSIONAL TIP You can create an AVI video file of the motion by clicking on the record bull's, eye in the **Drive Constraint** dialog box. With the increment value set to 4°, a file of about 10 megs will be created. This file can be played with any video player software, such as Windows Media Player.

The Rotation-Translation Motion Constraint

Make sure no files are open. Create a new *single user project* and name it Example-13-03.ipj. Make the *project (workspace) folder* the Chapter 13\Examples\Example-13-03 folder and click the Next button. Make sure that there are no libraries selected. Then pick **Finish** and **Apply** to make this the current project.

Figure 13-5.
The rack-and-pinion assembly.

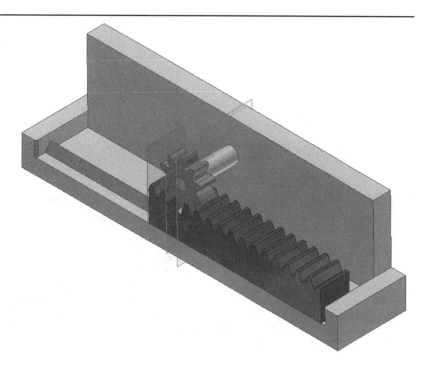

Open the file Example-13-03.iam. The pinion gear is inserted on the shaft and the rack is constrained to the slide on the frame with two mates. Three things need to be done to finish the example. These things are constrain the work planes to line up the teeth, apply a motion constraint between the pinion and rack, and apply a mate constraint between the rack and the frame to drive the motion.

The aligning of the teeth is first. Apply a flush mating constraint between the work plane on the pinion and the one at the center of the frame. See **Figure 13-6A**. Rename this constraint SUPPRESS ME 1. Apply a mate constraint between the work plane on the rack and the one on the frame. See **Figure 13-6B**. Rename this constraint SUPPRESS ME 2.

Figure 13-6.
A—Applying a flush constraint to the pinion and the frame. B—Applying a mate constraint to the rack and the frame.

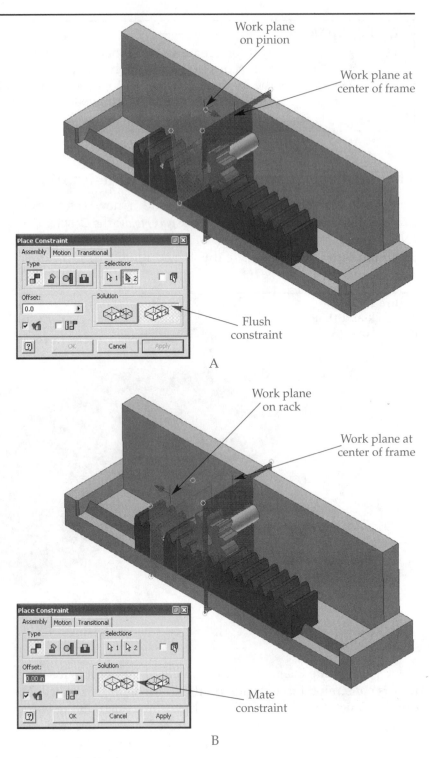

Work plane on pinion

Work plane at center of frame

Flush constraint

A

Work plane on rack

Work plane at center of frame

Mate constraint

B

Apply a rotation-translation motion constraint between the pinion and the rack. Once the **Place Constraint** dialog box is accessed, pick the two selections. With this motion constraint, one part rotates. This part is the pinion and is picked as the first selection. The other part translates and is the rack. This part is picked as the second selection. On the rack, select a face or work plane perpendicular to the direction of motion. This gives a vector (arrow) in the direction of the translation. These two vectors for the rack-and-pinion gear are shown in Figure 13-5. As the gear turns, the rack will slide on the green rail.

In the **Place Constraint** dialog box pick **Forward** in the **Solution** area. See **Figure 13-7**. The value for the relationship between the rotation and translation motions is entered in the **Distance** text box. This is the distance the second part (the rack) moves for one rotation of the first part (the pinion). This is equal to the pitch diameter times PI (2.5 × 3.1416), or 7.854. The calculation can be done in the dialog box by entering the equation (PI)*2.5. The value is bidirectional. This means if the rack is driven 7.854 units, the gear will rotate once. Pick **Apply** and then **Cancel** to exit the dialog box.

The constraints called SUPPRESS ME 1 and SUPPRESS ME 2 are shown in **Figure 13-8**. These constraints are used to line up the teeth, and prevent any motion of the rack and pinion. Both constraints have to be suppressed before the rack can be driven back and forth. Right-click on each constraint and pick **Suppress** in the pop-up menu.

Now the rack can be moved back and forth and the gear will rotate. To do this, put a mate constraint between the end of the rack and the frame. Rename this constraint DRIVE ME. Drive this constraint from 0 start to 8 end with the settings shown in **Figure 13-9**. Keep this file open for the next section.

Figure 13-7.
The distance value can be input as an equation.

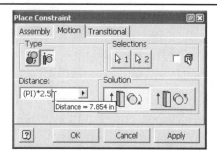

Figure 13-8.
The two constraints that need to be suppressed.

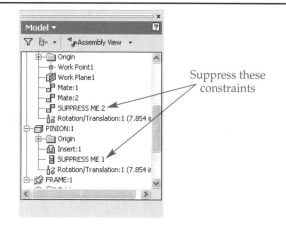

Suppress these constraints

The Collision Detector

When driving a constraint, the assembly does not know if the rack collides with the frame or if the gear teeth interfere with each other unless the **Collision Detection** check box is checked in the **Drive Constraint** dialog box. See **Figure 13-10**. In the rack and pinion example, the teeth do interfere. Drive the DRIVE ME constraint with the **Collision Detection** check box checked. The motion will stop after one increment and a collision detection message will appear. The rack and the pinion (the two interfering parts) will be highlighted on the screen. From the front view wireframe shown in **Figure 13-11**, the interference can be seen. In this case, the teeth are the wrong size.

Figure 13-9.
The **Drive Constraint** dialog box is used to determine the type of motion applied to the constraint.

Figure 13-10.
The **Collision Detection** check box is checked to determine if there are problems with assembly and the drive constraint.

Figure 13-11.
The interference is evident with the misalignment of the rack and pinion.

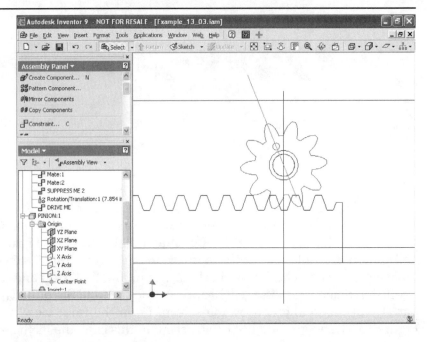

Figure 13-12.
The teeth are redesigned correctly so they do not interfere.

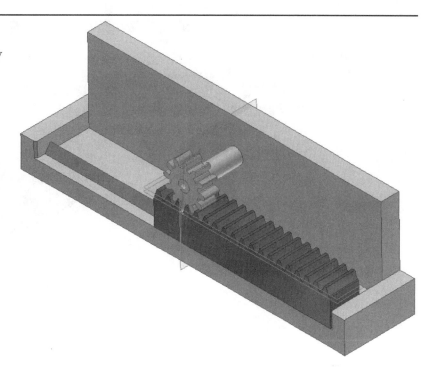

Make sure no files are open. Create a new *single user project* and name it Example-13-04.ipj. Make the *project (workspace) folder* the Chapter 13\Examples\Example-13-04 folder and click the Next button. Make sure that there are no libraries selected. Then pick **Finish** and **Apply** to make this the current project. Open the file Example-13-04.iam. See **Figure 13-12.** The teeth have been redesigned so they no longer interfere.

Drive the DRIVE ME constraint. Notice that the end is set at 10 inches. The collision detector will stop the motion when the rack runs into the left end of the frame.

The Rotation-Translation Constraint—Second Solution

Make sure no files are open. Create a new *single user project* and name it Example-13-05.ipj. Make the *project (workspace) folder* the Chapter 13\Examples\Example-13-05 folder and click the Next button. Make sure that there are no libraries selected. Then pick **Finish** and **Apply** to make this the current project.

Open the file Example-13-05.iam. See **Figure 13-13.** The threaded shaft has a right-hand thread and is supported in the frame. When it is rotated, the threaded nut moves. If the gear is rotated in a counterclockwise direction (looking at it from the right end of the assembly), the nut will move toward the gear.

Apply a rotation-translation constraint to the assembly, picking the gear as the first selection and then the block as the second selection. See **Figure 13-14.** Pick the first solution, Forward. The shaft has 12 threads per inch so set the **Distance** to 1/12. One rotation gives 1/12 inch translation. Remember, the direction of the rotation vector on the gear is meaningless. It is just an icon.

Drive the DRIVE ME angular constraint and the block will move toward the gear. Edit the Rotation/Translation constraint and pick the second solution, Reverse. This would be correct for a left-hand thread. If the gear is driven in the same direction, then the block will move away from the gear.

Figure 13-13.
An assembly with a
threaded shaft and
threaded nut.

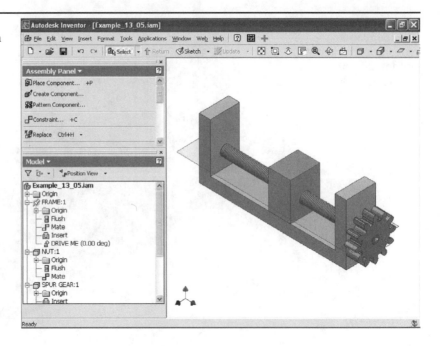

Figure 13-14.
Pick the first and
second selection for
the constraint.

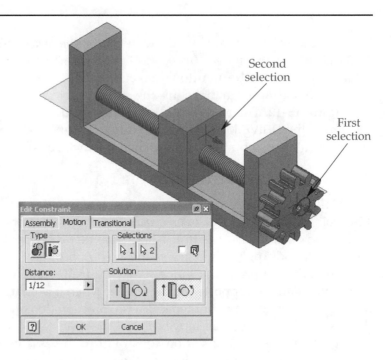

The Transitional Motion Constraint

The transitional constraint relates the motion of contacting faces on two parts such as the cam and follower shown in **Figure 13-15**. The dialog box is very simple with only two choices; the first being the moving part, the cam in the example. The second is transitioning face; the bottom of the follower.

Make sure no files are open. Create a new *single user project* and name it Example-13-06.ipj. Make the *project (workspace) folder* the Chapter 13\Examples\Example-13-06 folder and click the Next button. Make sure that there are no libraries selected. Then pick **Finish** and **Apply** to make this the current project.

Figure 13-15.
A cam and follower
assembly.

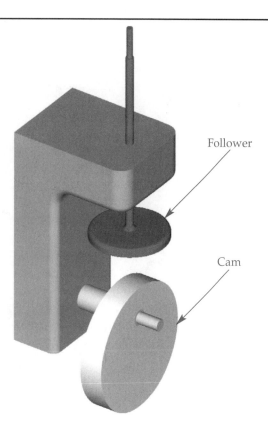

Follower

Cam

Open the file Example-13-06.iam. This simple cam is a circle with an offset center. The cam rotates on the shaft, pushing on the follower that slides up and down in the frame. Since there is no return spring, this would be a slow speed device. It would rely on the weight of the follower to maintain contact between the two surfaces. Use the **Rotate** tool to get an isometric view from the bottom showing both the cam surface and the bottom face of the follower. See **Figure 13-16**.

Apply a transitional constraint using the **Transitional** tab in the **Place Constraint** dialog box. Pick the contact face of the cam as the first selection and the bottom of the follower as the second. Once this is done, the follower will move down and contact the cam.

Drive the DRIVE ME Angle constraint on the cam and the follower will move up and down. Any pair of contacting faces on two parts can be used, as long as the face for the second choice is one face or contiguous multiple faces. For examples of this, see the exercises at the end of this chapter.

Figure 13-16.
Pick the first and
second selection for
the constraint.

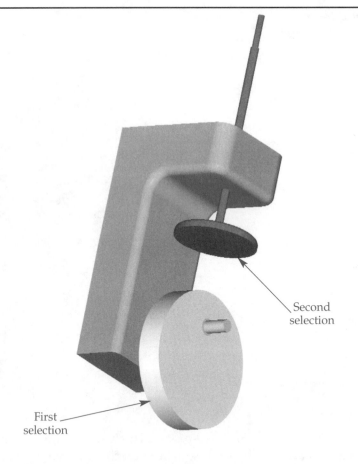

Second
selection

First
selection

Chapter Test

Answer the following questions on a separate sheet of paper.

1. What are the three motion constraints that relate relative motion of one component to another?
2. The _____ motion constraint is used on a rotating part to rotating part, such as a pair of gears.
3. The _____ motion constraint is used on a rotating part to translating part, such as a gear and rack.
4. The _____ motion constraint is used for surface contact, such as a cam and follower.
5. What can be used to determine if components in motion collide and interfere with each other?

Chapter Exercises

Exercise 13-1. Create a new project Exercise-13-01 with the path set to …Chapter 13/ Exercises/Exercise-13-01. Open file Exercise-13-01.iam. The pinion, the smallest gear, is 1″ in diameter, the idler is 2″, and the large gear is 4″ in diameter. Put rotational motion constraints between the three gears and drive the pinion to make the gears rotate.

Exercise 13-2. Create a new project Exercise-13-02 with the path set to …Chapter 13/ Exercises/Exercise-13-02. Open file Exercise-13-02.iam. Line up the gear teeth by constraining the work planes. Put rotational motion constraints between the four gears. Suppress the line up constraints and drive the angle constraint on the pinion gear.

Exercise 13-3. Create a new project Exercise-13-03 with the path set to …Chapter 13/ Exercises/Exercise-13-03. Open file Exercise-13-03.iam. This is a planetary gear train with the carrier (the red arm) grounded. In this configuration, it acts like a normal gear train. Use the work planes to line up the teeth. Put motion constraints between the 36 tooth sun (gold) and the 18 tooth planet (green) and between the planet and the 72 tooth ring gear. Suppress the constraints used to line up the gears. Place an angle constraint between the sun gear and the carrier work planes and drive it to make the gears rotate. Note that you can temporarily turn off the visibility of components to make selections easier.

Exercise 13-4. Create a new project Exercise-13-04 with the path set to …Chapter 13/ Exercises/Exercise-13-04. Open file Exercise-13-04.iam. Use the work planes to line up the teeth. This is a planetary gear train with the sun gear (gold) grounded. In this configuration the carrier arm (red), along with the shaft holding the planet gear, rotates as the input. For every rotation of the arm the planet rotates 3 times in the same direction. Put this motion constraint between the arm and planet. For every rotation of the arm, the ring gear rotates 1.5 times in the same direction. Put this motion constraint between the arm and the ring gear. Suppress the constraints used to line up the gears. Place an angle constraint between the ring gear and the sun gear work planes and drive it to make the gears rotate.

Exercise 13-5. Create a new project Exercise-13-05 with the path set to …Chapter 13/ Exercises/Exercise-13-05. Open file Exercise-13-05.iam. Place a transitional constraint between the two slides. Drive the lower slide (Slide_1) with respect to the frame causing Slide_2 to move up. Use collision detection to determine the limits of motion.

Exercise 13-6. Create a new project Exercise-13-06 with the path set to …Chapter 13/ Exercises/Exercise-13-06. Open file Exercise-13-06.iam. Place a transitional constraint between the pin (first selection) and one of the faces (second selection) in the contiguous slot. Drag and move the pin and it will follow the slot.

Exercise 13-7. Create a new project Exercise-13-07 with the path set to …Chapter 13/ Exercises/Exercise-13-07. Open file Exercise-13-07.iam. The shaft is threaded at 12 threads per inch with left-hand threads on the end with the drive pin and right-hand threads on the other end.
Place a rotation-transition constraint between the shaft (first selection) and the pivot (second selection) on each end using the opposite solutions. Suppress the SUPPRESS Angle constraint that lines up the top and bottom plates.
Put an angle constraint between the shaft and the green bottom plate and drive it to make the jack move up and down. Determine the limits for the drive angle.

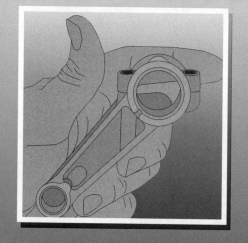

iParts and iFactories

Objectives

After completing this chapter, you will be able to:

* Explain the process of creating a part from an iPart using an iFactory.
* Create parts from iParts using an iFactory.
* Explain what iMates are and how to create and use them.
* Create an iPart from scratch.

User's Files

*The following is a list of files that you will need to work through this chapter. These files can be found on the **User's Files** CD included with this text.*

Examples

12 OD Flange.ipt
Cross.ipt
Elbow.ipt
Example-14-01.iam
Example-14-01.idv
Example-14-02.ipt
Pipe.ipt
PipeManifold.iam
PipeMainifold.idv
Reducer.ipt
Tank.ipt
Tee.ipt

Exercises

Exercise-14-01.ipt
Exercise-14-02.ipt

Inventor allows you to create and use iParts. An *iPart* is an "intelligent" part with parameters that are tied to a spreadsheet, allowing multiple sizes and configurations from a single part. All of the possible variations produce a group of similar parts called a *family of parts.* For example, a basic socket-head cap screw can be created. Then, its dimensions, such as length and diameter, can be tied to a spreadsheet that provides data for all variations in the group of similar parts. By using iParts, you can:

- Create parts with versions.
- Develop libraries of standardized parts.
- Create a basic design and use it in many different assemblies by varying data.
- Adhere to corporate specifications for design consistency in assemblies.

The process of placing an iPart into an assembly is called the *iFactory* because you assemble the specifications for, or "manufacture," the part being inserted from the iPart "raw material." There are two types of iParts—standard iParts and custom iParts. A *standard iPart* cannot be altered as it is inserted into an assembly. Parameters must be selected from a set list of options. A *custom iPart* has certain aspects of its definition that can be manually entered when the part is inserted into an assembly. For example, an iPart bolt may allow the designer to enter a custom value for the bolt length, rather than select from a list, as the part is inserted. This would be considered a custom iPart.

An *iMate* is a named, partial constraint definition that allows a part to be automatically constrained when inserted into an assembly. The constraint within an iMate will have an "i" added to the constraint name, such as iInsert, iAngle, and so on. In order to use an iMate, there must be a matching iMate in the assembly when the part is inserted. Multiple constraints can be defined in an iMate and applied as a group. This is called a *composite iMate.*

In this chapter, you will insert custom parts into an assembly by manufacturing them in an iFactory from iParts. You will also create your own iParts from scratch. Finally, you will create iMates, apply them to parts, and insert the parts into an assembly using the iMate feature.

PROFESSIONAL TIP  Before starting this chapter, if not already done copy the Chapter 14 folder from the CD to your hard drive. Copy all of the files and subfolders contained within the Chapter 14 folder. Keep the folder structure and file locations the same.

Assembling iParts

In this example, you will build a piping system off a storage tank. See **Figure 14-1.** Most of the various fittings you will add are iParts that have been created for you. Open Example-14-01.iam and zoom in on the flanged outlet on the side of the tank.

Placing an iPart

All iParts for this example are located in the Chapter14\Examples\Example-14-01 folder. Using the **Place Component** tool, place an instance of the iPart Pipe.ipt into the assembly. In the **Open** dialog box, do *not* check the **Use iMate** check box. See **Figure 14-2.**

Figure 14-1.
This is the completed assembly. You will add the components from the flange. Also, you will create the valves as iParts and then place them in the assembly.

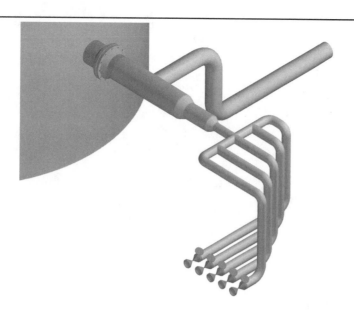

Figure 14-2.
Selecting an iPart for placement in an assembly.

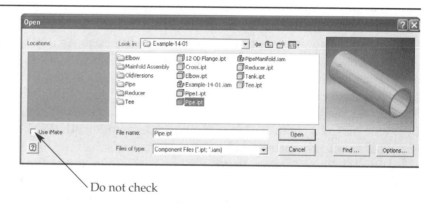

Do not check

Normally, when you pick the **Open** button in the **Open** dialog box, the dialog box is closed and a preview of the part appears attached to the cursor. However, since this is an iPart, an additional dialog box is displayed in which you can specify values for the part. See **Figure 14-3.** You can change the values by picking in the **Value** column for the appropriate property. Pick the value next to PipeNominal; a list box appears. See **Figure 14-4.** Select 12 from the list. Next, pick the value next to PipeLength. This time, a text box appears in place of the value. Type 10 and press [Enter]. Since you can manually enter a property, this is a custom iPart.

Figure 14-3.
Specifying the part parameters for a custom iPart.

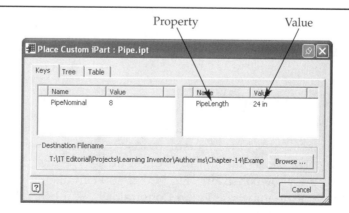

Property   Value

Figure 14-4.
Setting a parameter by selecting a value from a set list of options.

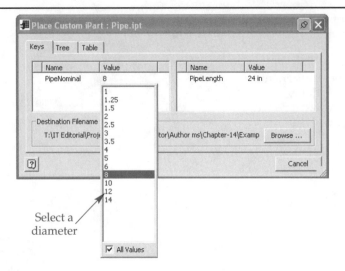

Select a diameter

As you make settings in the dialog box, you are actually defining a part for which there is no part file (IPT). You need to specify where to save the file. By default, the part file will be saved in the same folder where the assembly file (IAM) is located. However, a generic name, such as Part1, is specified. To change the name and/or location, pick the **Browse...** button. This displays the **Save As** dialog box.

For this example, pick the **Create New Folder** button in the **Save As** dialog box and create a new subfolder named Pipe. In a complex assembly with many parts, it is a good idea to create logically named subfolders, such as Pipe, Tee, Elbow, and so on. Then, navigate to the new subfolder, enter the name Pipe12x10 in the **File name** text box, and pick the **Open** button. The **Place Custom Part** dialog box is redisplayed and the path next to the **Browse...** button reflects the setting.

With the **Place Custom Part** dialog box still open, move the cursor in the graphics window. A preview of the part is attached to the cursor. Pick once near the flange to place one instance. Then, pick the **Cancel** button in the **Place Custom Part** dialog box or press [Esc].

The pipe is different from all the other fittings you will place in this example because it is a custom iPart. You entered a custom length for the part. This length can be any value. All other pipe fittings are standard iParts. You will select sizes from list boxes, as you did the pipe diameter. However, you are limited to the sizes provided in the list box.

PROFESSIONAL TIP

The process of saving a custom iPart as it is placed or creating an iPart in an iFactory is called "publishing" the part. Keep this in mind when using the Inventor documentation. The term is used often without much explanation.

Constraining the iPart

In *Chapter 11*, you learned how to remove degrees of freedom by applying constraints to parts. One of the methods you learned was to "drag and drop" the part into the correct location. In order to do this, the **Place Constraint** dialog box is open and set up for the constraint. However, you can also "drag and drop" to constrain a part without opening the **Place Constraint** dialog box. Hold down the [Alt] key and pick the outside surface of the pipe near one end. Drag the pipe toward the flange. As

Figure 14-5.
A—The **Insert** button in the **Place Constraint** dialog box. B—The insert icon next to the cursor.

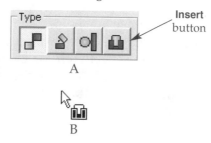

you drag, notice the icon next to the cursor. It appears the same as the **Insert** button in the **Place Constraint** dialog box. See **Figure 14-5.** This indicates that an insert constraint will be applied. When the cursor is over the end of the pipe stub on the flange, the inner circle is highlighted red and the red direction arrow points out of the pipe stub. Release the mouse button to place an insert constraint between the pipe and the flange.

The pipe is now properly aligned with the pipe stub on the flange. However, the pipe is still free to rotate about the centerline of the flange. To prevent the pipe from rotating, apply a mating constraint between the XZ plane of the pipe and XZ plane of the flange. To do this, expand the Origin branch in the **Browser** for both the flange and the pipe. Then, open the **Place Constraint** dialog box. Using the two buttons in the **Selections** area, pick the two planes in the **Browser**. Finally, pick the **Flush** button in the **Solution** area of the dialog box and apply the constraint. The pipe is now completely constrained and has zero degrees of freedom.

The Placed iPart

Now, take a look at the assembly tree in the **Browser**. See **Figure 14-6A.** The part is listed with the name you specified when placing the iPart, which is Pipe12x10. The :1 indicates that it is the first instance of this part file. Since this version of the part is now saved using the file name and path you specified earlier, you can place additional instances of the part by placing the IPT file into the assembly. Each additional instance will have a sequential number added to its name.

Figure 14-6.
A—The pipe in the assembly tree. B—The available part variations in the family of parts. The version selected has a check mark next to its name.

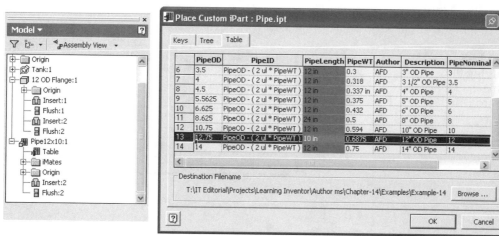

A   B

The Table branch displays the possible selections that appeared in the list box for the diameter when the part was placed. If you right-click on Table or any branch below it and select **Change Component** from the pop-up menu, the **Place Custom iPart** dialog box is displayed. On the **Table** tab the size you selected is highlighted, as shown in **Figure 14-6B.** You can then change the size or length of the pipe. However, if you do so, the part file will no longer reflect your logical file name.

Other branches below Pipe12x10:1 include Origin, iMates, and the two constraints that were applied. The Origin branch contains the three standard planes and axes for the part. The constraints can be edited in the same way as for other parts. The iMates branch contains the automated constraints defined by the iMate. Constraining with iMates is discussed in the next section.

Constraining iMates

Now, you will place a reducing tee into the assembly. Refer to **Figure 14-1.** Pick the **Place Component...** button in the **Assembly Panel**. In the **Open** dialog box, navigate to the Chapter14\Examples\Example-14-01 folder and select Tee.ipt. Even though you will be using iMates, do not check the **Use iMate** check box in the **Open** dialog box. When you pick the **Open** button, the **Place Standard iPart** dialog box is displayed, **Figure 14-7.**

Pick the value next to BasePipeNom to display a list box. Select 12 from the list box. This is the diameter of the pipe you inserted and constrained to the flange. Then, pick the value next to StubPipeNom to display a list box. Select 8 from the list box. This is the reduced diameter of the tee. Notice how both of the settings provided list boxes, not text boxes. Since the tee is a standard iPart, there are no user-defined settings. All settings must be selected from available values or options. Also, notice that you do not need to specify a file name and location for the part. However, Inventor will create a subfolder named Tee and save this version of the part, a 12″ × 8″ tee, in the folder under a descriptive name. Finally, pick once near the pipe in the graphics window to place one instance of the tee. Press [Esc] to close the **Place Standard iPart** dialog box.

The tee has defined iMates. You can verify this by looking at the assembly tree in the **Browser**. Notice that the Tee branch of the assembly tree indicates the selected values in its name. Also, there is an iMates branch below it. This indicates that iMates have been defined for the part. The pipe you inserted earlier also had an iMates branch because iMates were defined for it. If you expand the iMates branch for the Tee, you will see that three iMates have been defined. See **Figure 14-8.** These three iMates happen to be composite iMates, which are discussed later. If you expand the individual iMate branches, you can see which constraints are contained within the composite iMate.

Figure 14-7.
Specifying the parameters for a standard iPart.

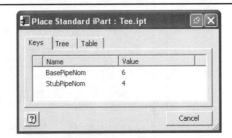

Learning Inventor

Figure 14-8.
The iMates for the
tee. These are
composite iMates.
The constraints
contained within a
composite iMate are
displayed below it
in the tree.

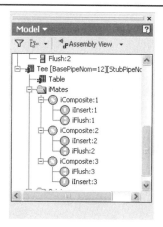

When a part that has iMates is selected, icons appear on the part in the graphics window indicating where the iMate constraints are located. See **Figure 14-9.** Inventor calls these icons *iMate glyphs.* By default, the iMate glyphs are only displayed for a part when it is selected. However, they can be displayed for a part when it is not selected. Right-click on the tee in the graphics window to display the pop-up menu. See **Figure 14-10.** Select **iMate Glyph Visibility** to display the iMate glyphs on the part, even when the part is not selected. A check mark appears next to the menu item when the display is turned on. Pick the menu item again to turn off the display of the iMate glyphs except when the part is selected. For this example, turn on the iMate glyph display for both the tee and the pipe.

Now, you are ready to constrain the tee to the pipe using the iMates set up for each part. Hold down the [Alt] key and pick and hold on the iMate glyph near the large-diameter opening of the tee. The iMate glyph turns green. You may need to move the cursor slightly for the iMate glyph to be displayed in green. Drag the tee by the iMate glyph to the iMate glyph on the pipe that is near the edge. When the two iMate glyphs meet, they turn red and the tee is properly aligned. Release the mouse button to constrain the tee to the pipe. Once the tee is in place, release the [Alt] key.

Figure 14-9.
The iMate glyphs
indicate where
iMates are located.

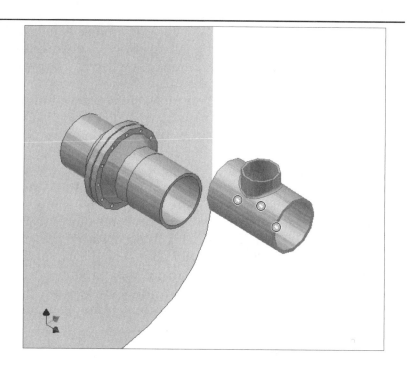

Figure 14-10.
Displaying iMate
glyphs for a part.

Copy	Ctrl+C
Delete	
Promote	Shift+Tab
Demote	Tab
Infer iMates...	
Replace Component	Ctrl+H
Create Note	
Component Selection	▶
Isolate	
Undo Isolate	
✓ Visibility	
✓ iMate Glyph Visibility	
✓ Enabled	
Grounded	
Contact Set	
Find in Browser	
Previous View	F5
Isometric View	F6
Properties	

A big advantage of using iMates is that all six of the degrees of freedom can be eliminated on a part in a single step. Because of the composite iMates set up for the tee and the pipe, the tee is now fully constrained. All six degrees of freedom were eliminated simply by dragging the tee into the correct location using the iMates. If you try to move the tee now, you will not be able to do so.

PROFESSIONAL TIP

As you drag using iMates, only the valid iMates on the mating part are displayed.

Placing the Remaining Parts

There are several more parts to be added to the assembly. These include a series of reducers, elbows, pipes, and a subassembly of the manifold. You will add these parts in this section. In the next section, you will create the valves as iParts and insert them into the assembly to complete it.

First, you will add elbows and pipes to the reduced-diameter pipe run from the tee. Zoom and pan as needed as you work. Place one instance of the part Pipe.ipt, which is the same part file you used earlier. This time, set PipeNominal to 8 and PipeLength to 24. Specify the filename as Pipe8x24 and the location as the Pipe subfolder. If you get a message asking to save the file and its dependents, pick **OK**. Press [Esc] to close the dialog box. Using iMates, constrain the pipe to the small diameter connection on the tee. See **Figure 14-11**.

Next, place an instance of the part Elbow.ipt. Set the PipeNominal to 8 and ElbowRad to 12. Since this is a standard iPart, you do not need to name it. However, keep in mind that Inventor will create a subfolder named Elbow and save this version of the part in it. Using iMates, constrain either end of the elbow to the pipe you just placed. There will be only one iMate glyph displayed on the pipe. Pick it as the target even if the iMate glyph is displayed at the opposite end of the pipe. The elbow will automatically

Figure 14-11.
Constraining the
pipe to the small
diameter connection
on the tee.

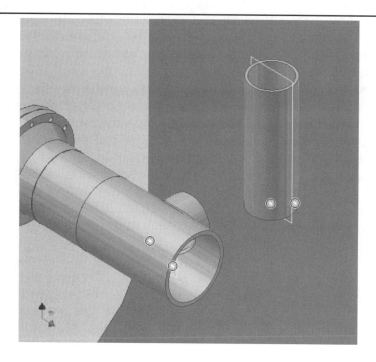

point down when constrained. If the preview points up, drag the elbow away from the pipe and release the mouse button. Then, repeat the process but select the iMate glyph on the opposite end of the elbow. Continue as follows.

1. Place another instance of the part Pipe.ipt. Set PipeNominal to 8 and PipeLength to 20. Specify the filename as Pipe8x20 and the location as the Pipe folder. Using iMates, constrain the pipe to the elbow.
2. Place another instance of the part Elbow.ipt. Set the PipeNominal to 8 and ElbowRad to 8. Using iMates, constrain the elbow to the pipe you just placed.
3. Place another instance of the part Pipe.ipt. Set PipeNominal to 8 and PipeLength to 72. Specify the filename as Pipe8x72 and the location as the Pipe folder. Using iMates, constrain the pipe to the elbow.

The reduced-diameter pipe run is complete. Now, you need to assemble the manifold. Most of the manifold is created for you as a subassembly that you will place into the tank assembly. However, you need to add a couple of reducers to decrease the diameter of the pipe run to 4″ so that the manifold subassembly can be attached.

1. Place an instance of the part Pipe.ipt. Set PipeNominal to 12 and PipeLength to 12. Specify the filename as Pipe12x12 and the location as the Pipe folder. Using iMates, constrain the pipe to the tee.
2. Place an instance of the part Reducer.ipt. Set LargePipeNom to 12 and SmallPipeNom to 8. Using iMates, constrain the reducer to the pipe you just placed. Pick the iMate glyph near the large-diameter end of the reducer.

Again, Inventor will create a subfolder named Reducer and save this version of the part in it. Continue as follows.

3. Place another instance of the part Pipe.ipt. Set PipeNominal to 8 and PipeLength to 12. Specify the filename as Pipe8x12 and the location as the Pipe folder. Using iMates, constrain the pipe to the reducer.
4. Place another instance of the part Reducer.ipt. Set LargePipeNom to 8 and SmallPipeNom to 4. Using iMates, constrain the reducer to the pipe you just placed. Pick the iMate glyph near the large-diameter end of the reducer.
5. Place another instance of the part Pipe.ipt. Set PipeNominal to 4 and PipeLength to 8. Specify the filename as Pipe4x8 and the location as the Pipe folder. Using iMates, constrain the pipe to the reducer.

Now, the manifold subassembly can be placed into the tank assembly. Using the **Place Component** tool, place one instance of the assembly file PipeManifold.iam into the tank assembly. Placing an assembly into another assembly makes the first assembly a subassembly.

You cannot make use of iMates on a subassembly, even if iMates are set up. Therefore, use "traditional" constraints to locate the manifold. Place an insert constraint between the end of the 4" pipe and the inlet of the manifold. Then, apply a mating constraint between the YZ plane of the part Cross contained within the subassembly and the XZ plane of the last section of pipe. Select the planes in the **Browser**. Also, pick the **Flush** button in the **Solution** area of the **Place Constraint** dialog box.

PROFESSIONAL TIP

The manifold subassembly contains many work planes. These may clutter your screen. Turn off the display of all work planes by selecting **Object Visibility** from the **View** pull-down menu. Then, select **User Work Planes** in the cascading menu to uncheck it. All work planes will be hidden. However, you can still select them in the **Browser** if needed.

Creating an iPart from Scratch

The iParts you have used so far were created for you. However, you can create your own iParts. You will create the gate valves for the tank assembly as iParts. The basic process is:

1. Create the part.
2. Constrain the part.
3. Select the iParts authoring tool.
4. Build the spreadsheet for the family of parts.
5. Publish the iPart.

When creating the part, pay close attention to the relationship between part features. Also, fully constrain everything. Try to have all dimensions relate back to one or two key features. For example, the thread diameter on a bolt may be a "base" dimension for a family of bolts. A pipe OD may be used as the "base" dimension for a family of pipe fittings. Once the part is modeled and constrained, the spreadsheet built, and the part is published, you have an iFactory for that iPart.

Modeling the Part

Open Example-14-02.ipt. The gate valve has been modeled for you. Otherwise, you would start with a design, create sketches, and then turn the sketches into features. Dimensions and constraints have also been added to the part. However, you will need to complete the iPart definition. The first thing you need to do is rename several dimension parameters to names that are logical for gate valves. Select **Parameters** from the **Tools** pull-down menu to display the **Parameters** dialog box. To rename a dimension parameter, simply pick on its name to display a text box. Then, type the new name and press [Enter]. Comments are added by picking in the **Comment** column to display the text box. Rename dimensions, alter equations, and add comments as indicated in **Figure 14-12**. Also, add a user parameter named PipeNominal with a nominal value of 1. Then, close the **Parameters** dialog box. Now, these parameters will make sense to anyone using this iPart.

Figure 14-12.
Change the dimension parameters as indicated in this chart.

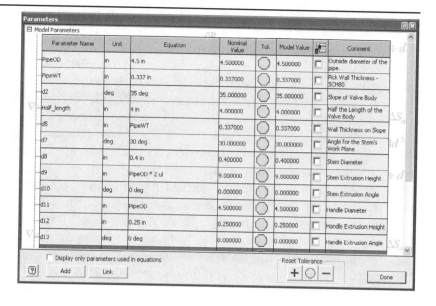

Creating iMates

The next step is to create iMates on the valve. It is not required for an iPart to have iMates, but iMates can greatly simplify the assembly process, as you saw earlier. Pick the **Create iMate** button in the **Part Features** panel. The **Create iMate** dialog box is opened, **Figure 14-13.** This dialog box is similar to the **Place Constraints** dialog box. However, there are two major exceptions. The **Transitional** tab is not included. Also, in the **Selections** area there is only one button. Remember, an iMate is one-half of a constraint definition.

Pick the **Insert** button in the **Type** area of the **Assembly** tab. Then, with the button in the **Selections** area on, pick the inside cylindrical edge of the valve body. Refer to Figure 14-13. The centerline should be highlighted in red. Pick the **Apply** button to apply the constraint and leave the dialog box open. An iMate glyph now appears at the end of the valve.

Figure 14-13.
Creating an iMate constraint.

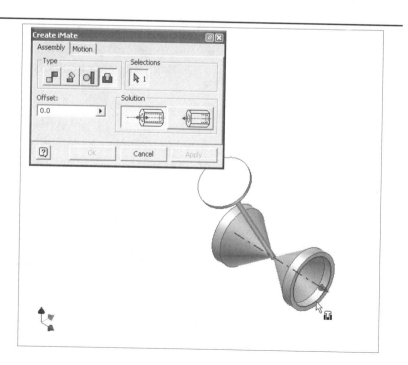

Rotate the view to the opposite end of the valve. Then, apply an insert constraint to the other end in the same manner. Pick the **OK** button to apply the constraint and close the dialog box. An iMate glyph is displayed on this end of the valve as well. Expand the iMates branch of the part tree in the **Browser**. There should be two constraints listed, each named iInsert and a sequential number.

You will now add a mating constraint to fully constrain the valve when the iMates are used to constrain it within an assembly. Select the **Create iMates** tool again. Pick the **Mate** button in the **Type** area of the **Create iMate** dialog box. Also, pick the **Flush** button in the **Solution** area. Select the XZ plane in the Origin branch and then pick the **Apply** button. Now, exactly repeat this step and then close the dialog box. There should be two constraints named iFlush in the **Browser**. The flush constraint was applied twice so one can be used on each end of the valve.

Now, you must create the composite iMates. Select iInsert:1 in the **Browser**, hold down the [Ctrl] key, and select iFlush:1. The two glyphs on the part turn green. Right-click on the names in the **Browser** to display the pop-up menu, and select **Create Composite**. The two separate constraints are replaced by one named iComposite. If you expand this branch, the two constraints are listed below it. Create another composite constraint from iInsert:2 and iFlush:2. Now, there should be two iComposite constraints in the iMates branch, one for each end of the valve.

PROFESSIONAL TIP

You can have Inventor create iMates (composite or not) from the assembly constraints on a part already in the assembly. After the part is inserted and constrained, right-click on the name of the part in the assembly tree and select **Infer iMates...** from the pop-up menu. Then, make appropriate settings in the **Infer iMates** dialog box.

iPart Authoring

So far, you have prepared the part to be an iPart. However, you have not yet *created* an iPart. If you save the file now, the valve can be placed into an assembly. However, you cannot "manufacture" a series of valves in an iFactory. In order to "manufacture" different valve sizes for use in assemblies, you must author or publish the iPart. Select **Create iPart** from the **Tools** pull-down menu. The **iPart Author** dialog box is displayed, **Figure 14-14.** This dialog box can be resized by dragging the lower-right corner of the dialog box.

Parameters tab

The upper-left pane of the **Parameters** tab lists all of the features that make up the part. Dimension parameters are listed underneath each feature's name. The upper-right pane lists the parameters that will be used to manufacture the part in the iFactory. Currently, the four named parameters are listed. Inventor assumes that renamed parameters will be used to manufacture the part. To add a parameter to the upper-right pane, highlight it in the upper-left pane and pick the **>>** button. To remove a parameter from the upper-right pane, highlight it on the right and pick the **<<** button.

The bottom pane is the spreadsheet containing all the values needed to generate the family of parts. Notice the four parameters listed in the right-hand pane are the column headings. As you add parameters to the right-hand pane, they will appear as column headings. Each row represents a different version of the iPart. If there are five different versions, there will be five rows. Since there is currently only one version of the part, there is only one row. An intersection between a row and a column is called a cell. The cell contains the data corresponding to the column. A cell can contain a value, Inventor equation, or Microsoft Excel formula. The current cell is indicated by a thick, black outline around the cell.

Figure 14-14.
The **Parameters** tab
of the **iPart Author**
dialog box.

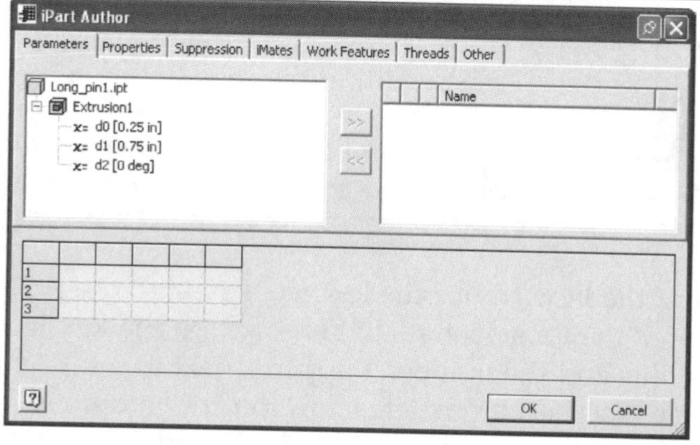

Entering data for the family of parts. For this family of valves, there will be 14 possible versions. Therefore, you need to have a total of 14 rows. However, there is currently only one row. Right-click anywhere on the row, except in the current cell, and pick **Insert Row** from the pop-up menu. A new row is added. Continue doing this until there is a total of 14 rows; notice the sequential numbers on the left of the spreadsheet.

Notice how the data in the cells is identical for each row. You now need to edit the data to reflect the family of parts. Pipe and related components, such as valves, are specified by the nominal diameter. The actual dimensions of a pipe—the inner diameter (ID) and outer diameter (OD)—are based on the nominal size. For example, a 4″ nominal pipe fitting has a 4.5″ OD. Refer to the chart in **Figure 14-15**. Enter the information shown in the chart for the PipeOD, PipeWT, Half_Length, and PipeNominal columns. You will add the other columns shown in the figure later.

Figure 14-15.
The final spreadsheet in the **iPart Author** dialog box should look like this. Refer to this chart when entering values.

	PipeOD	PipeWT	Half_Length	Pipe Nominal	Author	Stem	Handle	Description
1	1.315	0.179	1.5	1	AFD	Compute	Compute	1″ OD Gate Valve
2	1.66	0.191	1.875	1.25	AFD	Compute	Compute	1 1/4″ OD Gate Valve
3	1.9	0.2	2.25	1.5	AFD	Compute	Compute	1 1/2″ OD Gate Valve
4	2.375	0.218	2.5	2	AFD	Compute	Compute	2″ OD Gate Valve
5	2.875	0.276	3	2.5	AFD	Compute	Compute	2 1/2″ OD Gate Valve
6	3.5	0.3	3.375	3	AFD	Compute	Compute	3″ OD Gate Valve
7	4	0.318	3.75	3.5	AFD	Compute	Compute	3 1/2″ OD Gate Valve
8	4.5	0.337	4.125	4	AFD	Compute	Compute	4″ OD Gate Valve
9	5.5625	0.375	4.875	5	AFD	Compute	Compute	5″ OD Gate Valve
10	6.625	0.432	5.625	6	AFD	Compute	Compute	6″ OD Gate Valve
11	8.625	0.5	7	8	AFD	Compute	Compute	8″ OD Gate Valve
12	10.75	0.594	8.5	10	AFD	Compute	Compute	10″ OD Gate Valve
13	12.75	0.6875	10	12	AFD	Suppress	Suppress	12″ OD Gate Valve
14	14	0.75	11	14	AFD	Suppress	Suppress	14″ OD Gate Valve

Setting the key. Notice the key icons next to the parameter names in the right-hand pane of the **Parameters** tab. Left-click on the key icon next to PipeNominal. The key turns blue and the number 1 appears next to the key. The column in the spreadsheet for PipeNominal parameter is now the key column. A *key column* is used to select which version of the part will be "manufactured." Since the PipeNominal column is the key column, you will select the nominal diameter of the pipe when placing the valve into an assembly. The pipe fittings you placed into the tank assembly had the nominal pipe diameter set up as the key. There may be multiple keys (or key columns), one primary and up to eight secondaries. For example, the tee you inserted into the tank assembly had a primary key for the base nominal OD and a secondary key for the allowable sizes based on the primary key. The primary key *filters* the available selections for the secondary keys.

Setting the default version. You can set up a default version for the iPart. For example, if most of the gate valves you will be using are for 4" pipe, you can have the settings for this version displayed in the "place iPart" dialog box. Right-click on the row that contains the value 4 in the PipeNominal column. Then, select **Set as Default Row** from the pop-up menu. The background of the row is changed to a light green, indicating this row defines the default version.

Setting custom parameters. There are two basic types of iParts—custom and standard. After a custom iPart is placed into an assembly, additional features may be added. Additional features cannot be added to a standard iPart. To create a custom iPart, right-click on a column and select **Custom Parameter Column** from the pop-up menu. This indicates that the value for this column can be changed as the part is placed into an assembly. For example, the pipe iPart allowed you to enter a pipe length. The pipe iPart has a length column set up as a custom parameter column. A custom iPart can have multiple custom parameter columns. Since you are creating a standard iPart, no columns should be set up as custom parameter columns.

Properties tab

The properties on an Inventor model can be viewed by selecting **iProperties** from the **File** pull-down menu. The properties listed in the **Summary**, **Project**, and **Physical** tabs of the **Properties** dialog box can be added to an iPart definition. If you accessed the **Properties** dialog box, close it now and access the **iPart Author** dialog box. The properties listed in the **Properties** dialog box appear in a tree in the left-hand panel of the **Properties** tab of the **iPart Author** dialog box. If you add any of these properties to the iPart definition, they will be available when a drawing (IDW) is created. Highlight Author in the Summary branch in the left-hand pane. Then, pick the **>>** button to place it in the right-hand pane. Notice that a column has been added to the spreadsheet. See **Figure 14-16.** The column is named Author and each cell in the column displays the author of the IPT file you have open. These cells can be edited, if needed.

Figure 14-16.
The **Properties** tab of the **iPart Author** dialog box.

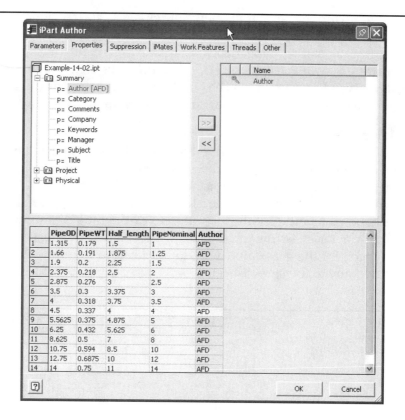

Suppression tab

The **Suppression** tab is used to set up certain features to be excluded from certain versions of the part. For example, the 12″ and 14″ valves do not have stems and handles. Highlight Stem in the left-hand pane of the **Suppression** tab and pick the **>>** button. Do the same for Handle. The two features are now listed in the right-hand pane and two columns have been added to the spreadsheet. See **Figure 14-17.** The

Figure 14-17.
The **Suppression** tab of the **iPart Author** dialog box.

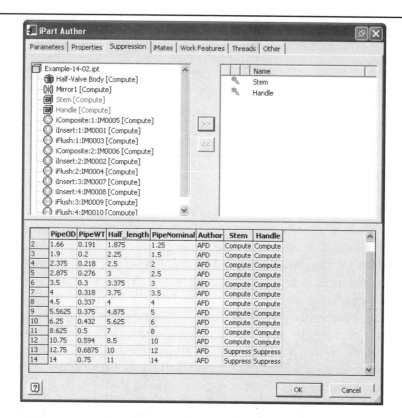

default text in the cells of the new columns is Compute, which means to generate the feature. Valid entries for Compute include Compute, U, u, C, c, ON, On, on, and 1. To suppress a feature, type Suppress in the cell. Valid entries for Suppress include Suppress, S, s, OFF, Off, off, and 0. To suppress the stem and handle for the 12″ and 14″ valves, type Suppress in the cells in the Stem and Handle columns for the 12″ and 14″ nominal sizes. Refer to the chart in **Figure 14-15.**

iMates tab

The **iMates** tab is used to fine-tune the iMates set up for the part. See **Figure 14-18.** You can include or exclude individual iMates for different part versions. In addition, you specify the offset values for the iMates, matching names, and sequence numbers. The iMates set up for the part are listed in the left-hand pane. Highlight a name in the left-hand pane and pick the **>>** button to display it in the right-hand pane and create a column for it in the spreadsheet. The values in the cells can then be edited. However, for the gate valve, you do not need to fine-tune the iMates. Therefore, do not add any to the spreadsheet.

Figure 14-18.
The **iMates** tab of the **iPart Author** dialog box.

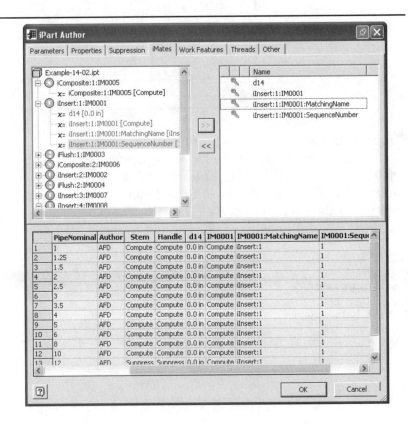

Work Features tab

Work features are useful in iParts to constrain parts in assemblies and to create pins in electrical parts. Work features are created in a part before you transform it into an iPart factory. You then can determine which work features to include or exclude in iPart instances.

Work features have default Include or Exclude settings. You can override the setting by selecting work features to include or exclude in the iPart table. Each row, which represents an instance of the iPart, can have work features included or excluded. Default settings that are work features constrained with iMates are Included, Pins (work points) in electrical parts are Included, and all other work features are Excluded (except those constrained with iMates).

Each row in the iPart table represents an instance. A column for each work feature indicates whether it is included or excluded. You can modify the setting for each row in the table. You do not need to make any settings in this tab.

Threads tab

If the part has any thread features, the **Threads** tab is used to add the feature to the spreadsheet and provide parametric control over the thread specifications. For example, if you were creating a family of threaded valves, you could specify the threads per inch, class, and orientation for each version of the part in the family. However, this is a family of welded valves so you do not need to make any settings in this tab.

Other tab

The **Other** tab is used to add custom columns to the spreadsheet. For the valve, add a Description column. Refer to **Figure 14-19.** First, pick the entry Click to add value in the pane at the top of the tab. A text box appears in the **Name** column of the pane. Type the name Description and press [Enter]. A new custom column named Description is added to the spreadsheet. Next, edit each cell in the column to the values shown in **Figure 14-15.**

Figure 14-19.
The **Other** tab of the **iPart Author** dialog box.

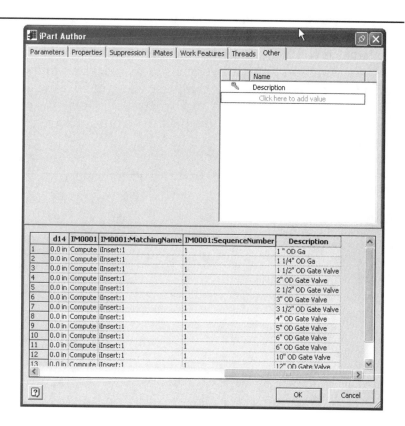

You can set up a default file name for the part that is saved when the iPart is placed into an assembly. The default file name is based on a column. For the valve, the Description column will be used. Right-click on that column heading and select **File Name Column** from the pop-up menu. Now, the part that is placed into an assembly will be saved to an IPT file with a name based on the description and the file name of the iPart. Since this is a standard part, the file name is automatically created by Inventor. For a custom part, you need to specify the file name as you place the part.

Testing the iPart in the iFactory

The iPart is now completely set up. Close the **iPart Author** dialog box. Then, save the part as Valve.ipt in the Chapter14\Examples\Example-14-02 folder using **Save Copy As...** in the **File** pull-down menu. Now, you need to "test build" the part to simulate the iFactory.

First, expand the Table branch of the part tree in the **Browser**. There should be 14 listings of PipeNominal = and a number corresponding to the nominal size. See **Figure 14-20.** Also, notice that the PipeNominal = 4 entry has a check mark next to it. This indicates that the size is the default part.

To simulate the iFactory, right-click on the first entry (PipeNominal = 1) and select **Compute Row** from the pop-up menu. The part should update in the graphics window to reflect the settings in the spreadsheet. Also, the name in the **Browser** has a check mark next to it. Repeat this procedure for each part in the list. For the 12" and 14" valves, check that the handle and stem are suppressed.

If you receive an error message, right-click on Table in the **Browser** and choose **Edit Table** from the pop-up menu. The **iPart Author** dialog box is displayed. Verify the data against the chart in **Figure 14-15.** Look for typos or missing data. Edit the spreadsheet as needed and pick the **OK** button to close the **iPart Author** dialog box. If you have Microsoft Excel installed and prefer to correct the spreadsheet in that software, select **Edit via Spread Sheet...** from the pop-up menu when you right-click on Table. The spreadsheet is opened in Excel. When you save and exit Excel, the spreadsheet is updated as if you had edited it in the **iPart Author** dialog box.

Once you have tested all versions of the part by simulating an iFactory, and all versions are correct, the iPart is complete. Close the IPT file. If you made any changes since you saved the copy, save another copy replacing the earlier copy. Be sure that the correct version is set up as the default before you save the file. The iPart is ready to be placed into an assembly.

Figure 14-20.
The Table branch of the part tree contains all of the part variations set up in the spreadsheet.

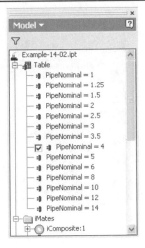

Inserting the Valves into the Tank Assembly

The valves now need to be "manufactured" and placed at the open ends of the manifold subassembly. This can be done in the tank assembly. However, you cannot make use of the iMates on a subassembly. The parts within the manifold have iMates set up. Right-click on the manifold in the tank assembly and select **iMate Glyph Visibility** from the pop-up menu. Since this is a subassembly and not a part, no iMate glyphs are displayed. In order to take advantage of the iMates, you must place the valves into the manifold assembly file.

Leave the tank assembly open. Then, open PipeManifold.iam. Place a 4″ gate valve into the assembly. Use the iMate drag method to constrain the valve to the end of one of the lower pipes. You may need to rotate the view and zoom to better see the iMate glyphs. Remember, if you drag to the wrong iMate glyph, move the part off the glyph, release the mouse, and try again. Do this for each pipe end so there is a total of five gate valves. Redisplay the isometric view, save the assembly, and close the file. In the tank assembly file, the subassembly should be updated. If not, pick the **Update** button on the **Inventor Standard** toolbar. Your assembly should now look like **Figure 14-1.**

Chapter Test

Answer the following questions on a separate sheet of paper.

1. What is an *iPart?*
2. Define *family of parts.*
3. What is an *iFactory?*
4. What is the difference between a *standard iPart* and a *custom iPart?*
5. What is an *iMate?*
6. What is a *composite iMate?*
7. List three advantages of using iParts.
8. What are the icons called that are used with iMates?
9. What is the basic process for constraining parts using iMates?
10. How can you use iMates on a subassembly?
11. What are the five basic steps in creating an iPart from scratch?
12. What is a *key column?*
13. How many key columns, or keys, can an iPart have?
14. How do you specify that the iPart you are creating is a custom iPart?
15. How can you test the iPart in a simulated iFactory?

Chapter Exercises

Exercise 14-1. *300lb. Raised Face Weld Neck Flange (RFWN) SCH80.* In this exercise you will create your own iPart using the specific values shown below. Open Exercise-14-01.ipt and pick **Create iPart** from the **Tools** pull-down menu. Enter the part's parameters and edit the D*n* values to those shown.

Parameters

Model Parameters

Parameter Name	Unit	Equation	Nominal Value	Tol.	Model Value		Comment
Flange_OD	in	9 in	9.000000		9.000000		Flange diameter
Flange_thk	in	1.063 in	1.062500		1.062500		Flange thickness
d2	in	(BoltCircle_dia - BoltHole_dia) - 0.25 in	6.250000		6.250000		Diameter of the raised face
d3	in	d2 - 1.25 in	5.000000		5.000000		Diameter for start of angled side
Flange_length	in	3.1875 in	3.187500		3.187500		Overall length of flange including raised face
d5	in	Pipe_OD - (wall_thk * 2 ul)	3.364000		3.364000		Flange ID for this schedule 80 flange
d6	in	0.5 in	0.500000		0.500000		Length of straight cylindrical section
d8	in	0.0625 in	0.062500		0.062500		Distance face is raised
d10	in	0.0625 in	0.062500		0.062500		Weld throat distance
d11	in	BoltHole_dia / 2 ul	0.375000		0.375000		Fillet on side of flange
d12	deg	37.5 deg	37.500000		37.500000		Angle of weld prep
Pipe_OD	in	4 in	4.000000		4.000000		The main key dimension
d15	deg	360 deg / 12 ul / 2 ul	15.000000		15.000000		Angle to locate the bolt hole properly
BoltCircle_dia	in	7.25 in	7.250000		7.250000		Bolt circle diameter
BoltHole_dia	in	0.75 in	0.750000		0.750000		Bolt Hole diameter
Bolt_count	ul	8 ul	8.000000		8.000000		Count of bolts
d26	deg	360 deg	360.000000		360.000000		Angle to fill for the pattern
d27	in	0.0625 in	0.062500		0.062500		Fillet radius

User Parameters

Parameter Name	Unit	Equation	Nominal Value	Tol.	Model Value		Comment
wall_thk	in	0.318 in	0.318000		0.318000		
Nominal_dia	in	3.5 in	3.500000		3.500000		

Display only parameters used in equations

Add | Link | Reset Tolerance | Done

You will need to create two user parameters to be used in the iPart. The wall_thk parameter will be used to calculate the pipe's inside diameter based on the wall thickness from the industry standard tables for a *schedule 80 bore flange*. The Nominal_dia parameter will be used as a key column in the iPart's table. When you place the iPart in an assembly, the specific flange will be chosen by this key column. Fill in the iPart table using the supplied values shown at the top of the next page. Please note that a default file name has been specified as a column as well. To do this use the **Other** tab and define a custom column named Filename. Once created and it is displayed in the table in the bottom half of the iPart table, then right-click on it and choose **File Name Column** from the pop-up menu. This will ensure that when the iPart is used in an assembly, the user will not be prompted for a file name to which to save the iPart version. It will also ensure that the file name will be standardized every time this iPart is used. When done entering in the values, pick **OK** to exit the dialog box and your iPart should be ready to use. Notice in the **Browser** the

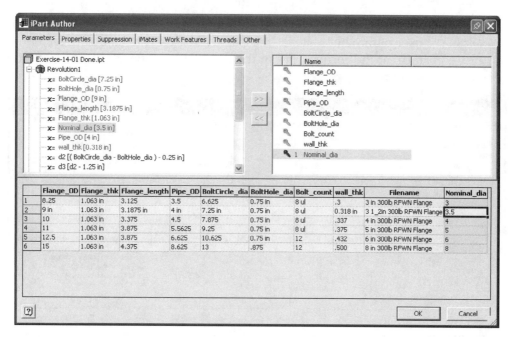

addition of the Table feature. Expand it and right-click on any version. Choose **Compute Row** from the pop-up menu to examine that version. Do this for all six versions to test their validity. Save it as Exercise-14-01_*user*.ipt, where *user* is your name. Begin a new assembly and place it into the assembly to test out its versions.

Exercise 14-2. *Mounting bracket for pressure vessel.* In this exercise you will create your own iPart using the values shown below. Open Exercise-14-02.ipt and choose **Create iPart** from the **Tools** pull-down menu. Enter the part's parameters and edit the D*n* values to those shown.

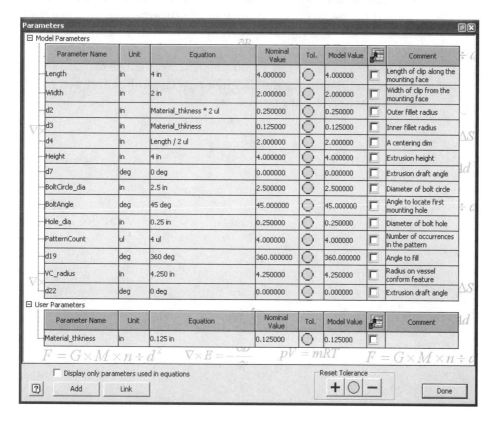

You will need to create a user parameter to be used in the iPart. The user parameter Material_thkness will be used to calculate the bend radii at the corners. Fill in the iPart table using the supplied values shown.

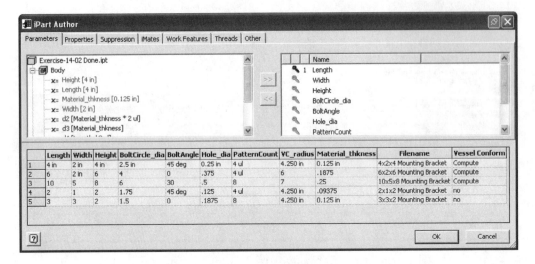

Please note that a default file name has been specified as a column as well. To do this, use the **Other** tab and define a custom column named Filename. Once created and it is displayed in the table in the bottom half of the iPart table, then right-click on it and choose **File Name Column** from the pop-up menu. This will ensure that when the iPart is used in an assembly, the user will not be prompted for a filename to which to save the iPart version and that the filename will be standardized every time this iPart is used. When done entering in the values, pick **OK** to exit the dialog box and your iPart should be ready to use. Notice in the **Browser** the addition of the Table feature. Expand it and right-click on any version. Choose **Compute Row** from the pop-up menu to examine that version. Do this for all the versions to test their validity. Save it as Exercise-14-02_*user*.ipt, where *user* is your name. Begin a new assembly and place it into the assembly to test out its versions.

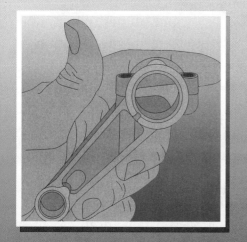

CHAPTER 15

Parameters in Assemblies

Objectives

After completing this chapter, you will be able to:

* Understand the use of parameters to control an assembly.
* Create your own spreadsheet to control the parameters to revise an assembly.
* Comprehend the underlying principles of controlling these parameters.

User's Files

*The following is a list of files that you will need to work through this chapter. These files can be found on the **User's Files** CD included with this text.*

Examples	Practices	Exercises
Example-15-01a.ipt	Practice-15-01.iam	Exercise-15-01.iam
Example-15-01b.ipt	Practice-15-01.xls	MasterSketch.xls
Example-15-01.xls		Body.ipt
Example-15-01.iam		MasterSketch.ipt
Example-15-02.xls		Plug.ipt
Example-15-02.iam		Plug_Sketch.ipt
		Exercise-15-02-10in.xls

The Process of Working with Parameters

In this chapter a part will be modeled as usual, but special attention will be paid to any parametric relationships needed to provide control. Ordinarily, when working with parameters the dimension parameters are renamed to meaningful nouns in the **Parameters** dialog box. The equation relationships between the dimensions are set up here also. However, to have a spreadsheet control the dimensions the parameters' names must be specified in the spreadsheet. Microsoft **Excel**'s math functions can also be used to create relationships between the parameters.

User parameters are added that may be necessary, such as *force* or *wall thickness*. User parameters are typically only used for a variable that is not dimensioned on the part. If no dimension parameter exists and another variable is needed, then a user parameter must be added.

Excel can then be used to create a spreadsheet with the names of the parameters that will be controlled in the assembly. Be careful to avoid any duplicate use of a parameter name. If duplicate parameter names are used, an error will occur when the spreadsheet is linked to the model.

Once the spreadsheet is linked and loaded, the dimension parameters may then be edited and the named parameters from the spreadsheet assigned to them. This must be done for all parts in the assembly that are to be controlled parametrically. Finally, the spreadsheet must be linked to the assembly, thereby affecting the parametric control of the assembly.

Using a Spreadsheet to Control an Assembly

This example illustrates the principle involved in controlling an assembly by a spreadsheet. The key thing to remember is that the spreadsheet is linked to every part that is to be controlled. The spreadsheet itself takes on a special status, because it has all the values needed for all parts contained within it. Open the Example-15-01a.ipt part file and select the **Parameters** button in the **Part Features** panel. This accesses the **Parameters** dialog box. See **Figure 15-1**. Review the dimension parameters that have been defined. They have been defined from the previous chapters' work. As features were created, Inventor created these dimension parameters.

At the bottom of the dialog box, find the **Link** button and pick it. Navigate to the Example folder for *Chapter 15* and select Example-15-01.xls and make sure the **Link** radio button is selected. Notice the **Start Cell** text box indicates A1. If titles and other information are in the spreadsheet, a different cell for Inventor to start loading values from should be indicated. See **Figure 15-2.** Change the value to A3 because we need Inventor to start reading in values from that cell.

Figure 15-1.
The **Parameters** dialog box.

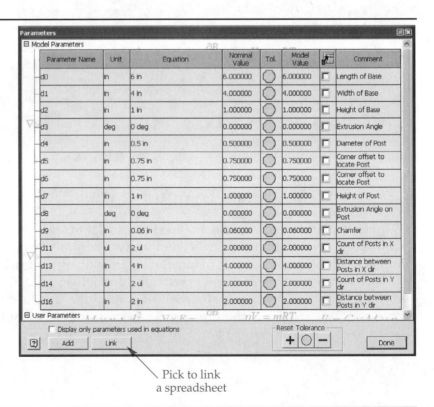

Pick to link a spreadsheet

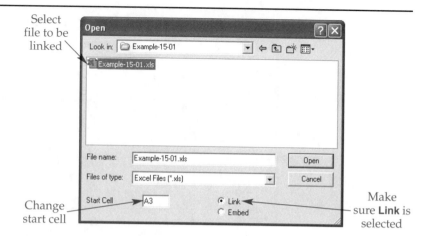

Figure 15-2. Opening a spreadsheet to be linked.

Select file to be linked

Change start cell

Make sure **Link** is selected

Click **Open** and the spreadsheet should now be linked to the part. Take a moment to scroll down to see the named parameters that have been imported. The **Parameters** dialog box can also be resized by clicking and dragging its edges.

The final step is to assign the named parameters in the spreadsheet to the dimension parameters defined in the part. The dimension parameter D0 controls the length of the part. Click once in the **Equation** field for D0. See **Figure 15-3.** Then pick the arrow and select **List Parameters** from the menu.

Choose BaseLength from the list, thereby assigning its value to control the dimension parameter D0. Continue doing this with all of the parameters. Use the parameter values listed in **Figure 15-4.** Any dimension parameter skipped will not be driven by the spreadsheet. This can be avoided when initially creating the part the spreadsheet is linked *first*. Then its parameters will be available for use as each of the features is built. When finished, pick **Done.** Update the part and save your work.

Open the Example-15-01b.ipt part file and do the same thing. Select the **Parameters** button in the **Part Features** panel. Using the **Parameters** dialog box, link the same spreadsheet, Example-15-01.xls. This time specify the starting cell to be E3. There are two sets of values defined within the spreadsheet—one for each part. Refer to **Figure 15-5** for the dimension parameters with the named parameters from the spreadsheet. When finished, pick **Done.** Update the part and save your work.

Figure 15-3. A—Select **List Parameters** to access the available parameters. B—The available parameters are listed in this dialog box.

Measure
Show Dimensions
Tolerance...
List Parameters

A

Parameters
BaseHeight
BaseLength
BasePostCountX
BasePostCountY
BasePostDepth
BasePostDia
BasePostDistX
BasePostDistY
BasePostOffset
BaseWidth

B

Figure 15-4.
The parameters used for this example.

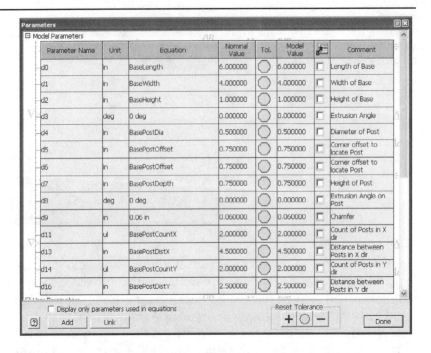

Parameter Name	Unit	Equation	Nominal Value	Tol.	Model Value		Comment
d0	in	BaseLength	6.000000	○	6.000000	□	Length of Base
d1	in	BaseWidth	4.000000	○	4.000000	□	Width of Base
d2	in	BaseHeight	1.000000	○	1.000000	□	Height of Base
d3	deg	0 deg	0.000000	○	0.000000	□	Extrusion Angle
d4	in	BasePostDia	0.500000	○	0.500000	□	Diameter of Post
d5	in	BasePostOffset	0.750000	○	0.750000	□	Corner offset to locate Post
d6	in	BasePostOffset	0.750000	○	0.750000	□	Corner offset to locate Post
d7	in	BasePostDepth	0.750000	○	0.750000	□	Height of Post
d8	deg	0 deg	0.000000	○	0.000000	□	Extrusion Angle on Post
d9	in	0.06 in	0.060000	○	0.060000	□	Chamfer
d11	ul	BasePostCountX	2.000000	○	2.000000	□	Count of Posts in X dir
d13	in	BasePostDistX	4.500000	○	4.500000	□	Distance between Posts in X dir
d14	ul	BasePostCountY	2.000000	○	2.000000	□	Count of Posts in Y dir
d16	in	BasePostDistY	2.500000	○	2.500000	□	Distance between Posts in Y dir

☐ Display only parameters used in equations

Reset Tolerance + ○ −

Add   Link   Done

Figure 15-5.
The dimension parameters with the named parameters from the spreadsheet.

Parameter Name	Unit	Equation	Nominal Value	Tol.	Model Value		Comment
d0	in	TopLength	6.000000	○	6.000000	□	Length of Top
d1	in	TopWidth	4.000000	○	4.000000	□	Width of Top
d2	in	TopHeight	1.000000	○	1.000000	□	Height of Top
d3	deg	0 deg	0.000000	○	0.000000	□	Extrusion Angle of Top
d4	in	TopHoleOffset	0.750000	○	0.750000	□	Corner offset to locate hole
d5	in	TopHoleOffset	0.750000	○	0.750000	□	Corner offset to locate hole
d6	in	TopHoleDia	0.500000	○	0.500000	□	Diameter of hole
d13	ul	TopHoleCountX	2.000000	○	2.000000	□	Count of holes in X dir
d15	in	TopHoleDistX	4.500000	○	4.500000	□	Distance between holes in X dir
d16	ul	TopHoleCountY	2.000000	○	2.000000	□	Count of holes in Y dir
d18	in	2 in	2.000000	○	2.000000	□	Distance between holes in Y dir

☐ Display only parameters used in equations

Reset Tolerance + ○ −

Add   Link   Done

Editing the Part

The dimension parameters are using the named parameters from the spreadsheet. So, to change the shape of the features that make up the part, the spreadsheet must be edited. First, access the Example-15-01a.ipt file using the **Windows** pull-down menu. Now in the **Browser**, right-click on the spreadsheet entry underneath the 3rd Party entry and choose **Edit** from the pop-up menu. See **Figure 15-6.**

This is a special spreadsheet that we will use to drive the assembly of this part and another. Notice the two sets of values for Example-15-01a and Example-15-01b. They are linked using Excel's native capability of referencing values used in existing cells. Double-click on cell F3 and notice in the formula line that it is actually equal to the value in cell B3. See **Figure 15-7A.** Find the 6.00 value for BaseLength in cell B3 and change it to 8.00. Notice that the value in cell F3 also changes to 8.00, thereby affecting the length of both parts. See **Figure 15-7B.** They will both then update accordingly in the assembly.

Figure 15-6.
Pick **Edit** to make changes to the spreadsheet.

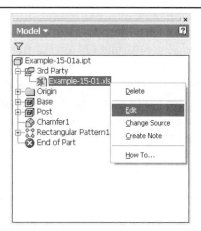

Figure 15-7.
A—The relationship between cell B3 and F3. B—Changing the value in cell B3 changes the value in cell F3 as well.

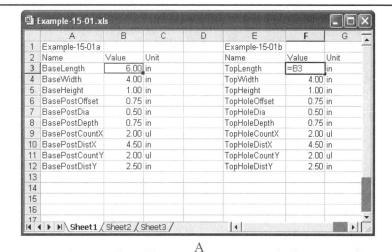

Take a moment to study the format of the cells for each part. Each part in the assembly that is to be controlled needs its own entry in the spreadsheet. Part entries are separated by a blank column for readability. Three columns are used—Name, Value, and Unit. Name is obviously for the name of the parameter, Value is for the numeric value, and Unit is for the type of units specified. Distances use inches (in) or millimeters (mm). Counts of items in a pattern are denoted by the special unit, ul. This stands for *unitless* and is used wherever something other than a distance is implied. The unit column is optional. If the unit column is not included, then all values will use the default unit value of the part file, such as inches or millimeters.

Save the spreadsheet and exit Excel. In Inventor, update both parts to see the changes take effect. Save both parts and exit both of them.

Open the Example-15-01.iam file and notice that there is no spreadsheet linked yet. Select the **Parameters** tools and the **Link** button within the dialog box. Locate the spreadsheet as before and specify A3 as the starting cell (perhaps you will recall that the first set of values were the drivers). Pick **Done** and update the assembly. In the **Browser**, right-click on the spreadsheet and choose **Edit**. Change the value for BaseWidth to be the same as for BaseLength, which is 8.00. Save the spreadsheet, exit Excel, and update the assembly to see the revisions. Take a moment to review in the **Browser** that each part has an entry for the linked spreadsheet under 3rd Party.

Flanged Pipe Run—An Advanced Example

This example is similar to the previous example, except this one demonstrates how to apply the named parameters to control parametric values used in an assembly. Open the Example-15-02.iam file. Examine the assembly structure by reviewing the listing in **Browser**.

Right-click on Pipe and choose **Open** from the pop-up menu. Once the Pipe.ipt file is open, access the **Parameters** tool and link the spreadsheet Example-15-02.xls. Specify the starting cell to be A3 as the values for the Pipe start there. Click once on the equation for D0 and change it to use named parameter PipeOD. Do the same for D2 using Length. When you are finished, pick the **Done** button. Update the part and the length should change from 24 in. to 36 in. Save the part and then close.

Open the Flange 300lb part. Access parameters and link the spreadsheet starting at cell E3. Click once on the equation for D1 and assign PipeOD to it. Continue with all the others shown in **Figure 15-8.** When you are finished, pick the **Done** button. Update, save, and close the part.

Open the Elbow part. Access parameters and link the spreadsheet starting at cell I3. Click once on the equation for D0 and assign PipeOD to it. Then assign Radius to D2. When you are finished, pick the **Done** button. Update, save, and close the part.

Open the Hexhead Bolt part. Access parameters and link the spreadsheet starting at cell M3. Assign HeadDia to D0, BoltDia to D7, and BoltLength to D8. When you are finished, pick the **Done** button. Update, save, and close the part.

Figure 15-8.
The dimension parameters with the named parameters from the spreadsheet.

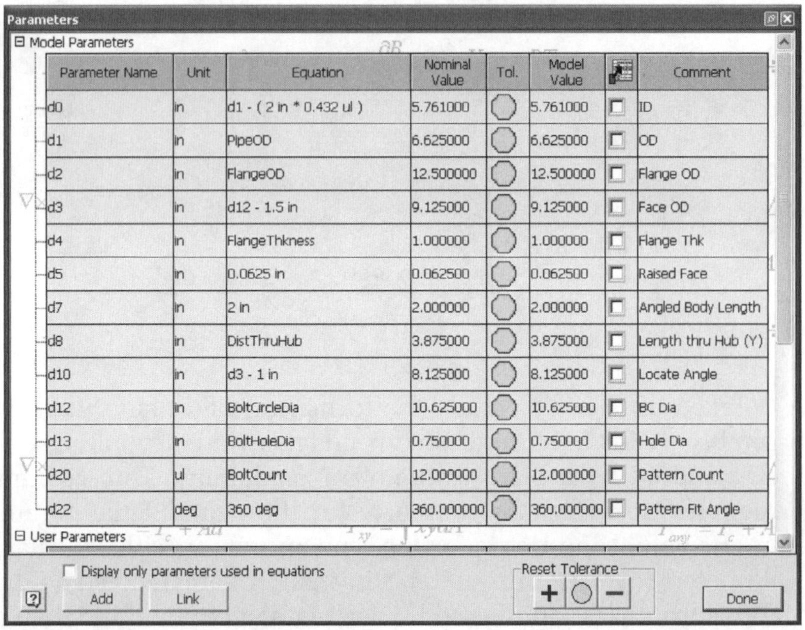

Parameter Name	Unit	Equation	Nominal Value	Tol.	Model Value		Comment
d0	in	d1 - (2 in * 0.432 ul)	5.761000	○	5.761000	□	ID
d1	in	PipeOD	6.625000	○	6.625000	□	OD
d2	in	FlangeOD	12.500000	○	12.500000	□	Flange OD
d3	in	d12 - 1.5 in	9.125000	○	9.125000	□	Face OD
d4	in	FlangeThkness	1.000000	○	1.000000	□	Flange Thk
d5	in	0.0625 in	0.062500	○	0.062500	□	Raised Face
d7	in	2 in	2.000000	○	2.000000	□	Angled Body Length
d8	in	DistThruHub	3.875000	○	3.875000	□	Length thru Hub (Y)
d10	in	d3 - 1 in	8.125000	○	8.125000	□	Locate Angle
d12	in	BoltCircleDia	10.625000	○	10.625000	□	BC Dia
d13	in	BoltHoleDia	0.750000	○	0.750000	□	Hole Dia
d20	ul	BoltCount	12.000000	○	12.000000	□	Pattern Count
d22	deg	360 deg	360.000000	○	360.000000	□	Pattern Fit Angle

Open the Hexhead Nut part. Access parameters and link the spreadsheet starting at cell Q3. Assign BodyDia to D0 and BoltDia to D7. When you are finished, pick the **Done** button. Update, save, and close the part.

The final step is to link the spreadsheet to the assembly. This allows for full control of the assembly. Open Example-15-02.iam and update it if necessary. Access parameters and notice that there is an assembly parameter for the count of the bolts used in the pattern. Link the spreadsheet starting at cell E3. Why start at cell E3? If you recall, this is where the values for the Flange begins. One of the parameters used in the flange controls the number of bolt holes in the bolt circle. If the flange's parameters are imported into the assembly, then they can be used to control the assembly's parameters, such as the count of hex head bolts used in the assembly pattern.

Once the spreadsheet is linked, assign BoltCount to d31 and edit the equation for d30 to read 360/BoltCount. In Inventor, assembly pattern angles are specified explicitly as opposed to the alternate method used in part feature patterns, where they are specified by an included angle—usually 360°. Update and save the assembly file under a different name.

Under the **Browser** entry 3rd Party is the linked spreadsheet. Right-click on it and choose **Edit** from the pop-up menu. In the spreadsheet, change the values as shown below:

- Change cell B3 to 4.50. This is the pipe OD.
- Change cell B4 to 24.00. This is the length of the straight section of pipe.
- Change cell F4 to 10.00. This is the flange OD.
- Change cell F5 to .75. This is the flange thickness.
- Change cell F6 to 3.375. This is the depth of flange.
- Change cell F7 to 7.875. This is the bolt circle diameter.
- Change cell F8 to 8. This is the count of bolts in the pattern.
- Change cell J4 to 6.00. This is the long radius for elbow.

Save the spreadsheet and update the assembly. The assembly should change from a 6" pipe run to a 4" pipe run with the correct number of bolts. If there are any problems, go into the individual part files and investigate the spreadsheet links.

Why are not more of the values parametrically related to the pipe OD so you would only change the pipe OD to change everything? All of the values are taken from standard tables for *Schedule 80 Pipe*. You must change each of the values because they are not functions of the pipe OD. For example, the flange OD is not always 1.5 times the pipe OD. There is no relationship between the dimensions controlling the parts.

NOTE	Alternatively, the pattern of bolts in this assembly could have been controlled by using the powerful Inventor capability of controlling an assembly pattern. This can be done by linking it to a feature pattern on one of the assembly's parts (for example, the pattern of bolt holes on one of the flanges).

❏ In this practice a spreadsheet will be manipulated and used to drive an assembly. Open the assembly Practice-15-01.iam.

❏ Right-click the Practice-15-01.xls spreadsheet in the **Browser** as shown below.

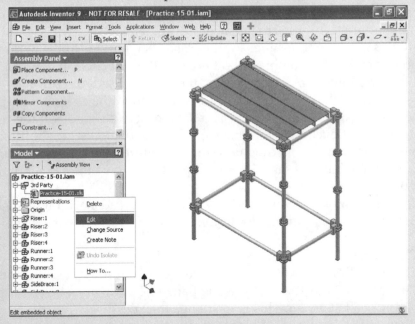

❏ Edit RunnerLengthNominal and SideBraceLengthNominal in the cells shown in orange. The riser height of the scaffold bay is not adjustable. Although the tiepoint pattern count is calculated (cell N4), Inventor gives an error when the riser height is changed. This is because the assembly constraints between the runners and the tiepoints cannot find the geometry that they were originally based on if the number of tiepoints changes.

❏ This practice demonstrates the ability to manipulate a main assembly composed of subassemblies with a spreadsheet. Make note that the spreadsheet is not linked to some of the subassemblies, such as Runner.Iam, yet it is linked to that subassembly's part, Runner.Ipt.

Summary

You have seen how important the proper use of parameters in an assembly can be for providing control and making revisions a simple process. In essence, this ability provides a central storage of named parameters and a means of importing them for each part in the assembly. Once the links are established, simply change the spreadsheet to revise the assembly. Without this capability a designer would have to use the tedious method of changing each part dimensionally to revise the assembly.

Chapter Test

Answer the following questions on a separate sheet of paper.

1. Why use parameters to control an assembly?
2. Why use a spreadsheet to control the parameters?
3. Are spaces allowed in a parameter's name?
4. What is the difference between the terms *link* and *embed* displayed on the file select dialog box when you have chosen to link a spreadsheet?
5. If you have inspected this Chapter's spreadsheets carefully, you would have noticed that some of the numeric values have units expressed in something other than IN. (Inches). The notation "UL" is used. What does this signify and what does the abbreviation stand for?
6. Is it necessary to link the spreadsheet into *every* part used in the assembly?
7. If you right-click on the spreadsheet entry in the **Browser,** a pop-up menu is accessed. If **Delete** is selected, is the spreadsheet entry and its parameters removed or does it simply sever the link to the spreadsheet?
8. If you right-click on the spreadsheet entry in the **Browser,** a pop-up menu is accessed. The **Change Source** option allows a link to be redirected to another spreadsheet. *True or False?*
9. Can any spreadsheet program, such as Quattro Pro, be used to control an assembly?
10. How would you apply this chapter's technology to your own design work?

Chapter Exercises

Exercise 15-1. *Flanged Pump Head—Parameters and Derived Parts.* Make sure no files are open. Create a new *single user project* and name it Exercise-15-01.ipj. Make the *project (workspace) folder* the Chapter 15\Exercises\Exercise-15-01 folder and click the Next button. Make sure that there are no libraries selected. Then pick **Finish** and **Apply** to make this the current project. Open Exercise-15-01.iam. In this exercise the spreadsheet will be linked to each of the parts in the assembly. There is a derived part used in each of the parts—Master Sketch.ipt is used in the Body.ipt and Plug_Sketch.ipt is used in the Plug.ipt. The spreadsheet must also be linked to these derived parts. Finally, the spreadsheet is linked to the assembly to enable control of the number of bolts in the pattern. As with previous work in this chapter, after linking the spreadsheet the numeric values of the dimension parameters must be replaced with the named parameters brought in with the spreadsheet. Each of the part's parameter tables are listed below:

Parameters for the Body.ipt

Parameters for the Master Sketch.ipt (derived part)

Parameter Name	Unit	Equation	Nominal Value	Tol.	Model Value		Comment
d1	in	Flange_Thick	1.000000	◯	1.000000	☐	
d2	in	Bore_Depth	2.68750	◯	2.687500	☐	
d3	in	Body_ID	1.750000	◯	1.750000	☐	
d4	in	Bore_Dia	3.000000	◯	3.000000	☐	
d5	in	Flange_OD	6.000000	◯	6.000000	☐	
d6	in	Body_OD	4.500000	◯	4.500000	☐	
d7	in	Body_Length	2.250000	◯	2.250000	☐	
d8	deg	90 deg	90.000000	◯	90.000000	☐	
d9	in	Bolt_Circle	5.250000	◯	5.250000	☐	
d10	deg	HoleCL	36.000000	◯	36.000000	☐	

Parameters for the **Plug.ipt**

Parameter Name	Unit	Equation	Nominal Value	Tol.	Model Value		Comment
d1	ul	1.0000 ul	1.000000	◯	1.000000	☐	
d3	in	Hole_Dia	0.406000	◯	0.406000	☐	Hole dia
d5	in	Counter_Bore	0.609000	◯	0.609000	☐	Counterbore dia
d6	in	CB_Depth	0.375000	◯	0.375000	☐	Counterbore depth
d10	ul	Num_Holes	10.000000	◯	10.000000	☐	Hole count
d12	deg	360 deg	360.000000	◯	360.000000	☐	Pattern included angle
d13	in	0.025 in	0.025000	◯	0.025000	☐	Chamfer

Parameters for the **Plug-Sketch.ipt** (derived part)

Parameter Name	Unit	Equation	Nominal Value	Tol.	Model Value		Comment
d1	in	Flange_Dia	6.000000	◯	6.000000	☐	Flange Diameter
d2	in	Plug_Dia	2.992500	◯	2.992500	☐	Plug Diameter
d3	in	Flange_Thk	0.875000	◯	0.875000	☐	Flange Thickness
d4	deg	15 deg	15.000000	◯	15.000000	☐	
d5	in	0.25 in	0.250000	◯	0.250000	☐	
d6	in	Oring_Gv	0.125000	◯	0.125000	☐	O-ring groove
d7	in	d6	0.125000	◯	0.125000	☐	
d8	in	d6	0.125000	◯	0.125000	☐	
d9	in	Plug_Length	2.562500	◯	2.562500	☐	Plug length
d10	deg	90 deg	90.000000	◯	90.000000	☐	
d11	in	BC	5.250000	◯	5.250000	☐	Bolt circle dia
d12	deg	Hol_CL	60.000000	◯	60.000000	☐	Pattern angle

Exercise assembly parameters

Parameter Name	Unit	Equation	Nominal Value	Tol.	Model Value		Comment
d0	in	0.0 in	0.000000	◯	0.000000	☐	
d1	in	0.00 in	0.000000	◯	0.000000	☐	
d2	in	0.0 in	0.000000	◯	0.000000	☐	
d3	deg	HoleCL	60.000000	◯	60.000000	☐	Pattern Angle
d4	ul	No_Holes	6.000000	◯	6.000000	☐	Hole Count

Once all parameters are updated, experiment with changing the values in the spreadsheet to explore different configurations of this assembly.

Exercise 15-2. *10^2 Pipe Run.* Create a spreadsheet for Example 15-2 to use 10″ pipe. Modify the spreadsheet to use the values for 10″ SCH80 pipe. Save it under a different name, such as Exercise-15-02-10in.xls. The correct values are shown below.

	A	B	C	D	E	F	G
1	Pipe				Flange 300lb		
2	Name	Value	Unit		Name	Value	Unit
3	PipeOD	10.75	in		PipeOD	10.75	in
4	Length	48.00	in		FlangeOD	17.50	in
5					FlangeThkness	1.25	in
6					DistThruHub	4.63	in
7					BoltCircleDia	15.25	in
8					BoltHoleDia	1.00	in
9					BoltCount	16.00	ul

	I	J	K	L	M	N	O	P	Q	R	S
1	Elbow LR				Hexhead Bolt				Hexhead Nut		
2	Name	Value	Unit		Name	Value	Unit		Name	Value	Unit
3	PipeOD	10.75	in		BoltDia	1.00	in		BoltDia	1.00	in
4	Radius	15.00	in		HeadDia	1.75	in		BodyDia	1.75	in
5					BoltLength	3.50	in				

Open up each of the example assembly's part files and use the **Change Source** option to the revised spreadsheet. Update the parts and save each of them. The spreadsheet name on the **Browser** may not change to reflect the new name. However, if you access the **Parameters** tool and look for the spreadsheet entry at the bottom, you will notice that it does indeed reference the correct new spreadsheet. It is simply a matter of renaming the spreadsheet's name in the **Browser**.

Open the assembly and update it. Finally, link it also to the revised spreadsheet by using the **Change Source** option.

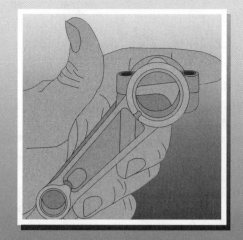

CHAPTER 16

Surfaces

Objectives

After completing this chapter, you will be able to:

* Explain the differences between a surface and a solid.
* Adjust the display of a surface.
* Explain the basic process for creating a surface.
* Offset a surface from a solid or surface.
* Replace a face on a solid with a face that matches the shape of a surface.
* Use surfaces as construction geometry.
* Stitch surfaces into a quilt.
* Visually analyze the faces of a part for draft angle or continuity of surface topology.

User's Files

*The following is a list of files that you will need to work through this chapter. These files can be found on the **User's Files** CD included with this text.*

Examples
Example-16-01.ipt
Example-16-02.ipt
Example-16-03.ipt
Example-16-04.ipt
Example-16-05.ipt
Example-16-06.ipt
Example-16-07.ipt
Example-16-08.ipt
Example-16-09.ipt
Example-16-10.ipt

Exercises
Exercise-16-01.ipt
Exercise-16-02.ipt
Exercise-16-03.ipt
Exercise-16-04.ipt

A surface defines the form and shape of an object but does not have volume. A solid also defines the form and shape of an object, but a solid has volume. Inventor is primarily used to create solid objects. As you will learn in *Chapter 19*, even thin, sheet metal parts are created as solids. However, you can create surfaces to use as construction geometry. You used this method in Exercise 10-4 in *Chapter 10*. See **Figure 16-1.** In addition, there are some instances where you may want to create the final product as a surface model, such as when sharing a model with software that does not support solid modeling.

This process for creating a surface is basically the same as for creating a solid. First, sketch and constrain the geometry. The sketch for a surface does not need to be closed. Then, finish the sketch and select the feature creation tool you wish to use, such as **Extrude** or **Revolve**. In the feature creation dialog box, such as the **Extrude** dialog box, pick the **Surface** button in the **Output** area. See **Figure 16-2.** If the sketch is open, the **Solid** button will be disabled and the **Surface** button will be on. Then, select the profile, if needed, and create the feature.

Figure 16-1.
In Exercise 10-4, you used a construction surface to terminate the extrusion when creating the slots.

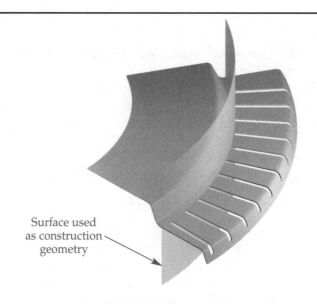

Surface used as construction geometry

Figure 16-2.
Creating a surface extrusion.

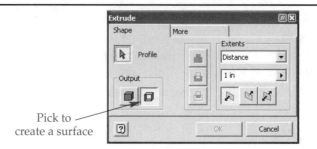

Pick to create a surface

Surface Display

By default, surfaces are displayed translucent in Inventor. This can be changed so that they are displayed opaque. Select **Application Options** in the **Tools** pull-down menu. The **Options** dialog box is displayed. Select the **Part** tab. In the **Construction** area of the tab, check the **Opaque Surface** check box. Then, pick the **OK** button to save the setting and close the dialog box. This setting is only applied to new surfaces you create. Existing surfaces are unaffected.

You can also turn translucency on and off for an individual surface feature. Right-click on the surface feature name in the **Browser** and select **Translucent** from the shortcut menu. A check mark appears next to the menu item when translucency is on.

Surfaces are always displayed in yellow, either translucent or opaque. You cannot change the material for a surface feature, as you can a solid feature. The material set up for the part is not applied to surfaces in the part.

Extruded Surfaces

Open Example-16-01.ipt. This file contains an open sketch composed of a line and an arc. It is fully constrained. Pick the **Extrude** button in the **Part Features** panel. In the **Extrude** dialog box, select the **Profile** button in the **Shape** tab. Then in the graphics window, pick the sketch (which is made up of the arc and tangent line).

Now, notice in the dialog box that the **Surface** button in the **Output** area is automatically selected. This is because the sketch is open. A solid cannot be created from an open sketch. If the **Solid** button is selected, the **Examine Profile Problems** button (a red cross) is displayed at the bottom of the dialog box. Also, notice that the operation buttons—**Join**, **Cut**, and **Intersect**—are grayed out when the **Surface** button is selected.

Select **Distance** in the **Extents** drop-down list. Then, enter 1 in the text box below the drop-down list. Finally, pick the **OK** button to extrude the surface. Notice in the **Browser** that the feature is named ExtrusionSrfx. The Srf suffix indicates that the feature is a surface, not a solid.

Revolved Surfaces

Open Example-16-02.ipt. This file contains a sketch that you will revolve into a surface. Like the previous example, this is an open sketch. Pick the **Revolve** button in the **Part Features** panel. In the **Revolve** dialog box, pick the **Profile** button in the **Shape** area. In the graphics window, select all of the lines in the sketch except for the long, vertical line. The **Surface** button in the **Output** area of the dialog box is automatically turned on. With the **Axis** button on in the dialog box, pick the long, vertical line as the axis of revolution. Select **Angle** in the **Extents** drop-down list. Then, enter 180 in the text box below the drop-down list. Finally, pick the **OK** button to create the surface.

Lofted Surfaces

Open Example-16-03.ipt. This file contains two closed sketches that will be used to create a lofted surface. Pick the **Loft** button in the **Part Features** panel. In the **Loft** dialog box, pick the **Surface** button in the **Output** area of the **Curves** tab. Pick Click to add in the **Sections** column. Then, select both sketches in the graphics window. Pick the **OK** button to create the loft. A surface is generated between the two sketches. However, notice that the ends—the area of each closed sketch—are open. This is more apparent if translucency is turned off for the surface.

Swept Surfaces

Open Example-16-04.ipt. This file contains two sketches. The curved line will be used as the path for a swept feature. The circle sketch will be the cross section. Pick the **Sweep** button in the **Part Features** panel. In the **Sweep** dialog box, pick the **Surface** button in the **Output** area of the **Shape** tab. Then, pick the **Profile** button and select the circle in the graphics window. Pick the **Path** button in the dialog box and select the curved line in the graphics window. Finally, pick the **OK** button to create the swept surface. If you turn the translucency for the surface off, it is apparent that this is a tube, not a solid part.

Offsetting a Surface

You can offset a surface from an existing surface or solid. This can be used to create a surface for use as a construction feature or to simulate a thin object such as a bottle. Open Example-16-05.ipt. This file contains a loft surface created from a series of cross sections. You will use the **Thicken/Offset** tool to create a parallel surface.

Pick the **Thicken/Offset** button in the **Part Features** panel. The **Thicken/Offset** dialog box is displayed, **Figure 16-3**. In the **Output** area of the **Thicken/Offset** tab of the dialog box, pick the **Surface** button. Then, with the **Select** button on, pick the surface in the graphics window. The selected surface is displayed in cyan. Since this part is one continuous surface, the entire part is selected. In the **Thicken/Offset** tab, select the **Direction** button so the previewed surface is to the outside of the original surface. Finally, enter .5 in the **Distance** text box and pick the **OK** button to create the offset surface. A copy of the surface is scaled so that the distance between it and the original is .5″. See **Figure 16-4**. If you turn off the translucency for the original surface, you can better see the offset.

Figure 16-3.
Creating an offset surface. By setting the output to a solid, you can thicken a solid using this dialog box.

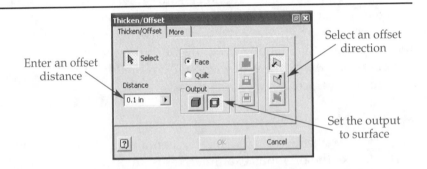

Enter an offset distance

Select an offset direction

Set the output to surface

Figure 16-4.
An offset surface and the original surface.

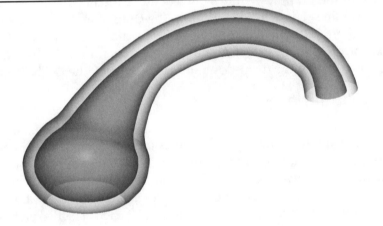

You can also offset a surface from a solid part. Open Example-16-06.ipt, which contains a solid part. Next, open the **Thicken/Offset** dialog box. In the **Output** area, select the **Surface** button. Pick the **Select** button and the **Face** radio button next to the **Select** button. Then, pick any face on the part in the graphics window. A single face is selected and previewed offset. You can continue to pick individual faces to offset as a surface. However, the selected faces must be contiguous and form a single area. To create the offset surface, pick the **OK** button.

Suppose you want to offset all faces on the solid that are contiguous. You can individually select all faces. However, there is an easier method. Undo the offset operation you just performed. Then, open the **Thicken/Offset** dialog box. Pick the **Surface** button and the **Face** radio button in the **Thicken/Offset** tab. Then, pick the **More** tab.

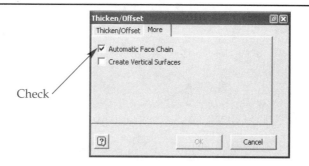

Figure 16-5.
All faces in a chain can be selected with one pick when this check box is checked.

Check the **Automatic Face Chain** check box, **Figure 16-5.** Now, pick the **Select** button in the **Thicken/Offset** tab and select any face on the part other than the top or bottom face. All of the faces around the perimeter of the part are selected because they are a chain. The top and bottom faces are not part of the chain because of the sharp edge. Next, pick the top face and create the offset surface. A surface "shell" is generated at the specified offset distance from the solid.

Now, suppose you want to offset the surface you just created. The **Face** and **Automatic Face Chain** options work the same when selecting a surface as they do on a solid. However, you can select *all* faces on a contiguous surface with one pick using the **Quilt** option. A *quilt* is a single continuous surface consisting of two or more faces. Open the **Thicken/Offset** dialog box and pick the **Quilt** radio button. Then, with the **Select** button on, pick anywhere on the surface in the graphics window. The entire surface is selected and displayed in cyan. Do not try to pick on the solid. You cannot select faces on a solid with the **Quilt** option.

PROFESSIONAL TIP

If the output for the **Thicken/Offset** tool is set to solid, the result is a thickened part with the feature's name in the **Browser** reflecting this. By setting the output to **Solid**, you can create a solid feature from an offset surface.

Replace Face

In this example, you will create a loft surface and use it to replace the planar face on top of a rectangular solid. Open Example-16-07.ipt. This file contains a rectangular solid box and four unconsumed sketches. Two of the sketches will be used as the cross section. The other two sketches will be used as rails, or guides.

Pick the **Loft** button in the **Part Features** panel. In the **Curves** tab of the **Loft** dialog box, pick the **Surface** button in the **Output** area. It is important to select this button before selecting cross sections or rails. Next, pick Click to Add in the **Sections** column. Then, pick the first cross section sketch. See **Figure 16-6.** In this particular case, two separate picks are required for each cross section. The first pick selects the sketch. The second pick specifies the curve to be used as the cross section; select the arc. Select the second cross section as well. Once the second cross section is selected, a preview appears in the graphics window. Then, pick Click to Add in the **Rails** column and select one rail. Two picks are required for each rail—one to select the sketch and one to select the curve. Finally, pick the **OK** button to create the surface. See **Figure 16-7.**

Next, pick the **Replace Face** button in the **Part Features** panel. The **Replace Face** dialog box is displayed, **Figure 16-8.** With the **Existing Faces** button in the dialog box on, pick the top planar face on the solid in the graphics window. Then, pick the **New Faces** button in the dialog box. Select the loft surface you just created in the graphics window. Pick the **OK** button to replace the face, **Figure 16-9.**

Figure 16-6.
Creating a loft
surface.

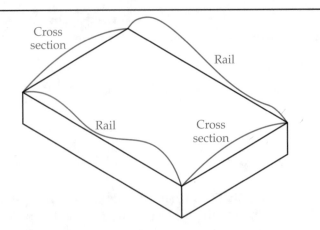

Figure 16-7.
The completed loft
surface.

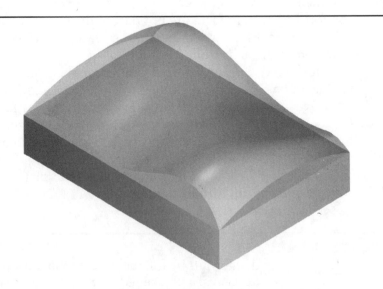

Figure 16-8.
Replacing a face.

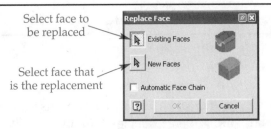

Figure 16-9.
The original flat face
is replaced by a face
that matches the
shape of the loft
surface.

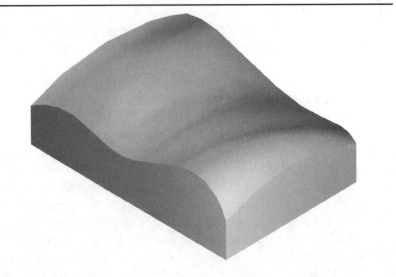

Since the top face was replaced, the solid still has six faces. If the loft had been created as a solid that was joined to the existing solid, additional faces would have been created. This is one of the benefits of using this procedure.

Surfaces as Construction Geometry

As mentioned earlier, surfaces are often used as construction objects. For example, a surface can be used to terminate an extrusion or revolution. A surface can also be used to split a part, which is discussed later in this chapter. In this example, a surface will be used to terminate an extrusion. Open Example-16-08.ipt. This file contains a rectangular part and two unconsumed sketches. One sketch will be used to create the construction surface. The other sketch will be extruded as a solid.

Using the **Extrude** tool, extrude the sketch named Sketch for Construction Surface 4.00". See **Figure 16-10A.** Select the **Surface** button in the **Output** area of the **Extrude** dialog box, and then select the sketch. Pick the **Midplane** option and then pick the **OK** button to create the surface. The surface should project above and below the top face of the solid part. Notice that the surface is slightly curved.

Next, extrude the sketch named Sketch for Subtraction Solid. Pick the **Solid** button in the **Output** area of the **Extrude** dialog box. Then, select the profile in the graphics window. Select **To** in the **Extents** drop-down list and, with the "select" button below the drop-down list on, pick the surface in the graphics window. Finally, pick the **Cut** button and then pick **OK** to create the feature. See **Figure 16-10B.** Notice how the back of the subtracted feature matches the curvature of the surface.

Figure 16-10.
A—The part showing the two unconsumed sketches. B—The curved back of the cutout feature was used as the terminator for the extrusion.

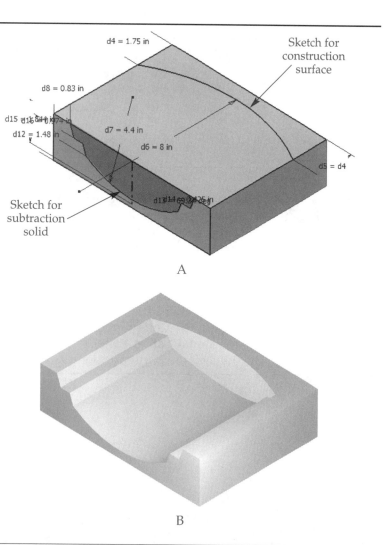

Stitched Surfaces

As described earlier, a quilt is a single surface consisting of two or more surfaces that are connected as a single unit. The **Stitch Surface** tool allows you to "sew" separate surfaces into a quilt. In order for the tool to work, the mating edges of the surfaces to be stitched need to be identical.

Open Example-16-09.ipt. Using the **Revolve** tool, revolve Sketch1 90 degrees about the X axis. See **Figure 16-11**. Next, use the **Extrude** tool to extrude Sketch2 2″ extending away from the revolution. One edge on each surface touches the other surface. Additionally, those two edges are identical.

Pick the **Stitch Surface** button in the **Part Features** panel. The **Stitch** dialog box is displayed, **Figure 16-12**. With the **Surfaces** button in the dialog box on, pick both surfaces in the graphics window. Then, pick the **OK** button in the dialog box to stitch the surfaces into a quilt.

Figure 16-11.
These two sketches will be used to create two surfaces, which will then be stitched into a quilt.

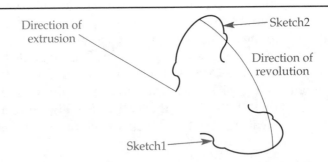

Figure 16-12.
Stitching surfaces into a quilt.

PROFESSIONAL TIP

When certain files are imported, such as IGES files, they may come in as surfaces. In order to be used as a solid feature or part, these surfaces must be promoted to the part environment. By default, this is automatically done when importing the file. Once imported, a Construction branch is added to the part tree. To manually promote the surfaces, right-click on the branch and select **Edit Construction** from the shortcut menu. Then, stitch all surfaces into a quilt using the **Stitch** tool in the **Construction Panel**. Then, use the **Promote** tool in the **Construction Panel** to promote the stitched surface to a solid in the part environment.

Inventor provides a tool for visually analyzing the faces on a part. The part can be a solid or a surface. There are two analysis modes. One mode is for analyzing a draft angle and the suitability of the part for removal from a mold. The other mode analyzes the continuity of the part's surface topology.

The analysis can be toggled on and off using the **Analyze Faces** button on the **Inventor Standard** toolbar. The current settings are used for the analysis. To change the current settings, or to switch between modes, open the **Analyze Faces** dialog box by selecting **Analyze Faces** from the **Tools** pull-down menu. See **Figure 16-13.**

To display an analysis of the face draft, pick the **Draft** button in the **Style** area of the **Analyze** dialog box. The remaining settings in the dialog box change to reflect the draft selection. In the middle of the dialog box are two text boxes in which you enter the minimum and maximum draft angles. The color spectrum shown at the bottom is applied to the surface of the part. This helps the designer to determine whether or not the part can be removed from the mold. If not, then the draft angle will have to be changed and a new analysis run. In the **Selection** area of the dialog box, you can choose to have the analysis applied to the entire part or to selected faces. To save the settings, pick the **OK** button in the **Analyze** dialog box.

Open Example-16-10.ipt. This part will need to be pulled from a cavity of a mold. See **Figure 16-14**. Because the draft angles vary, each shape will present a different degree of difficulty when pulling the part out of the mold. Open the **Analyze Faces** dialog box and pick the **Draft** button in the **Style** area. Set the **Draft Start Angle** to –2 degrees and the **Draft End Angle** to 10 degrees. Pick **OK** to see the draft analysis applied to the part. The cylinder was extruded without any taper, therefore, we know that any face that is the same color as the cylinder, purple, has 0 degrees of draft. Any area that contains a color that is left of the purple on the scale has a negative taper and could be difficult to remove from the mold. The bottom of the bullet-shaped feature is colored blue, which indicates an undercut situation that may be difficult to pull out of the mold.

The analysis of surface topology is called a zebra analysis because black and white stripes are displayed on the surface. See **Figure 16-15.** To display the zebra analysis, pick the **Zebra** button in the **Style** area of the **Analyze** dialog box. The remaining settings in the dialog box change to reflect the zebra setting. The zebra stripes are parallel on areas of flatness on the part. The zebra stripes will converge where there are areas of radical departure from the curvature. One thing to keep in mind when displaying a zebra analysis is that the stripes are not "fixed" to the part. As you rotate the view, the lines will "move around" on the part.

Figure 16-13.
Setting the analysis mode to examine the draft angle.

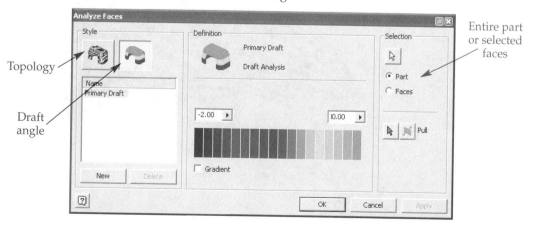

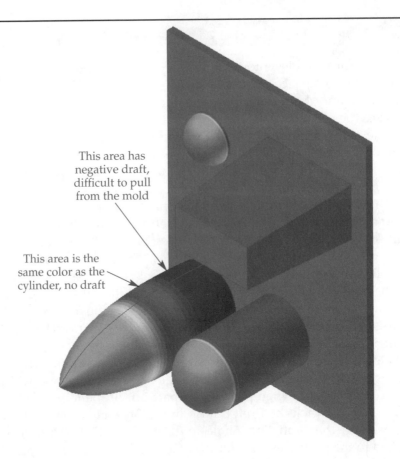

Figure 16-14.
The draft angle is represented by a corresponding color.

This area has negative draft, difficult to pull from the mold

This area is the same color as the cylinder, no draft

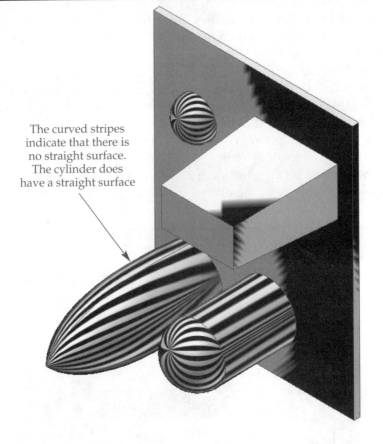

Figure 16-15.
Surface topology can be analyzed using zebra stripes.

The curved stripes indicate that there is no straight surface. The cylinder does have a straight surface

Chapter Test

Answer the following questions on a separate sheet of paper.

1. Explain the differences between a surface and a solid.
2. How does the process for creating a surface differ from the process used to create a solid?
3. Briefly describe how to change the display of an existing surface from translucent to opaque.
4. Which tool is used to create a new surface that is offset from the original surface?
5. Define *quilt.*
6. Which tool is used to create a quilt from separate surfaces?
7. What must be true of two surfaces that are going to be joined using the tool in Question 6?
8. Give two examples of how a surface can be used as construction geometry.
9. In which dialog box are settings made for the analysis of the faces of a part?
10. What are the modes available for analyzing the faces of a part?

Chapter Exercises

Exercise 16-1. Open Exercise-16-01.ipt. Create an offset surface that is .1" from the original and to the outside. There are six faces on the original solid; offset all six. After the offset surface is created, use the **Fillet** tool and apply rounds to the four vertical corners. Use a radius of .25".

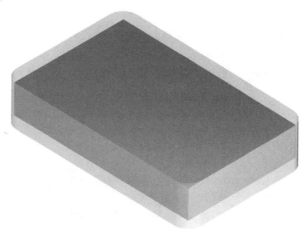

Exercise 16-2. Open Exercise-16-02.ipt. Create a revolved surface using the sketch named Sketch for Revolution. Revolve the sketch 180 degrees through the existing solid. Finally, use the **Split** tool to remove the solid material inside of the bottle-shaped surface.

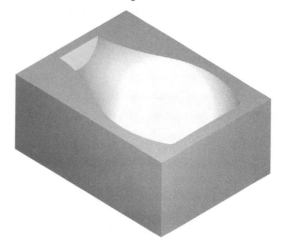

Exercise 16-3. Open Exercise-16-03.ipt. Create a surface by extruding the sketch 10". Create a new sketch on the YZ plane. Project the top and bottom edges of the surface onto the sketch plane. Then, draw a rectangle with its corners on the endpoints of the projected lines. Apply rounds to the corners using a radius of your choice. Finally, extrude the sketch into a solid using the surface as the terminator.

Exercise 16-4. Open Exercise-16-04.ipt. Create a loft surface using the two "section" sketches and the two "rail" sketches. Note: These sketches do *not* require two picks as were required in the example in the text. Then, replace the top face of the solid with the surface. Create an offset surface that is .5" above the new top face. Next, extrude the sketch named Sketch for Extrusion using the offset surface as the terminator. Finally, apply .375" radius fillets and rounds as shown below.

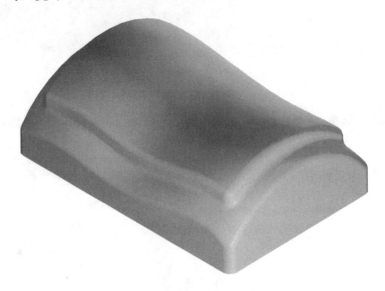

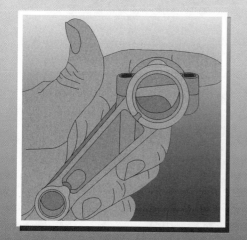

CHAPTER 17

Assembly Drawings

Objectives

After completing this chapter, you will be able to:

* Create 2D orthographic, section, and break out views from assemblies.
* Annotate assembly drawings.
* View, modify, and insert parts lists.
* Create drawing views using design view representations.

User's Files

*The following is a list of files that you will need to work through this chapter. These files can be found on the **User's Files** CD included with this text.*

Examples	Exercises	Practices
Example-17-01.iam	Exercise-17-01.iam	Practice-17-01.iam
Example-17-02.iam	Exercise-17-02.iam	
Example-17-03.iam	Exercise-17-03.iam	
	Exercise-17-04.iam	

Views

Make sure no files are open. Create a new *single user project* and name it Example-17-01.ipj. Make the *project (workspace) folder* the Chapter 17\Examples\Example-17-01 folder and click the Next button. Make sure that there are no libraries selected. Then pick **Finish** and **Apply** to make this the current project. Open Example-17-01.iam and examine the model to familiarize yourself with the various parts and subassemblies used in the assembly. See **Figure 17-1**. Notice that several of the work planes are visible—these will not appear in the drawing views. Pay attention to the orientation of the coordinate system. This plays a crucial part in specifying the viewing direction for the creation of the drawing views. Create a new drawing (.IDW) and save it as Example-17-01.idw. Before you begin to create views, choose an E-size sheet to use. Right-click on the sheet in the **Browser** and choose **Edit Sheet** from the pop-up menu. In the **Edit Sheet** dialog box, select **E** from the **Size** drop-down list, and pick **OK** to exit.

Figure 17-1.
Notice the appearance of the work planes and the orientation of the coordinate system.

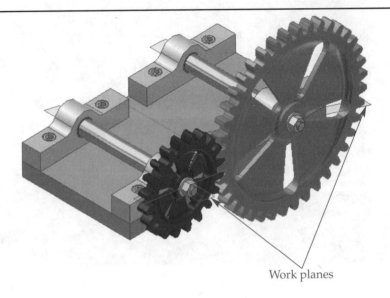

Work planes

Using the **Base View** tool, create a base view from the bottom orientation of the assembly. Pick in the lower-left corner of the drawing to place the base view. Refer to **Figure 17-2** for the correct orientation and location. Even though the bottom view orientation of the assembly was selected, this will be the front view for this drawing. You cannot always count on an assembly being oriented properly for the creation of a given view, so you must become comfortable with using the various options available.

You started with the front view, as most drawings are developed, and now are able to create three additional views projected from the base view. Right-click on the base view (VIEW1:Example-17-01.iam) in the **Browser** and choose **Projected** from the **Create View** cascading menu. See **Figure 17-3**. When the cursor is moved around the drawing, different views appear. Pick on the drawing, above the base view, to the right of the base view, and diagonally to the upper-right of the base view. When all three points have been specified, right-click and choose **Create**. Inventor takes a moment and produces three projected views. The drawing's top view and side view are known as orthographic projections and the last view is isometric.

Figure 17-2.
The correct orientation of the base view.

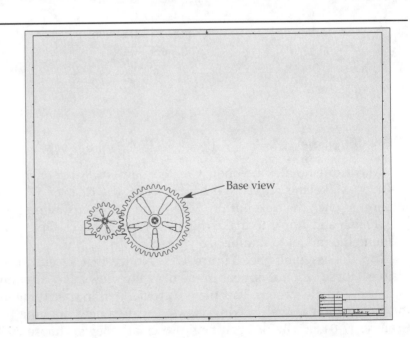

Base view

Figure 17-3.
Creating projected
views.

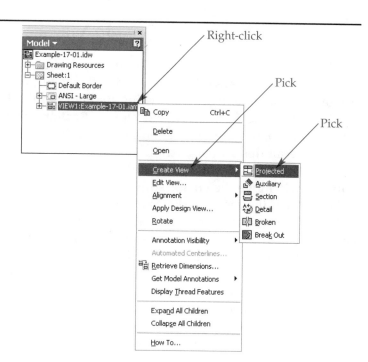

The gears in the front view appear to be missing some geometry. There are some fillets that are not shown. Edit the view by double-clicking on it and picking the **Options** tab in the **DrawingView** dialog box. Place a check in the box labeled **Tangent Edges**. Pick **OK** and observe the effect on the gears in the front view.

Notice in the **Browser** that the views are not referred to as Front or Top, but instead have numbers assigned to them. Furthermore, dependent views are sublisted underneath their parent views. In **Figure 17-4**, View 2 and View 3 are dependent on View 1 (the base view).

All of the views currently inherit their rendering style from the front view, upon which they are based. If you change the style for the front view, all others should follow suit. Try it now—change the style of the front view to **Shaded** (yellow cylinder). After you review the drawing, pick **Undo** to change the front view style to **Hidden Line**. Now double-click on the isometric view and change its style to use the **Shaded** method.

Figure 17-4.
View names and the
layout of dependent
views.

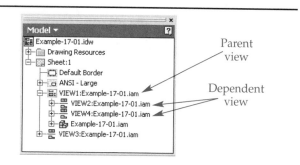

The side view that was created does not convey many details of our assembly. A *section view* would reveal the interior details by cutting away some of the assembly. Delete the side view by right-clicking its name in the **Browser** and picking **Delete**. Right-click on the front view and choose **Section** from the **Create View** cascading menu. To properly describe the large gear, an *aligned section view* needs to be used. You may recall this type of section view and how to create it from Chapter 8. Hover over the center of the large gear and "acquire" the center point—do not click at this point. Slowly move the cursor straight down and ensure that a dotted line is being projected from the center of the gear. Pick a point outside the gear and in-line with its center. Now, bring the cursor back to the center and click that point, and then slowly move the cursor toward the upper-right. While doing this, try to "acquire" the center of the web cutout's smaller arc(s). Refer to **Figure 17-5**. Once the point shows up green, proceed outside of the gear and pick a point. Finally, right-click and choose **Continue**. Drag the section view to the right of the front view.

If you zoom in on Section A-A, you will notice that all of the parts in the assembly are sectioned. The drafting convention for sectioned assemblies specifies that shafts, fasteners, and most standard components are not sectioned. We need to tell Inventor not to section these. Expand the plus signs in the **Browser** for the section view and you should be able to see a listing of all of the parts that make up this assembly.

It is difficult to link the part and its corresponding label in the **Browser**. For example, there are two shafts, but you want the shaft that is being used to support the Large_Gear. If you hover over the names in the **Browser,** you will notice that the corresponding part is highlighted red in the drawing view. If you left-click on one of them, the highlight color changes from red to green. Zoom in on Section A-A for the next step. Right-click on the second instance of the shaft part and uncheck **Section**. There will be a pause as Inventor updates the section view to reflect the change. Repeat these steps to remove the section of the second instance of the hex nut and washer. Now, Section A-A is a view that follows the drafting convention.

If you would like to remove the listing of parts in the **Browser**, simply right-click on View1 in the **Browser** and choose **Collapse All Children**. Save your work before continuing on.

Figure 17-5.
Creating an aligned
section view.

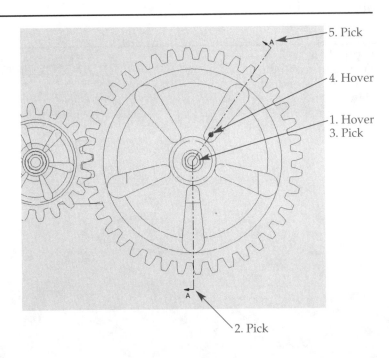

5. Pick

4. Hover

1. Hover
3. Pick

2. Pick

❑ Make sure no files are open. Create a new *single user project* and name it Practice-17-01.ipj. Make the *project (workspace) folder* the Chapter 17\Practices\Practice-17-01 folder and click the **Next** button. Make sure that there are no libraries selected. Then pick **Finish** and **Apply** to make this the current project.

❑ Open Practice-17-01.iam and create a D-size drawing as shown below. All views are half scale and the top and side views are sections. Refer to the figure as you practice creating section views for this assembly. Centerlines and dimensions are not required in this practice and will be covered in the next section. When you are finished, save the drawing as Practice-17-01.idw.

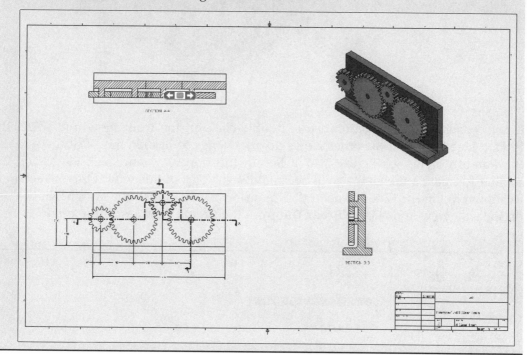

Break Out Views

Inventor has the capability of removing a portion of an assembly and revealing what is underneath. This could be used to remove part of an outer cabinet, for example, to reveal the machinery inside. The first step in creating a *break out view* is to create a sketch inside the view. The sketch determines the boundary of the break out. **Figure 17-6** is the final projected view you will create in this section.

Make sure no files are open. Create a new *single user project* and name it Example-17-02.ipj. Make the *project (workspace) folder* the Chapter 17\Examples\Example-17-02 folder and click the **Next** button. Make sure that there are no libraries selected. Then pick **Finish** and **Apply** to make this the current project. Open Example-17-02.iam and create a new drawing file. Save the drawing as Example-17-02.idw. Change the sheet to a D-size and select the **Base View** tool. The scale is .25 and the orientation is front. Scale is typically in the format 1/4 or 1:4, but Inventor only accepts scale in decimal format. Pick a point in the lower-left corner of the drawing to place the view. You will see an empty square because the outer sheet metal cabinet is hiding the internal parts. Create a top view above and a side view to the right of the base view.

Figure 17-6.
The final projected view with a break out.

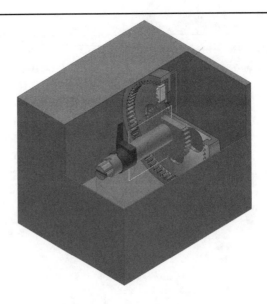

In order to see the parts inside, double-click on the front view and select the **Hidden Line** button. If the other views do not change to hidden line, edit them to use the **Parent's** style. See **Figure 17-7**. Click in the side view and pick **Sketch** on the toolbar to create a sketch that will be considered a part of this view. Draw a rectangle as shown in **Figure 17-8**. No need to dimension it—just finish the sketch. Right-click on the side view and choose **Break Out** from the **Create View** cascading menu.

Figure 17-7.
Hidden lines are shown.

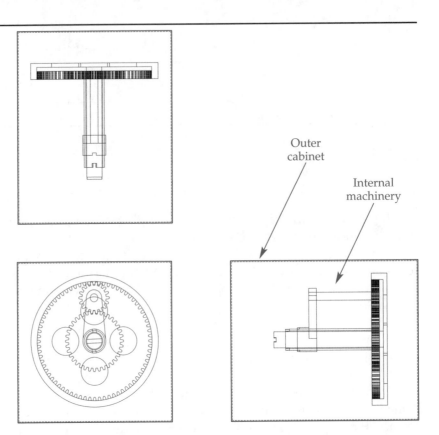

Outer cabinet

Internal machinery

Figure 17-8.
The rectangle defines the break out.

Rectangle that will define the break out

Sketch is part of the view

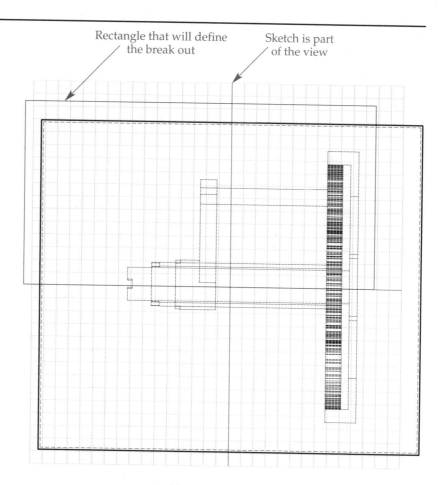

In the **Break Out View** dialog box, **Figure 17-9**, **Profile** will not be selected because Inventor has already found the sketch profile. In order to create a break out view, you must specify the depth of the break out. Make sure that the method of specifying the depth is set to **From Point** and the **Show Hidden Edges** button is not selected. With the **Depth** button depressed, change the depth value to 10.00. Now pick the start point of the break out view in the top view. In other words, from where do you want Inventor to measure the 10.00"? In the top view, pick the lower-right corner of the rectangle representing the sheet metal cabinet. Pick **OK** to exit the dialog box and the break out is created in the right side view.

The break out view does not convey much information. We need to create an *isometric view* projected from the break out view. Right-click on the view and choose **Create View**, then **Projected** and pick a point near the upper left, then **Create** the isometric view. Change the view style to **Shaded** and the drawing should look like **Figure 17-10.**

Figure 17-9.
The **Break Out View** dialog box with the **Depth** button selected.

Unselected when only one profile is available

Selected

Unselected

Change to 10.00

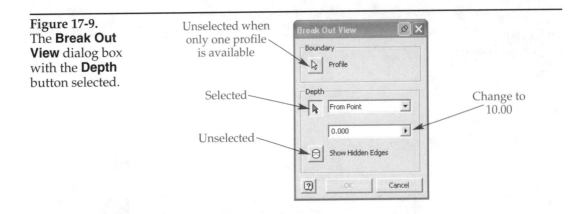

Figure 17-10.
The completed views with a shaded isometric break out view.

Projected from the break out view

Break out view

Break out starts here and goes left 10.00"

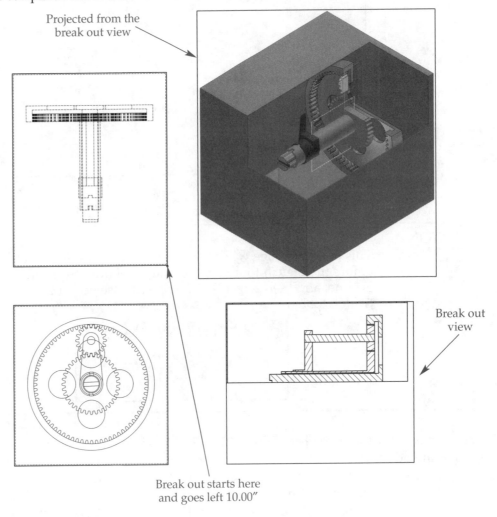

This technically is not a break out view—only an isometric view of the break out view, which is, of course, the side view. Always remember, the break out view is generated *into* the view perpendicular to the "paper."

There are other methods of specifying the depth of the cut:

- **To Sketch.** Uses an existing sketch in another view to specify depth.
- **To Hole.** Uses an existing hole feature in the assembly or part to specify depth.
- **Through.** Break out cuts all the way through the assembly or part.

As you learned in *Chapter 9*, you can apply various types of annotation to the 2D drawings. All of the annotation that you could apply to a part drawing is available for an assembly drawing. We will reuse Example-17-01.idw for the following instruction. Open this file now.

Centerlines

Adding centerlines to an assembly drawing is the same as adding them to a part drawing. Using the skills that you learned in the beginning of *Chapter 9*, apply centerlines to Example-17-01.idw. In the top view, apply center marks to the socket head cap screws. Also in the top view, apply centerline bisectors to the two shafts. After applying the centerline bisectors, you will have to drag their endpoints to make them longer as shown in **Figure 17-11**. In the front view, apply center marks to the two gears and lengthen these as well.

Dimensions

Dimensioning an assembly is similar to dimensioning a part. After picking the **Dimensioning** tool, select the part edges or centerlines. Once the proper geometry is selected, the dimension can be placed on the drawing. Refer to **Figure 17-12** as you place the dimensions on the top view. On the front view, place a horizontal dimension between the two gears. This documents the distance between the two gear shafts. Use the endpoints of the center marks you placed previously. On the section view, place the dimensions shown in **Figure 17-13**.

After placing a dimension, you may need to change its numeric value or alter the appearance of the dimension text. There are several techniques available. Choose the appropriate one.

To add text to the default dimension text

At times you will want to add text to a dimension, such as TYP, REF, or 6x. Right-click on the dimension text and choose **Text...** from the pop-up menu. In the dialog box, the default numeric value is **<<>>**. You cannot delete it or edit it—you can only add to it by placing the cursor at the end of the symbol and typing in the necessary text.

Figure 17-11.
Select and drag the centerline ends to make them longer.

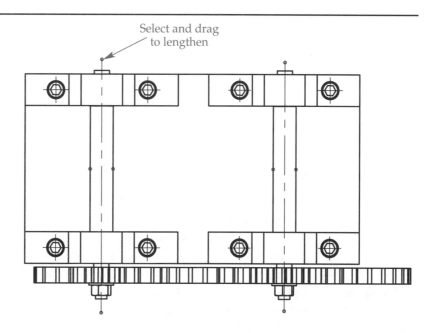

Select and drag to lengthen

Figure 17-12.
The top view with dimensions.

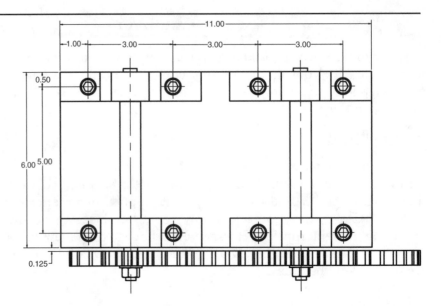

Figure 17-13.
The section view with dimensions.

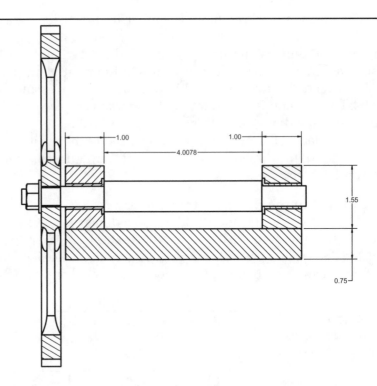

Do this for the two dimensions in the top view shown in **Figure 17-14**. To add TYP to the 1.00 dimension, you will have to place the cursor at the end of the **<<>>** symbol and press [Enter] to start a new line. For the 0.50 dimension, just start typing after the symbol. The TYP designation signifies that those dimension values are typical for like features in the assembly. In this case, the distance from the corner of the base plate to the location of the hole is typical for all similar holes.

To add tolerancing information

Double-click on the dimension and choose the type of tolerance desired from the dialog box. Notice that as you pick the options, they are applied immediately to the selected dimension. Change the 6.00 dimension between the gears in the front view to that shown in **Figure 17-15**. Use the **Precision** text box for a handy way of changing the decimal places used in the numeric value. The **Nominal** text box in the upper-middle of the dialog box is for typing in a numeric value that is different than the model value.

Figure 17-14.
Adding text to dimensions.

Cannot edit or delete

Text added to next line

Text added to next line

Text added to same line

Click and drag end of centerline to change its length

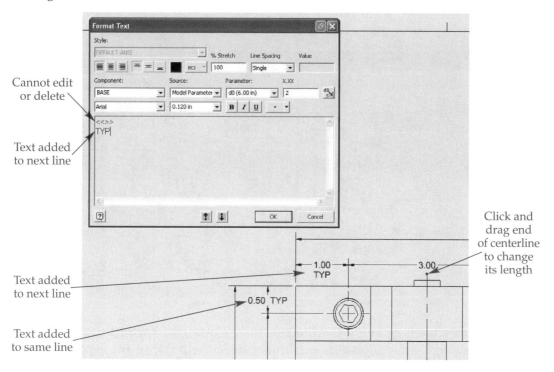

Figure 17-15.
Adding a tolerance to a dimension.

Pick to change dimension

The **Model Value** cannot be edited

Pick to change decimal places

Choose the tolerance type here

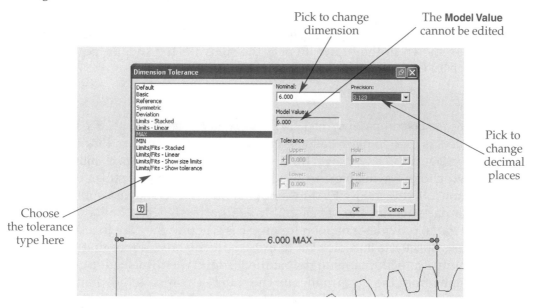

To completely replace the dimension text

Add a thickness dimension to the large gear shown in the top view, **Figure 17-16**. Right-click on the dimension text and choose **Hide Value**. The dimension's text will be replaced by a symbol **<TEXT>**. In order to enter your own text, right-click on the dimension and choose **Text...** from the pop-up menu. In the ensuing dialog box, strike out the <TEXT> entry and enter SEE VENDOR PRINT. Pick **OK** to exit the dialog box when finished typing. This dimension now has no value, so double-clicking on it produces no result. However, right-clicking on it and unchecking **Hide Value** will make the default numeric value reappear along with your text appended.

Figure 17-16.
Replacing the
dimension with
a note.

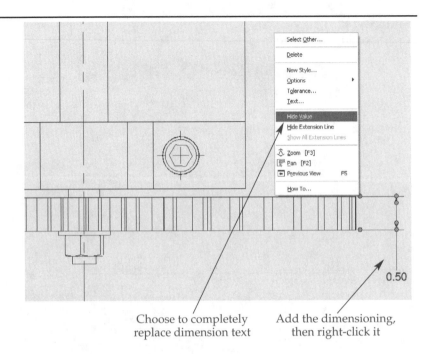

Choose to completely
replace dimension text

Add the dimensioning,
then right-click it

Assembly drawings are not bidirectionally associative. This means, unlike part drawings, if you change the value of a dimension it will not result in the assembly model changing.

Leader Text

Leader text is added to an assembly drawing in the same manner as it is added to a part drawing. See **Figure 17-17**. Select the **Leader Text** tool, pick on a tooth of the large gear, pick on the drawing to locate the text, right-click, and pick **Continue**. Type in the text, as shown in the figure. Repeat the steps for the smaller gear. This time type: 20T PINION GEAR.

Balloons

Locate the **Balloon** drop-down menu in the **Drawing Annotation Panel**. See **Figure 17-18**. The **Balloon** tool is used to place balloons one-at-a-time on each part. Pick this tool and select any component in any view. This accesses the **Parts List-Item Numbering** dialog box. The two choices available in the **Level** area determine if subassemblies will receive balloons. If you choose **First-Level Components**, each subassembly is given a balloon—the parts within the subassembly are not assigned a balloon. The other choice, **Only Parts**, provides a balloon for all parts whether they are in a subassembly or not. Choose **Only Parts** and pick **OK** to exit the dialog box. Notice that the arrow is started. Continue to pick points to define the leader. When finished, right-click and choose **Continue** from the pop-up menu. The number will be applied automatically. Press [Esc] when done placing balloons. Undo any balloons that you have placed.

The **Auto Balloon** tool is also accessed from the **Balloon** drop-down menu. It is used to place all the balloons necessary by picking a view. Pick the **Balloon All** tool to access the **Auto Balloon** dialog box. See **Figure 17-19**. Pick the **Select View Set** button in the **Selection** area and then pick the isometric view. This will then enable the **Add / Remove Components** button. This button allows you to select the components that will be included in this auto-ballooning operation. Use the **Window** selection method to pick everything in the isometric view.

In the **Item Numbering** area there are the **Level Components** and **Only Parts** options. These are the same options just covered. Select the **Only Parts** option.

Figure 17-17.
Adding leader text.

Leader text tool

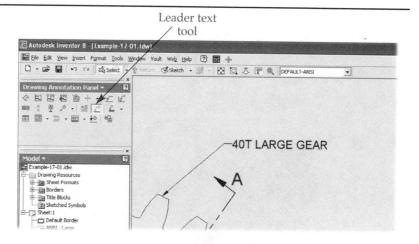

Figure 17-18.
Balloon drop-down menu.

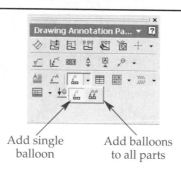

Add single balloon

Add balloons to all parts

Figure 17-19.
Dialog box for adding balloons automatically.

Pick to place balloons for all parts except those within a subassembly

Pick to place balloons for all parts

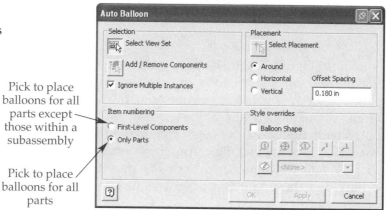

The **Placement** area of the dialog box provides more control over how the balloons are applied to the view. They can be placed all around the view with the **Around** option, or lined up horizontally or vertically by selecting **Horizontal** or **Vertical**. Pick the **Select Placement** button and try each selection. Move your cursor around the isometric view and a visual representation of the balloons and their placement will be presented. The **Around** selection is the most dramatic. As you move the cursor, the balloons rearrange themselves dynamically. The numerical value for **Offset Spacing** controls the distance between circles' edges. Use the **Select Placement** tool to place a row of horizontal balloons and then change offset spacing value. When finished, make sure **Around** is selected and pick a point to the right of the isometric view.

The **Style override** area is where you may elect to override the default balloon style defined in the **Styles and Standards Editor**. This is rarely done at this time, because it can be changed using the **Styles and Standards Editor**. At this point, choose **Apply** and the balloons will be placed. Pick **Cancel** to exit the dialog box. Your view

Figure 17-20.
View after the **Balloon All** tool was used.

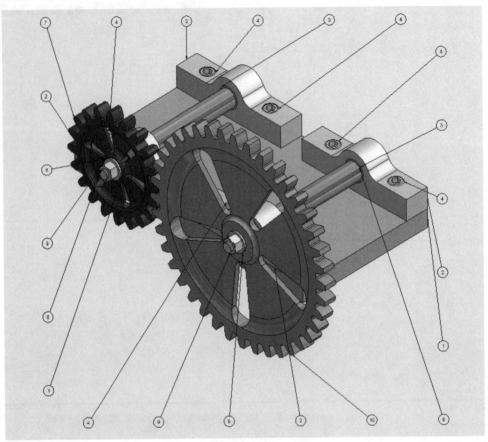

should be similar to **Figure 17-20.** The parts are numbered in the same order that the assembly file was created. This may not be the correct order for the drawing. Reordering the balloon numbers will be covered in the parts list section of this chapter.

As previously mentioned, the appearance of the balloons is controlled by the **Styles and Standards Editor**. Before continuing, zoom into the isometric view so the balloons are visible. Pick **Styles Editor** from the **Format** pull-down, or right-click on a balloon and pick **Edit Balloons Style** from the pop-up menu. Find **Balloon** in the listing. The plus sign should be expanded and Balloon (ANSI) selected. If not, do so now.

In the **Sub-styles** area, there are three sub-styles that control the appearance of the balloons. See **Figure 17-21.** They are, in turn, controlled by **Styles**. Pick on the pencil point graphic next to **Leader Style**. This will take you to the page controlling that feature. For example, you can change the arrowhead type. Click on the **Back** button along the top to return to the **Balloon Styles** page. If at any point you pick **Save** to write the changes to the style, the balloons will update on the drawing screen. Now, change the **Text Style** to Label Text (ANSI) and pick the **Save** button. The balloons will resize to accommodate the size of the text. Below the **Sub-styles** area is the **Offset Spacing** setting. This controls the distance between balloon edges when the **Auto Balloon** tool is used. In the **Balloon Formatting** area, the shape of the balloon and what property is to be shown in the balloon can be changed. It is usually Item, but it could be changed. The **Symbol Size** area controls the balloon size to accommodate any changes in text size. Uncheck the **Scale to Text Height** checkbox and the diameter of the balloon can be changed using the **Size** text box. The **Stretch Balloon to Text** checkbox allows the balloon shape to become an oval in order to accommodate long text items. The **Comments** box is used to list the ANSI standard the balloons are based on.

Figure 17-21.
The appearance of the balloons is controlled by the **Styles and Standards Editor**.

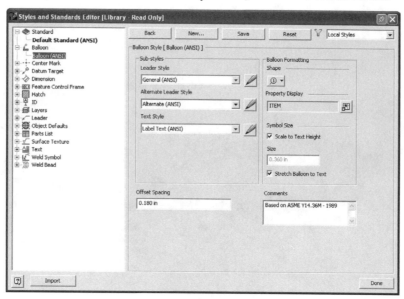

Now the balloons must be moved to create a legible drawing. The balloons are relocated by clicking on each one and dragging it to a new location. Move all balloons closer to the parts they identify. If you want to line up the balloons, either horizontally or vertically, drag a balloon over its neighbor until a dotted line appears. Keeping the dotted line visible, move the balloon in a straight direction. Release the left mouse button to place the balloon. If a balloon is placed on top of a leader, the leader is erased below the balloon. For drawing clarity, avoid overlapping balloons and leaders.

The drawing is looking good so far, but there are too many balloons. Ideally, there should be only one or two balloons per item number. To accomplish this, some of the balloons will have multiple leaders. Refer to **Figure 17-22** and look at item number 4 with four leaders pointing to four fasteners. To create multiple leaders, delete all item 4 balloons except for one—select the balloons and hit the **Delete** key. Right-click on the remaining balloon and choose **Add Vertex/Leader** from the pop-up menu. Pick another instance of the fastener for the origin of the leader. Lead it to the balloon and pick the center of the balloon. Continue doing this for all four fasteners visible in the isometric view.

Continue to refer to **Figure 17-22** as you clean up the balloons. Delete as many as necessary to match the figure. It is not necessary to assign a balloon to every part visible in the view. For parts that are grouped together logically, such as nuts, bolts, and washers, the balloons are often shown together with only one leader pointing to the group of parts. Notice that this has been done for items 8 and 9 in the figure. Right-click on balloon 8 and choose **Attach Balloon** from the pop-up menu. Pick the part that you want to group with balloon 8, in this case the hex nut—item 9. Repeat these steps for both sets of the washer and nut.

At times it may be necessary to use the **Single Balloon** tool to select just one part and balloon it. Check the resulting item number because you can inadvertently pick the wrong part. In these situations you may need to make the adjacent part invisible, as long as Show Contents is enabled.

To edit a balloon and override the item number, you could right-click the balloon, choose **Edit Balloon** from the pop-up menu, and type a different number. If you type the number in the **Item** cell, all balloons identifying this part and their reference in the parts list change to the new item number. If you type the number in the **Override** cell,

Figure 17-22.
Balloons have been relocated and modified.

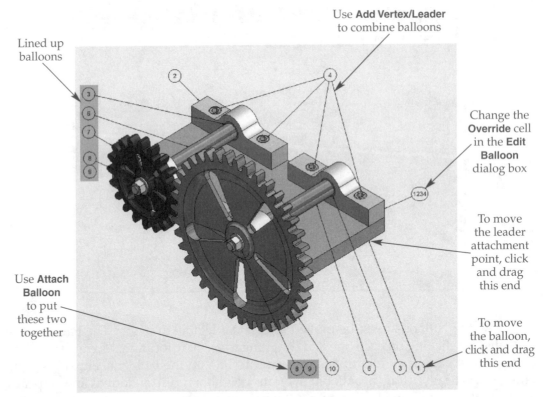

Lined up balloons

Use **Add Vertex/Leader** to combine balloons

Change the **Override** cell in the **Edit Balloon** dialog box

Use **Attach Balloon** to put these two together

To move the leader attachment point, click and drag this end

To move the balloon, click and drag this end

only the selected balloon is changed. Right-click on one of the item 2 balloons, choose **Edit Balloon...**, and change the number in the **Override** cell to 1234. What happened to the balloon circle? It expanded into an oval to accommodate the longer number. Also notice that the other item 2 balloon was unchanged. The **Edit Balloon** dialog box has an option to change the balloon type to a nonstandard one. Save your work before continuing on to the next section where a parts list for this assembly is created.

Parts Lists

A *parts list* is a table of information used to identify and describe the parts required for an assembly. See **Figure 17-23**. Do not confuse the parts list with the bill of materials. The bill of materials is the iProperties database of parts in the assembly, the parts list is the annotation of the information kept in the database. You will continue to use Example-17-01.idw for this section. Pick the **Parts List** tool on the **Drawing Annotation** tool panel. You are asked to pick a drawing view. This is because of the possibility of the drawing having views generated from different assemblies. This is a nice feature that will be covered later in the chapter. Select the **Isometric View**, although any of the views would do.

You do not have to list every part used in the assembly. The **Range** section of the dialog box allows you to specify which items will be included in the parts list. If **Items** is selected, the item numbers must be entered in the adjacent window. These can be entered in any of the following formats:

- **1,2,5.** Specific items are listed
- **1-4.** A range of items is listed
- **1-5,8,10-13.** A combination of specific and range of items is listed

Table Wrapping specifies the number of sections (columns) to divide the parts list into. More than one section may be necessary if the parts list is too tall for the drawing. Specifying **Left** or **Right** instructs the parts list to continue in the chosen

Figure 17-23.
Edit Parts List
dialog box.

Click and drag separator to change column width

Right-click column header to view options

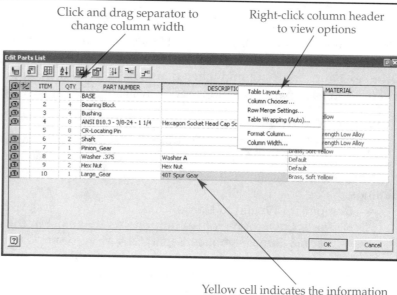

Yellow cell indicates the information does not match information within part

direction if additional rows require another section. Pick **OK** to exit the **Parts List-Item Numbering** dialog box. Pick a point on the drawing to locate the lower-right corner of the parts list. If you slowly move the cursor near the border or title block, the parts list will snap into position.

The parts list may be moved, but only by selecting the green dots at the corners. To widen the columns in the parts list, double-click on the parts list to make the **Edit Parts List** dialog box appear. Hover over the column partition between **PART NUMBER** and **DESCRIPTION** until the visual indicator appears—click and drag the partition. This is similar to the method used in popular spreadsheet programs. Alternatively, right-click on the **PART NUMBER** column header, select **Column Width**, and type a larger value.

Many of the parts do not have descriptions. Parts lists are actually populated, in part, by the part file's iProperties. Open each part file and edit the iProperties to add descriptions. Save each IPT file. When you are done, review the assembly's parts list.

- **Base.ipt.** 11 × 6 3/4″ Plate
- **Bearing Block.ipt.** 5 × 1 x 7/8″ Bearing Block
- **Bushing.ipt.** 1/2″ Bronze Bearing
- **CR Locating Pin.ipt.** 1/4″ dia. Pin
- **Shaft.ipt.** 7 1/4″ × 3/4″ dia. Round Bar
- **Pinion_Gear.ipt.** 20 Tooth Pinion Gear
- **Large_Gear.ipt.** 40 Tooth Spur Gear

Only the top-level parts' descriptions have been updated. The parts contained within the Subassembly Bearing Assembly are not updated. You will have to delete the parts list and place a new one to see the new descriptions. Do that now to update the parts list. To edit cells in the parts list, double-click on the parts list itself. Certain cells cannot be edited, such as **Quantity**. When information is added or changed with this method, the cell turns yellow. This indicates that the information does not match the part file.

The **Edit Parts List** dialog box offers many ways to control the appearance of the parts list.

Compare

This is used to compare the **Parts List Item** values with the parts' current iProperties values. If the values differ, the cells are highlighted in yellow.

Column chooser

You may want to add more columns than are currently in the parts list. Click on the **Column Chooser** icon, and find **Material** from the **Parts List Column Chooser** dialog box. Pick the **Add->** button to include it in the list in the right-hand pane. Press **OK** to exit and a new column labeled **Material** should have appeared listing the material used in each part. This information was specified earlier during part creation.

Row merge settings

By default, same parts are grouped together. To merge rows based on other criteria, such as material or description, select the **Row Settings Merge** button. Select the **Merge Similar Components** radio button. The drop-down lists in the **Row Merge Settings** dialog box are:

- **Component Type(s) to Merge.**
  - **Parts Only:** Merge only components that exist in the top assembly.
  - **Parts & Subassemblies Separately**: Merge components that exist in the top assembly, and then separately merge subassemblies.
  - **Parts & Subassemblies Together:** Treat parts and subassemblies the same for merging.
- **First, Second, and Third Key.** Specifies a property to use for identifying a discrete part when two or more parts have the same property.

Sort

At times you may want to change the sorted display of the parts list. Typically, parts lists are sorted by the item number, but the capability exists to re-sort under different criteria. Sort the current parts list under **Quantity**, **Part Number**, then **Description**. The **Ascending** and **Descending** buttons control the order of the sorting, **A-Z** or **Z-A**.

Export

At times you may need to bring the parts list data into an external database or word processor. Several formats are supported. Export the parts list to a *tab-delimited* file named PartsList.txt.

Table layout

Heading and Table Settings and **Row Height** are adjustable within this dialog. The default heading for the parts list is **Parts List**. This heading is seen on the drawing at the top of the parts list table. You may elect to specify a different name with this capability.

Renumber

If you had sorted by quantity, as instructed in the **Sort** description (if not, do it now), the item numbers are not in order. Click on the **Renumber** tool and notice that all of the items have new numbers. Exit the dialog box and notice the balloons are also renumbered. Since the item numbers have changed, the only way to revert back to the item sorted display is to use **Undo**.

Add custom parts below

It is often necessary to include an item in the parts list that is not actually modeled in the assembly, such as paint or lubricant. The **Add Custom Parts** tool is used to make a new entry for a custom part below the currently selected row. Select the last row and pick this tool's icon. In the blank space provided, type Lubricant under **Part Number** and Triflow lubricant under **Description**. Exit the dialog box to see the results on the parts list.

Custom part above

Similar to the previous option, the **Add Custom Parts** tool creates a new entry for a custom part above the currently selected row.

Figure 17-24.
Many standards related to the parts list are controlled with the **Styles and Standards Editor**.

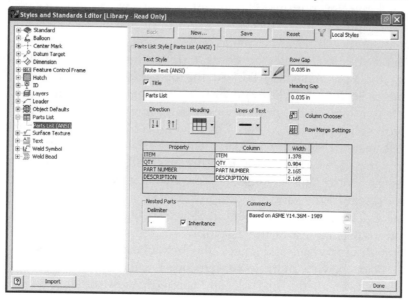

Changing standards

Controlling standards related to the parts list is done with the **Styles and Standards Editor**. This is accessed by picking **Styles Editor** from the **Format** pull-down, or right-clicking on the parts list and picking **Edit Parts List Style** from the pop-up menu. See **Figure 17-24**. Find **Parts List** in the listing, pick the plus sign, and select Parts List (ANSI). There are several control options only available here, such as the settings for nested parts in subassemblies.

Printing the Drawing

Many of the details of printing a drawing are controlled by Inventor. For example, lineweights are specified by the drafting standards and rarely have to be changed. Printing an assembly drawing is very similar to the procedure used to print a part drawing. Choose the output device, paper size, scale (if any) and print the drawing. For the most part, what you see on the graphics screen is what you will receive on the paper. Try it by printing Example-17-02.idw.

Creating Drawing Views Using Design View Representations

You may recall that Inventor saves design views in each assembly. You can save many design views within this file. These design views can also be used to make drawing views. If the design view is modified and saved, the drawing view automatically updates to show the current design view.

Make sure no files are open. Create a new *single user project* and name it Example-17-03.ipj. Make the *project (workspace) folder* the Chapter 17\Examples\Example-17-03 folder and click the Next button. Make sure that there are no libraries selected. Then pick **Finish** and **Apply** to make this the current project. Open Example-17-03.iam and examine the assembly. It uses the *contact solver*, a feature that stops part movement when contact is made. Enable the contact solver by going to **Tools, Application Options**, and then the **Assembly** tab. At the bottom of the dialog box, check the **Activate Contact Solver** box. Click and drag on the **Input Link** to animate the mechanism. Now take a look at the design views that have been saved. Click on each to familiarize yourself with their contents. You will use some of these design views: NE Isometric, SE Isometric, Locked down position, Open position, No fasteners, and Bottom view to create drawing views. The following three default design views: system.nothing visible, system.all visible, and User.default cannot be used to establish associativity between the drawing views and the design view.

Start a new drawing using the default C-size sheet. Create a base view generated from the bottom orientation of the assembly, set the scale to 1.0, select the **Locked down position** design view, and check the box marked **Associative** as shown in **Figure 17-25**. Place the view in the lower-left corner of the sheet. From that view, project three more views to create the top, side, and isometric views as shown in **Figure 17-26**.

All of the drawing views are associative to the linked design view. Put the drawing and the assembly on the screen at the same time by picking **Arrange All** from the **Window** pull-down menu. See **Figure 17-27**. Within the assembly, drag the red input link to a new position. Now click on the drawing window to make it active. The views in the drawing update to reflect the changes in the assembly file.

PROFESSIONAL TIP

Only public design views are associative If you are experiencing problems getting drawing views to update, make sure that each is associated to the design view. To do so, double-click on each and ensure that the **Associative** box is checked.

It is possible to change the design view associated with an existing drawing view, but with mixed results. Double-click on the top view and choose a different design view from the dialog box. Any one that you choose will not affect the drawing view's orientation, but will affect the visibility of components in the drawing view. Try it

---

Figure 17-25.
Select locked down position design view and check the box marked **Associative**.

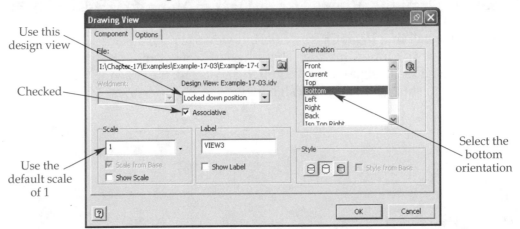

Figure 17-26.
Drawing with
projected views.

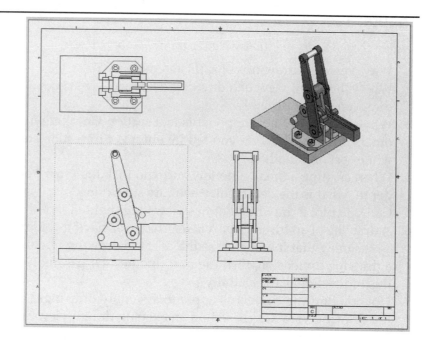

Figure 17-27.
Viewing two files
at once using
Arrange All.

3. Make this
window active
to view updates

2. Move input
link

1. Make this
window active

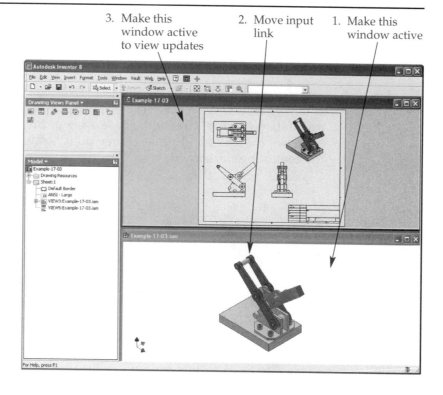

with the design view No fasteners. The fasteners disappear in the drawing view, so far
so good. Now change the design view to Bottom View. Notice that there is no change.
Since the orientation of the drawing view is changed, other drawing views projected
from it might be affected.

Summary

You have seen how creating assembly drawings is very similar to creating drawings for piece parts. These drawings are fully parametric and associated with the
assemblies they represent—if the assembly, or any of its parts, changes, the drawing
will update to reflect the changes. Assembly drawings task computers the hardest of
all of Inventor's file types, including assembly files.

Chapter Test

Answer the following questions on a separate sheet of paper.

1. Why create 2D drawings of a 3D assembly?
2. Why is the front view of a drawing *not* always created from the front viewing direction in the assembly?
3. Why are some of the components in a section view shown "sectioned" and others not? And how do you tell Inventor *not* to section some components while sectioning others?
4. When creating a break out view and you pick the **From Point** and enter the depth, what is the depth entry actually specifying?
5. Can you plot a drawing that has views highlighted in red in the **Browser**?
6. A drawing of an Inventor part is referred to as having "bidirectional associativity" — meaning that if you change the part the drawing updates and if you change a drawing dimension, then the part updates. Do assembly drawings possess this bidirectional associativity?
7. How do drafting standards apply to assembly drawings?
8. If a drawing view was based on an existing design view and that design view was changed and saved, does the drawing view update to reflect the revised design view?
9. How do you add a column to the parts list?

Chapter Exercises

Exercise 17-1. *Single Cylinder Engine.* Using Exercise-17-01.iam create the drawing as shown in the figure. Use a C-size sheet, a scale of .33, and an orientation of front. Create the base view in the lower left-hand corner of the sheet and project three other drawing views. Apply the centerlines as shown.

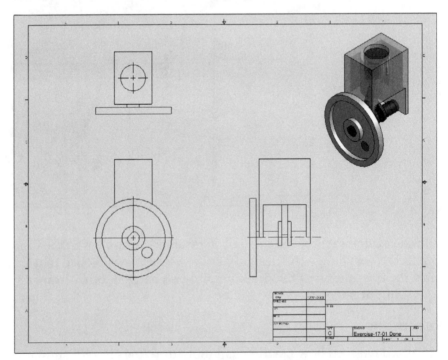

Exercise 17-2. *Adaptive Assembly*. Using Exercise-17-02.iam create the drawing as shown in the figure. Use a C-size sheet, a scale of 1.0 and an orientation of front. Create the base view in the lower left-hand corner of the sheet and project three other drawing views. The isometric view uses a scale of .75. Apply the centerlines and dimensions as shown.

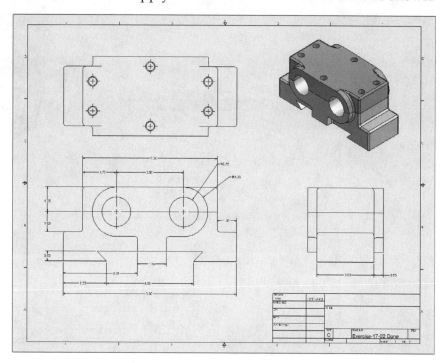

Exercise 17-3. *Sliding Clamp*. Exercise-17-03.iam create the drawing views as shown in the figure. Use a C-size sheet, a scale of .75, and an orientation of right. Create the base view in the lower left-hand corner of the sheet and project three other drawing views.

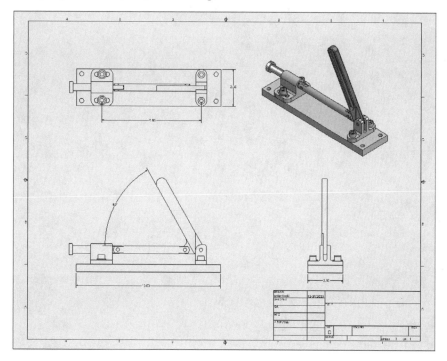

Exercise 17-4. *Scissors Lift*. Using Exercise-17-04.iam create the drawing views as shown in the figure. Use a C-size sheet, a scale of .25, and an orientation of left. Create the base view in the lower left-hand corner of the sheet and project three other drawing views. Apply balloons to the isometric view and place a parts list next to the title block.

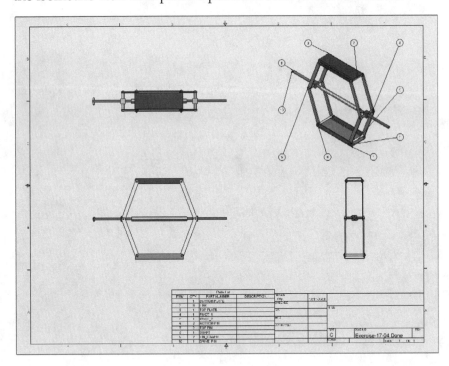

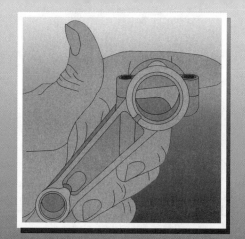

CHAPTER 18

Presentation Files

Objectives

After completing this chapter, you will be able to:

✳ Create automatic and manual explosions.
✳ Add linear and rotational tweaks to create and modify exploded views.
✳ Understand how trails should appear in drawings and animations.
✳ Modify animations to make the components move at the correct time and speed.
✳ Use design views to create multiple exploded views.
✳ Set different camera angles to emphasize specific areas of an assembly.
✳ Rearrange the sequences of an animation.
✳ Apply colors, styles, textures, and lights to create a refined presentation.

User's Files

*The following is a list of files that you will need to work through this chapter. These files can be found on the **User's Files** CD included with this text.*

Examples	Exercises
Example-18-01.ipj	Slider Clamp.iam
Drawer-box-top.iam	Explosion-Ex-18-01.avi
Drawer-box.iam	Explosion-Ex-18-03.avi
Drawer.iam	EX03-Clamp-Linkage.iam
Example-18-03.ipn	Explosion-Ex-18-04.avi
Example-18-04.ipn	EX04-Latch-Assembly.iam
Example-18-05.ipt	Explosion-Ex-18-05.avi
Example-18-05.ipn	EX05-VALET.iam
Example-18-06.ipt	
Example-18-07.ipt	
Example-18-08.ipt	

Presentation Files

Exploded views of an assembly are created in a separate file called a *presentation file.* A presentation file has an IPN extension and is related to one specific assembly file. A single presentation file can contain many exploded views of the same assembly. Each exploded view is based on a saved design view. Once exploded, the view can be animated to reassemble the components. Audio-video files with an AVI extension can be created from the animation and shown on computers without using Inventor. A wooden valet will be used as an example to study the features, principles, and steps in creating presentation files. The valet is designed to be built with simple hand tools, such as a saw, drill, and screwdriver. It has four subassemblies: the drawer, drawer front, drawer box, and drawer box top. The final assembly of all the components is shown in **Figure 18-1**.

Figure 18-1.
The valet assembly and components.

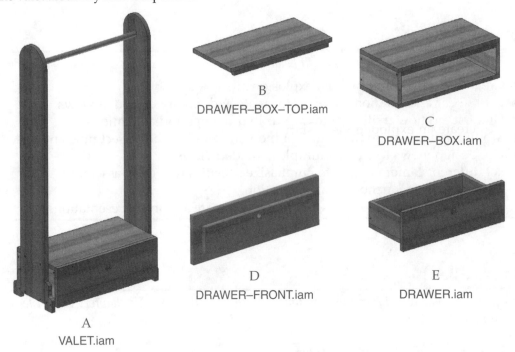

B
DRAWER–BOX–TOP.iam

C
DRAWER–BOX.iam

D
DRAWER–FRONT.iam

E
DRAWER.iam

A
VALET.iam

Figure 18-2.
Selecting design view representation.

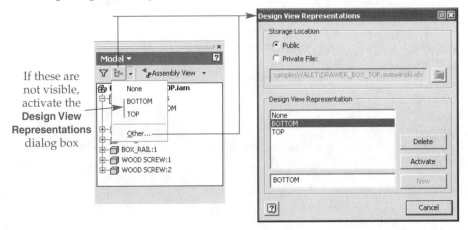

If these are not visible, activate the **Design View Representations** dialog box

Figure 18-3.
Choose an assembly
file and a design
view.

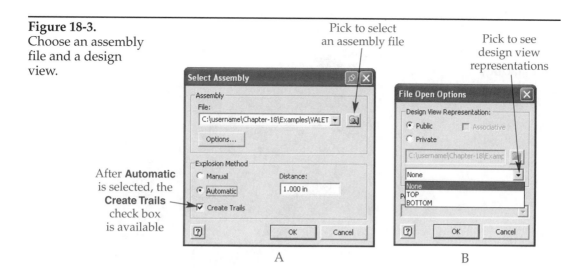

Pick to select
an assembly file

Pick to see
design view
representations

After **Automatic**
is selected, the
Create Trails
check box
is available

A

B

The first subassembly to work on is the drawer box top. Close Inventor and use Windows Explorer to find the project file Example-18-01.ipj in the Chapter-18\ Examples\VALET folder. Double-click on the IPJ file to open Inventor and Example-18-01.ipj is set as the current project. Now, open the assembly file DRAWER_BOX_TOP.iam by setting the **Files of type:** field to Assembly Files(*.iam) to make selection easier. The file opens with the BOTTOM design view representation active, but there is also a TOP view available. See **Figure 18-2.**

To create an exploded view, start a new presentation file, using the Standard.ipn template in English units. Pick the **Create View** tool in the **Presentation** panel to display the **Select Assembly** dialog box. See **Figure 18-3A**. The default assembly file is Drawer_box_top.iam since it is the only one open. Note that the assembly file does not have to be open to create a presentation file.

- Pick **Options...** and select the Bottom representation.
- The two selections for **Explosion Method** are Manual and Automatic. Select Automatic and set the distance to 2″.
- Check **Create Trails**. Trails are thin lines relating the assembled position to the exploded one.
- Pick **OK** to create the first exploded view. See **Figure 18-4**.

The automatic explosion applies only to mate constraints that involve a face or a plane, the tangent constraint, and some iMate constraints. See **Figure 18-5**. It does not apply to flush, angle, or insert constraints. In the example, the top is the grounded part so it does not move. The rail was mated face to face with the top. During the explosion, the rail moved 2″ perpendicular to the mate constraint. This is called a *tweak* in the **Browser**. The screws will not move automatically because they were put in with an insert.

Right-click on DRAWER_BOX_TOP in the **Browser** and select **Auto Explode**. See **Figure 18-6.** The distance entered in the **Auto Explode** dialog box will be added to the 2″. With 1.00 in the text box, pick **OK**. As you can see in the **Browser,** the rail now has two tweaks on it.

Adding Tweaks

To move the wood screws you have to pick the **Tweak Components** tool in the **Presentation** panel. **Tweak Components** is the only presentation tool that has a hot key, **T,** for easier selection. This tool will add translation or rotation tweaks. The **Tweak Components** tool is also used to edit existing trails. To translate, pick a direction, select one or more components, and drag-and-drop on the screen or enter a value in the current units. See **Figure 18-7.**

Figure 18-4.
An exploded view created automatically.

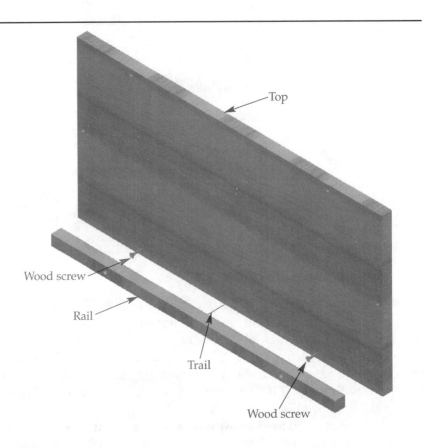

Top

Wood screw

Rail

Trail

Wood screw

Figure 18-5.
These are the only mating conditions that allow a move during an automatic explosion.

Geometry for mate	Result of auto explode
Face to face	Explodes normal to face
Face to plane	Explodes normal to face
Face to edge	Explodes normal to face
Face to point	Explodes normal to face
Face to axis	Explodes normal to face
Plane to plane	Explodes normal to plane
Plane to edge	Explodes normal to plane
Plane to point	Explodes normal to plane
Plane to axis	Explodes normal to plane
Insert (axis and face mate)	Explodes normal to face along axis
iMate—insert	Explodes normal to face along axis
Tangent	Explodes normal to planar face
iMate—tangent	Explodes normal to planar face
iMate to face	Explodes normal to face

- Pick the **Tweak Component** tool. In the **Tweak Component** dialog box, put a check mark in **Display Trails**.
- The **Direction** button and the **Z** button will both be on. Move the cursor around the assembly. When the cursor is on an edge, the Z axis aligns with that edge. When the cursor is on a face, the Z axis extends perpendicular to that face.
- Pick the face to set the direction of the Z axis, as shown in the figure.
- The **Components** button is now selected and the cursor icon has changed.

Figure 18-6.
A—Right-click on
the assembly in the
Browser. B—The
dialog box used to
adjust the tweaks.

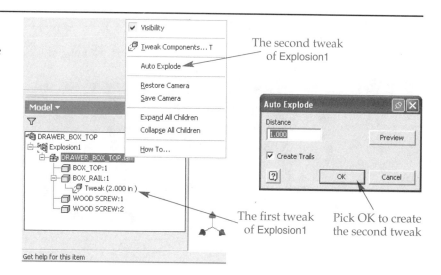

The second tweak
of Explosion1

The first tweak
of Explosion1

Pick OK to create
the second tweak

Figure 18-7.
Tweak the screws to
manually move
them.

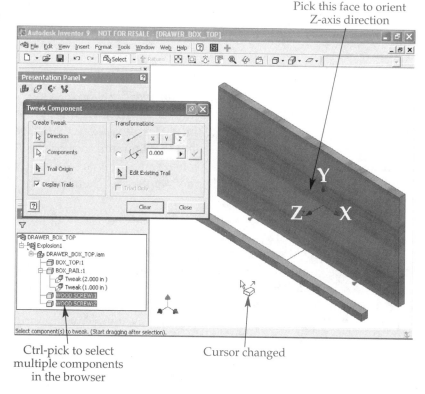

Pick this face to orient
Z-axis direction

Ctrl-pick to select
multiple components
in the browser

Cursor changed

- Pick the two wood screws, either in the assembly or in the **Browser.** Press the [Ctrl] button while picking both screws.
- Left-click and hold on the screws, and drag-and-drop them about 6″ away from the assembly. Note that the XYZ triad moves with the screws.
- Selecting **Clear** will accept the tweak and keep the dialog box open. Selecting **Close** will accept the tweak but end the command. Select **Close.**

You will practice using the **Tweak** tool and learn its other features in later examples. For now, save the file as Example-18-01.ipn. Start a new Standard.idw file. Pick the **Base View** tool and select the Example-18-01.ipn file in the drop-down list. Set the Orientation to Current and the Scale to 0.5. Click to place the view. The drawing with balloons and a parts list is shown in **Figure 18-8.**

Figure 18-8.
A drawing with an exploded view, balloons, and a parts list.

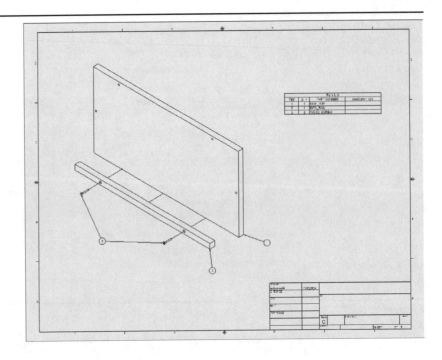

The Animate Tool and Editing Tweaks

The exploded view in this presentation file worked well in the drawing, but a flaw will be revealed when the view is animated. Animation temporarily assembles the components back together in the order they were exploded. In the presentation file (Window, Example-18-01.ipn), select the **Animate** tool in the **Presentation** panel. In the **Animation** dialog box, set the Interval to 15. *Interval* is the number of steps for each tweak. Leave the Repetitions at 1, pick **Apply** and then pick the **Play Forward** button. See **Figure 18-9**.

The speed of the screws and the rail, during the animation, is controlled by the distance each has to travel and the number of steps to complete the movement. The screws move quickly since their tweak distance is 6". The screws will move 6" and this has to be done in 15 steps, the interval setting. This means that each step will move the screws 6/15". Compare this with the distance the rail must travel during each step. The rail has two tweaks: the first is 1", therefore 1/15" per step and the second at 2/15" per step. Watch carefully as you play this and you can see the rail speed up during the second tweak.

The screws should assemble after the rail. Pick the **>>** button to expand the **Animation** dialog box and pick the **Reset** button to activate it. Pick WOOD_SCREW:1, and note that the Move Up and Move Down buttons are now active. Also note, since you picked both screws earlier for one tweak, the **Ungroup** button is active. See **Figure 18-10**.

Figure 18-9.
These are the controls to animate an explosion.

Play forward

Set to 15

Set to 1

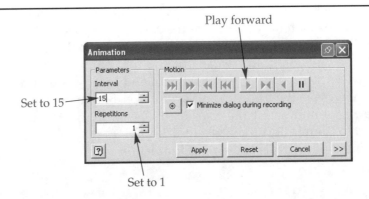

Figure 18-10.
The **Animation** dialog box is expanded to reveal the sequence of tweaks.

Expand/collapse button

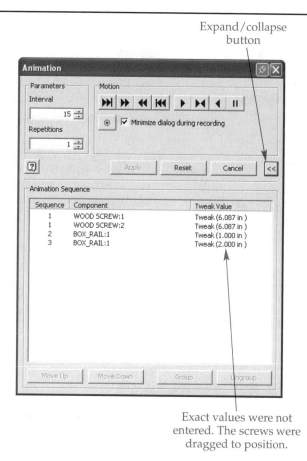

Exact values were not entered. The screws were dragged to position.

Pick the **Ungroup** button and move each of the now independent wood screws down to the bottom, one step at a time, so it looks like **Figure 18-11**. Pick **Apply** and then **Play** and the rail will assemble first and then the screws, one at a time.

Figure 18-11.
The screws have been moved to the bottom of the list.

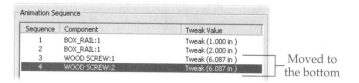

Moved to the bottom

You *cannot* edit the names or values of the tweaks in this dialog box. To change the value of an automatically created tweak, like the 2″ tweak on the rail, pick the tweak in the **Browser**. The tweak distance appears in a box at the bottom of the **Browser** panel and can be edited there. See **Figure 18-12**. Change the value for the rail to 3.0″ and hit [Enter]. Now right-click the 1-inch tweak for the rail and delete it.

The value of a manually applied tweak can be changed by picking the trail and dragging the green dot or using the **Browser** as previously shown. See **Figure 18-13A**. You can also right-click on a trail and select **Edit**. The **Tweak Component** dialog box appears and you can drag the component back and forth or enter an exact value in the Tweak Distance box. See **Figure 18-13B**.

In more complex assemblies, an automatic explosion may generate a view that will require many manual tweaks. A more efficient process is to apply all tweaks manually. This gives more control over the values, produces fewer steps in the animation, and is easier to edit. For example, close all files, open the Drawer-box.iam file, and create a new standard presentation file. Create a view of the assembly with the auto-

Figure 18-12.
The distance of an auto-exploded tweak can be changed only in the **Browser**.

Tweak value

Figure 18-13.
There are two additional methods of editing a manually applied tweak. A—Change the distance in the **Tweak Component** dialog box. B—Drag the trail's green dot.

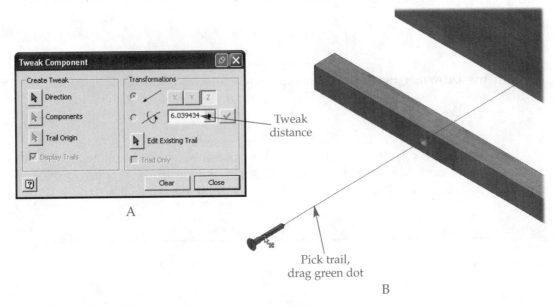

matic explosion method and a distance of 4.0". See **Figure 18-14**. Since this explosion was created automatically, any editing of the tweaks would have to be done through the **Browser**. Note the different directions of the trails. Think about how much work you would have to do to fix this exploded view and make the animation look right!

Figure 18-14.
Confusing views
can sometimes be
created when using
automatic
explosions.

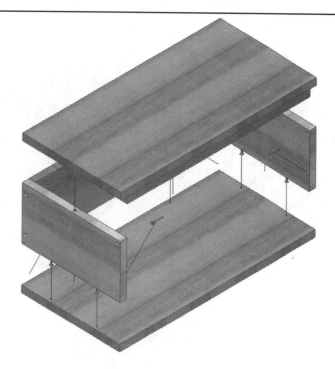

PRACTICE 18-1

❑ With the DRAWER_BOX.iam file open, review the parts in this assembly. The BOX_BASE is the grounded part. The two sides and back are fastened with ten screws. The top subassembly is located with four dowel pins and fastened with two screws, one from each side. What you need to think about is how you would assemble this and what you want to show in the exploded view. Remember, you are also going to both animate this assembly and create drawings. There will be conflicting requirements on the trails.

❑ Create a new standard presentation file, create a view of the assembly with the manual explosion method, and save it as Practice_18_01. Create the tweaks shown in the steps below by selecting the components from the **Browser**. Note: Pick the green check mark each time you enter a value. Pick **Clear** and keep the dialog box open.

   ❑ Tweak the drawer box top up 29".
   ❑ Within the drawer box top assembly, tweak the numbers 1 and 2 wood screws down 7".
   ❑ Within the drawer box top assembly, tweak the box rail down 4".
   ❑ Tweak the numbers 1, 2, 3, and 4 dowel pins up 11".
   ❑ Tweak the numbers 5, 6, 7, 8, 9, and 10 wood screws down 14".
   ❑ Tweak the numbers 1, 2, and 11 wood screws and the left box side up 6".
   ❑ Tweak the numbers 1 and 2 wood screws left 10".
   ❑ Tweak the number 11 wood screw left 12".
   ❑ Tweak the numbers 3, 4, and 12 wood screws and the right box side up 6".
   ❑ Tweak the numbers 3 and 4 wood screws right 11".
   ❑ Tweak the number 12 wood screw right 9".
   ❑ Tweak the box back up 4".

❑ See the figure on the next page. Now animate the assembly with the interval set to 10. Play it forward and reverse. Some of the screws will enter the box before the sides are assembled. You will rearrange the order, using the **Animate** tool, to correct this error. Also notice some of the components move at the same time. These components were grouped when they were selected together during the tweaking process. Leave the **Animation** dialog box open and pick the **>>** button to expand it for the following steps.

(Continued)

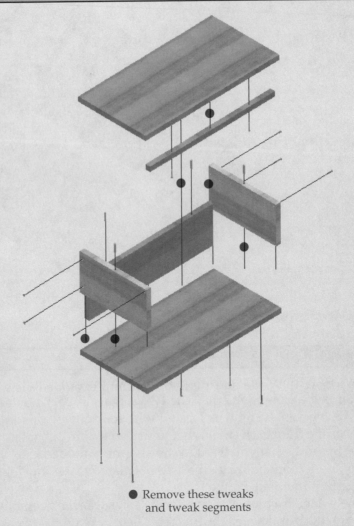

● Remove these tweaks
and tweak segments

❏ Select WOOD SCREW:12 Tweak(9.000 in) and pick **Move Down** ten times. This
should put it at the bottom of the list.
❏ Select WOOD SCREW:11 Tweak(12.000 in) and pick **Move Down** seven times.
❏ Pick the **Apply** and **Play** buttons. Notice WOOD SCREW:11 and WOOD
SCREW:12 moved after the top was positioned.
❏ Select WOOD SCREW:1 Tweak(6.000 in) and WOOD SCREW:2 Tweak(6.000 in).
Pick **Move Down** once.
❏ Select WOOD SCREW:3 Tweak(6.000 in) and WOOD SCREW:4 Tweak(6.000 in).
Pick **Move Down** once.
❏ Pick the **Apply** and **Play** buttons. Notice the screws are assembled after the
boards are in place.
❏ The presentation is acceptable for an animation, but this view will have to be
cleaned up if it will be added to a drawing. There are trails showing that are not
necessary for a drawing. You can remove an entire trail or remove a segment of
the trail. To remove the entire trail, right-click the trail and pick **Hide Trails**. To
remove a segment of a trail, right-click the segment and uncheck **Visibility**.

Creating Multiple Exploded Views

You can have many different views of the same assembly in one presentation file. Each view is based on a design view representation. Close all other files and open Drawer.iam. Check the three design view representations:

- TOP. All components visible, view from the top-left isometric.
- BOTTOM. All components visible, view from the bottom-left isometric.
- FRONT WITH LOCK. Only front subassembly and lock components visible, view from the top back isometric. The DRAWER_FRONT and the DRAWER_SUB_FRONT have **Enabled** unchecked so you can see inside them. See **Figure 18-15**.

The design view representation information is stored in the assembly file and contains the names of the views and information about each one including:

- Component, sketch, and work feature visibility.
- Colors and textures on components.
- Viewing angle and zoom.
- The default views: None and Administrator.

Create a new presentation file and save it as Example_18_02.ipn. Create a view using the TOP design view with:

- The DRAWER_FRONT subassembly and the five lock components moved out 7.5″ with no trails displayed.
- The four screws in the sides moved 6″ with trails.
- The three screws in the bottom moved down 12″ with trails. See **Figure 18-16**.
- Create a new IDW file and place the current exploded view as a base view with a 0.25 scale. Save the file as Example-18-02.

Figure 18-15.
Components with **Enabled** unchecked appear transparent.

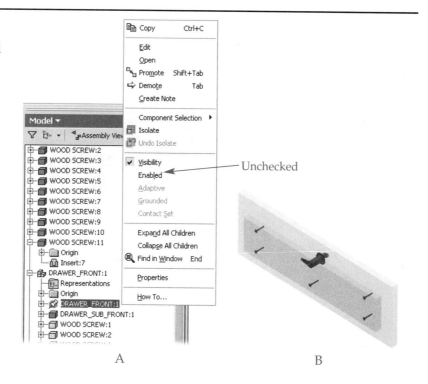

A

B

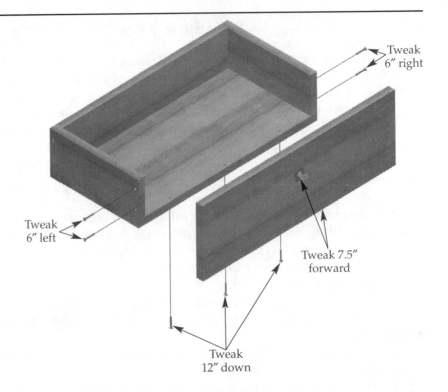

Figure 18-16.
An exploded view created with manual tweaks.

Tweak 6" right

Tweak 6" left

Tweak 7.5" forward

Tweak 12" down

Go back to the presentation file, pick **Create View**, set **Explosion Method** to Manual, and select the FRONT WITH LOCK design view representation. This creates a new view called Explosion2. Note that wooden parts are enabled even though this was unchecked in the assembly file. Create an exploded view similar to that shown in **Figure 18-17**. Place this view as another base view with a scale of 0.5 in the same drawing. See **Figure 18-18**.

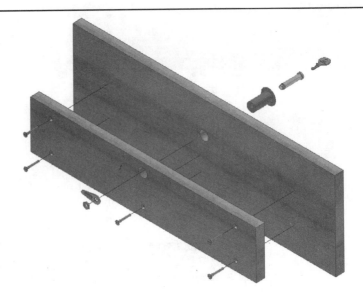

Figure 18-17.
Some trails are not visible in this manually exploded view.

Figure 18-18.
The exploded views have been placed in the drawing.

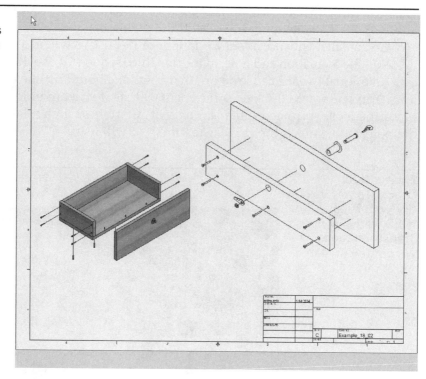

Complex and Rotational Tweaks

Open Example-18-03.ipn. You will create a complex linear tweak, a rotational tweak, and an animation with additional complexity, more controls, and different camera angles. First, you will add two linear tweaks to WOOD SCREW 5.

1. Zoom in on the area, see **Figure 18-19.** Pick the **Tweak Component** tool and select the right face of the DRAWER_SUB_FRONT to set the direction. Pick WOOD SCREW 5 and note the Z and X direction.
2. Drag the screw about 6" in the Z direction.
3. Do not pick **Clear**, but pick **X** in the **Transformations** section.

Figure 18-19.
Zoom in on this area of the view to add a complex linear tweak.

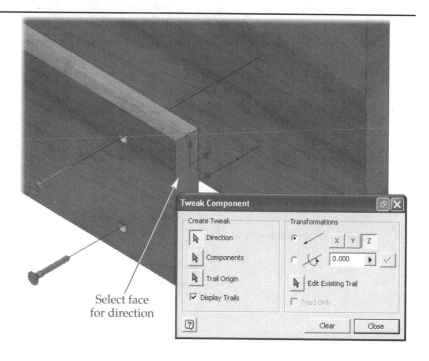

Select face for direction

4. Now drag the screw 3″ in the negative direction. See **Figure 18-20**.
5. Pick the **Animate** tool and expand the dialog box using the **>>** button.
6. Change the sequence as shown in **Figure 18-21**. This will require some ungrouping.
7. Run the animation and note, in the first three sequences, the drawer front is invisible.

Rotational tweaks cause an angular rotation about any one of the three principle axes. You will rotate the lock components 90° as if you are unlocking the drawer. To place the axis, you need either an axis of rotation or a circular feature. The inside diameter of the nut is the circular feature you will use.

Figure 18-20.
Drag the screw in the X direction while watching the **Transformations** window.

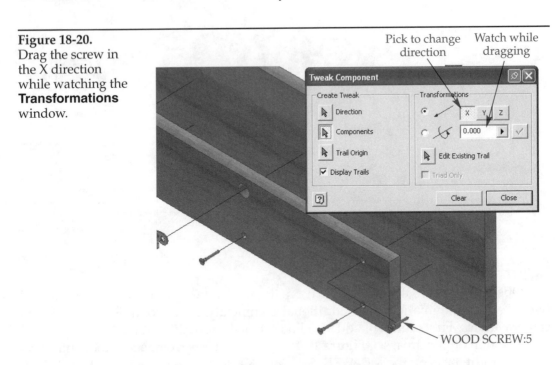

Figure 18-21.
You may have to use **Ungroup** to rearrange the tweaks so they are in this order.

These sequences are the two new tweaks to WOOD SCREW:5

This sequence is grouped so that it moves as a threaded object would

Move this group to the bottom of the list

Select sequence and **Move Up** or **Move Down**

While reordering, some sequences may need to be ungrouped

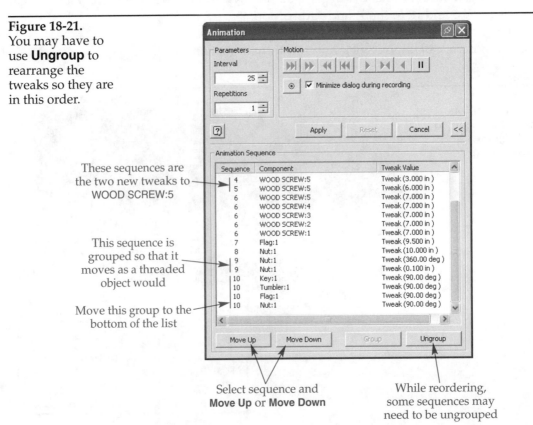

8. Zoom in on the nut.
9. Pick the **Tweak Component** tool and select the **Rotational Transformation** radio button.
10. Pick the inside diameter of the nut.
11. While the **Components** button is selected, pick the tumbler, flag, nut, and key in the **Browser**.
12. Input a 90° angle, pick the green check mark, and pick **Clear**. See **Figure 18-22**.
13. Pick the **Animate** tool and expand the **Animation** dialog box by picking the **>>** button.
14. Change the sequence so the rotational tweak is at the end of the list, as shown in **Figure 18-23**. It will help if you group the four rotational tweaks before moving them.
15. Run the animation to make sure you have the sequences in the right order.

Figure 18-22.
Rotational tweaks can be dragged or typed in, as shown.

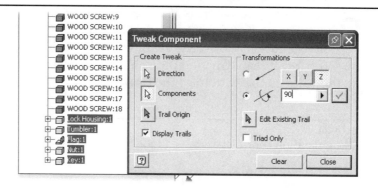

Figure 18-23.
Group the four rotational tweaks before moving them.

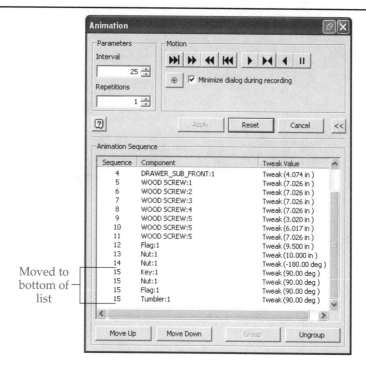

Moved to bottom of list

More on Animation

There are more animation sequence controls available within the **Browser**. See **Figure 18-24**. Pick the funnel icon at the top of the **Browser** and select **Sequence View**. This view shows an animation task, Task1, comprised of 11 sequences. Each sequence is a group of tweaks. Expand Sequence1 and Sequence11 as shown in the figure. Sequence1 is the translation of the lock housing and Sequence11 is the final rotation of the lock components. You can move the sequences up and down in the **Browser** using the drag-and-drop method. Be aware of the renaming of the sequences as you reorder them. When you move one, the list is renumbered so the sequences are always in numerical order. This is an easier method than moving them in the **Animation** dialog box.

Under each sequence there is a list of hidden components. In Sequence1 the drawer front is hidden, whereas it is not in Sequence11. To make the drawer front visible, right-click on it and delete it from the list. Do this for the first three sequences.

Run the animation and watch the nut closely. As it is installed, the nut looks as if it has a left-hand thread. Change the value in Sequence10 to –360° by picking it and editing the value at the bottom of the **Browser**.

Figure 18-24.
There are animation controls available in the **Browser**.

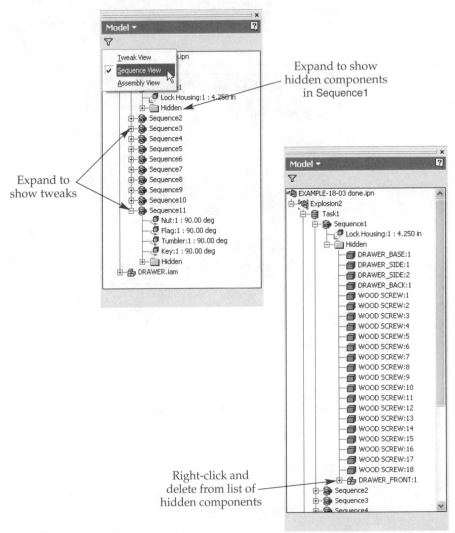

Expand to show hidden components in Sequence1

Expand to show tweaks

Right-click and delete from list of hidden components

In the **Browser**, right-click Task1 and pick **Edit**. This opens the **Edit Task & Sequences** dialog box, which lists the tasks. There is only one task in this example. Below the tasks list is an area where you can type in a description of the task. To the right is a **Play Forward** button so you can see the animation of the entire task.

Sequence1 is listed in the **Sequences** window. The drop-down list will show all 15 sequences. You can select a sequence and type in a description in the window below the sequences list. To the right is a **Play Forward** button that animates this sequence only (it puts any previous sequences in place first). At the lower left you can set the intervals for each sequence. Set this to 12. By setting the intervals for each sequence, you can make the animation flow smoothly.

In the lower right is the **Set Camera** button. This lets you change the viewpoint and zoom for a series of sequences. Change the viewpoint and zoom as shown in **Figure 18-25**. Pick **Set Camera**, and then pick **Apply**. Pick **Play Forward** on Sequence1 and pick **Reset**.

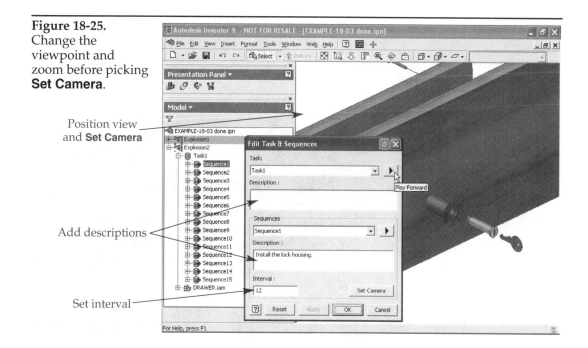

Figure 18-25. Change the viewpoint and zoom before picking **Set Camera**.

The following is a procedure of similar changes that will set up the camera position for specific sequences:

1. Set the interval for Sequence2 to 10 and Sequence3 to 8. Pick the sequence, **Play Forward**, and **Reset**.
2. For Sequence4, set the interval to 12, position the view as shown in **Figure 18-26**. Pick **Set Camera** and **Apply**. Run the task to see the results.
3. For Sequence12, position the view as shown in **Figure 18-27**. Pick **Set Camera** and **Apply**.
4. For Sequence14, position the view as shown in **Figure 18-28**. Pick **Set Camera** and **Apply**.

Figure 18-26.
View position for
Sequence4.

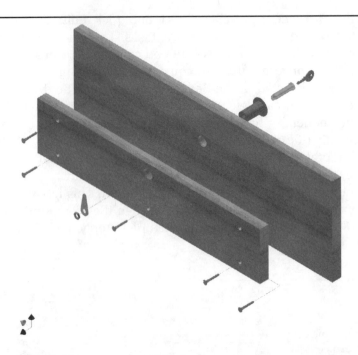

Figure 18-27.
View position for
Sequence12.

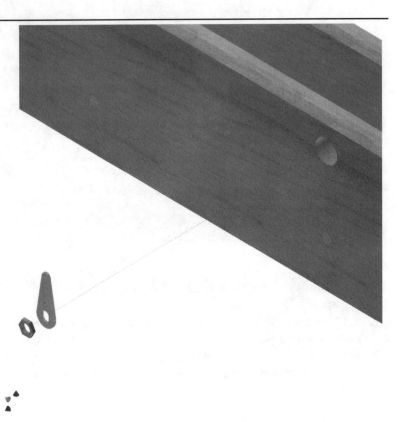

5. For Sequence15, position the view as shown in **Figure 18-29**. Pick **Set Camera** and **Apply**.

The changes you made in Task1 now apply to the **Animate** tool as well. Remember, this tool is used to group tweaks. If you wanted screws 1 through 4 to go in at once, group them. Grouping reduces the number of sequences, which may make larger animations more manageable. It is also the only way to show a nut rotating while it moves on a threaded shaft.

Figure 18-28.
View position for
Sequence14.

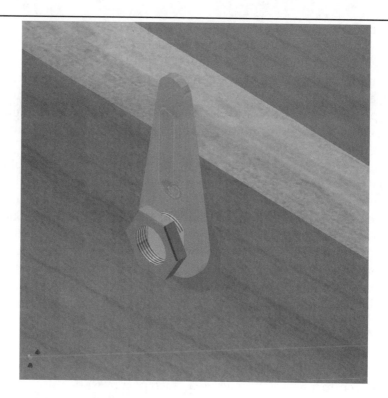

Figure 18-29.
View position for
Sequence15.

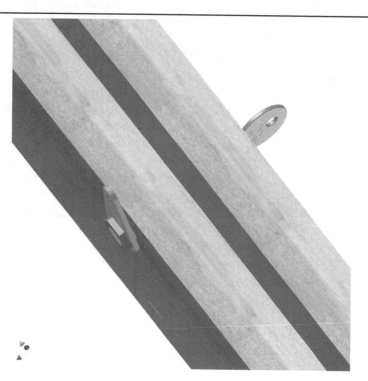

Styles

Applying styles to parts, modifying styles, and creating new styles are three skills developed in previous chapters that will now be utilized to enhance your presentation files. In this section you will open a presentation file and apply styles that will help highlight specific parts location and function. Next you will create custom lighting, also in a presentation file, to expose details that would otherwise be hidden in shadow. One benefit of performing these tasks within a presentation file is that the assembly and part files remain unchanged. The part should have the correct material assigned to it so that weight, cost, and performance calculations are accurate.

Part Colors in Assemblies

In Chapter 12 you created a design view representation of a single-cylinder engine. The engine block was made invisible in this representation so that the internal components could be viewed and highlighted. If you need to highlight various parts of an assembly without creating multiple design view representations, you could apply contrasting color styles to the parts that require highlighting. This method works well in a meeting where the discussion is too unpredictable to preplan multiple design view representations.

Open Example-18-04.ipn. In the **Browser**, expand the explosion and the assembly so that the six parts are available for selection. Now select those six parts by picking the first part and shift-picking the last part. See **Figure 18-30**. Look at the **Color Selection** drop-down box at the end of the **Inventor Standard** toolbar. If the drop-down box is grayed-out, then one of the three selections at the top of the **Browser** was mistakenly picked. Once the correct parts are selected, pick in the **Color Selection** drop-down box and pick **Lexan(Clear)**. All of the selected parts should be transparent. Now pick **Crank:1**, from the **Browser** or the graphics window, and change the color to **Lime**. Pick **Undo** and then practice highlighting other parts of the assembly.

Lighting

In a presentation file, lighting is the only style available when picking **Active Standard...** from the **Format** pull-down menu. This is because lighting can be applied to the entire presentation whereas each part must be selected to change its color in the presentation file. In Chapter 2 you learned what each lighting effect does and how to change it. In this section you will learn how to apply those skills to create a presentation that is easily understood.

Open Example-18-05.ipt. This is a plastic molded part that should look familiar; it is a model from Chapter 7. Since this part is made of black plastic, some of the inner details are difficult to see. Now open Example-18-05.ipn. This is a presentation file that contains an assembly of the plastic molded part. Assembly files typically contain more than one part but it is not necessary, and since a part file cannot be placed into a presentation file, an assembly file is required. As you can see, the details are easier to understand in the presentation file.

Pick **Active Standard...** from the **Format** pull-down menu. Pick the **Active Lighting Style** drop-down list to see the available styles. See **Figure 18-31**. Now, pick **Cancel** to close the dialog box and pick **Styles Editor...** from the **Format** drop-down menu. Expand the **Lighting** style in the left area of the **Styles and Standards Editor** dialog box to reveal the **My Lighting** style—pick it. See **Figure 18-32**. You will see three lights are used for this style. Pick each radio button in the settings area to view the location and color of each light. Notice the color swatch when the third light is selected. This light is colored gray so that it is not as bright as the other lights. This is the only way to control the brightness of individual lights.

Figure 18-30.
The six highlighted parts in the **Browser.**

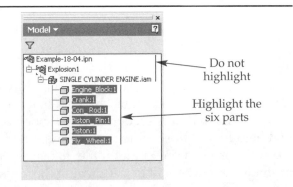

Figure 18-31.
Lighting is the only style that changes the entire presentation.

Pick to select a lighting style

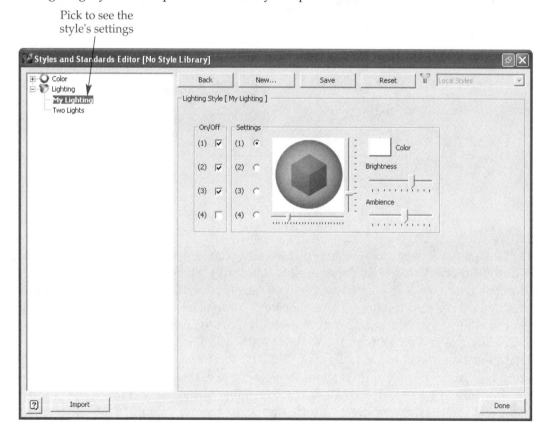

Figure 18-32.
New lighting styles can improve the visibility of a presentation.

Pick to see the style's settings

Other than Default, there are 112 standard styles available. There are 56 styles that are colors from Beige to Yellow. There are 53 styles that are materials from Aluminum to Zinc. There are 6 of the materials styles that are subtitled Texture. There are 14 of the materials styles that are subtitled Polished. There are seven of the materials styles that are subtitled Clear. There are three of the materials styles that are Chrome.

As you can see, there are many standard styles available. If needed, there are many more new styles that you can create. With practice, you should be able to create an impressive presentation that utilizes the proper combination of animation, materials, colors, textures, and lighting.

Figure 18-33.
The **Texture Chooser** dialog box reveals many texture options.

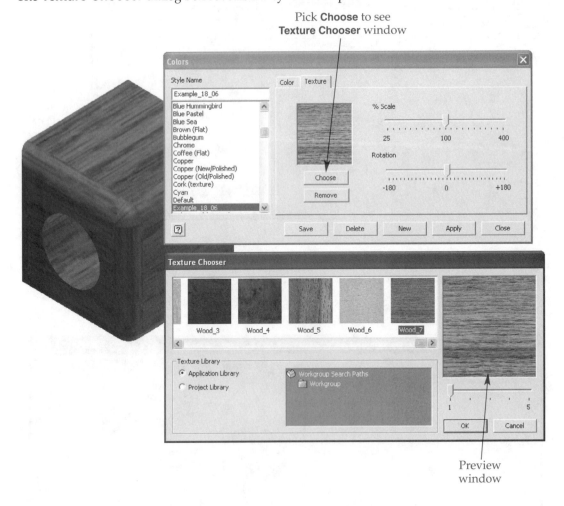

Pick **Choose** to see
Texture Chooser window

Preview
window

Texture

The last characteristic of a style is texture. This can be found by picking the **Texture** tab in the **Colors** dialog box. There are 98 standard textures. Open Example-18-06.ipt and notice a new style, called Example-18-06. See **Figure 18-33**. The preview window represents the selected texture when **Emissive** is set to black and all other color options are set to white. Change the **Ambient** color on this part to white and note the difference.

The texture is Wood_7 with the scale at 100% and the rotation at 0°. The preview image is just that and has no effect on how the texture is applied to the part. The texture is applied to the entire part but, as you can see, the angle of the pattern varies with the different faces. To see how rotation affects this part, change the rotation angle to 90° on the **Texture** tab and pick **Apply**. To see the effect of scale, change the scale to 25% and pick **Apply**.

A texture can be applied to just one face as long as it is saved as a style in the current part. Open the file Example-18-07.ipt. Open the **Colors** dialog box and review the style names. The one you saved in the previous example is not there. To bring the new style into this file:

1. Pick **Organizer** from the **Format** pull-down menu. Pick the **Color Styles** tab in the **Organizer** dialog box.
2. Pick **Browse**, find Example-18-06.ipt, and open it.
3. In the source document, find the style Example-18-06 and copy it into the active document. See **Figure 18-34**.
4. Close this dialog box.

Figure 18-34.
The **Organizer** allows you to copy lighting styles, materials, and color styles to another part file.

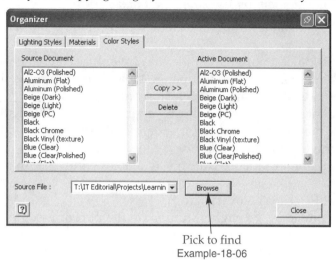

Pick to find
Example-18-06

5. Pick the top face of the part to highlight it. Right-click on the face and select **Properties** from the pop-up menu.
6. In the **Face Color Style** drop-down box, pick the style Example-18-06 and pick **OK**.
7. Apply the same style to the front face.

Lights

Lighting is an additional control of image quality. There can be one to four lights shining on a part or assembly. Each light has settings that allow you to control the direction, color, brightness, and ambience (ambient light).

Open Example-18-08.ipt and pick **Lighting...** from the **Format** pull-down menu. This accesses the **Lighting** dialog box, where the **Style Name** is set to TwoLights and lights 1 and 2 are on. The **Settings** area shows the current setting for each light. See **Figure 18-35**. The result is white light shining on the part from just above your head as you view the screen. It lights up the top face of the part. Move the horizontal slider to the right and then to the left. You can see the effect on the faces of the part as the light goes from over your right shoulder to over your left. The **Brightness** slider controls the brightness of all the lights in the style, not just light 1. The **Ambience** slider controls the total ambient light in the style. To control the brightness of a single light, change the color of the light to a darker color, in this case gray. As shown by the scroll bars, light 2 shines from below the horizon and to the right.

Figure 18-35.
Up to four lights can be set up and applied.

Pick slider to see current setting   Vertical slider

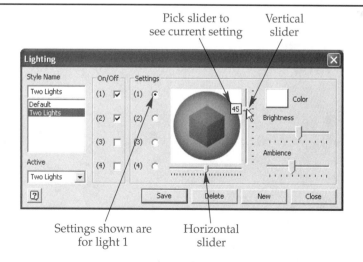

Settings shown are for light 1   Horizontal slider

Chapter Test

Answer the following questions on a separate sheet of paper.

1. Exploded views of an assembly are created in a separate file called a(n) _____ file.
2. The file extension of an animation file that Inventor can create is _____.
3. Does the assembly file have to be open to create a presentation file?
4. The two methods of creating an explosion are _____ and _____.
5. Each movement of a component during an explosion is called a(n) _____.
6. The speed of the components during an animation is controlled by the distance each has to travel and the number of _____.
7. There are animation sequence controls available within the **Animation** dialog box and the _____.
8. The four qualities each color has are _____ _____ _____ and _____.
9. There can be one to _____ lights shining on the part or assembly.
10. To make a part appear polished, change the _____ setting to 100%.
11. A texture can be applied to just one face as long as it is saved as a style in _____.

Chapter Exercises

Exercise 18-1. Create a new project Exercise_18_01 with the path set to …Chapter 18\ Exercises\Exercise-18-1. Run the AVI file Explosion-Ex-18-01.avi. Open the file Slider Clamp.iam. Change the color styles and create an exploded view without trails, similar to the AVI file. Create an AVI file similar to Explosion_Ex_18_01.avi.

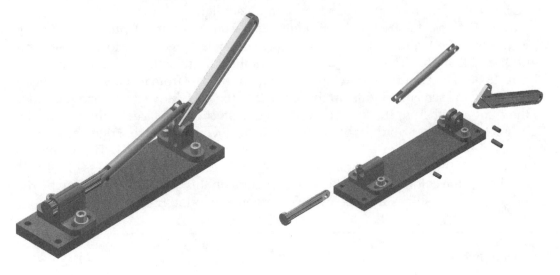

Exercise 18-2. Create a new project Exercise-18-02 with the path set to …Chapter 18\Exercises\Exercise-18-2. Create two exploded views in a drawing as shown in the figure.

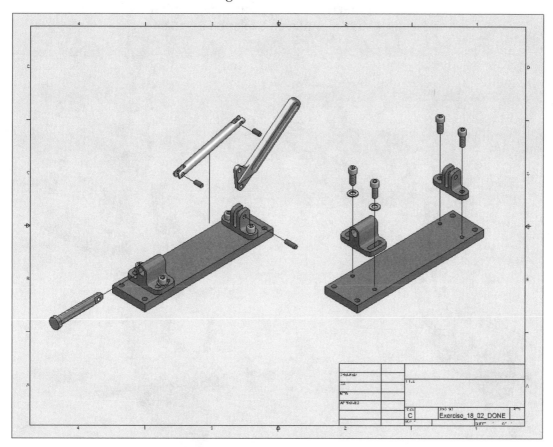

Exercise 18-3. Create a new project Exercise-18-03 with the path set to …Chapter 18\Exercises\Exercise-18-3. Run the AVI file Explosion-Ex-18-03.avi. Open the file EX03-Clamp-Linkage.iam. Change the color styles and create an exploded view, without trails, similar to the AVI file. Note that the cast support has several machined faces. Create an AVI file similar to Explosion-Ex-18-03.avi.

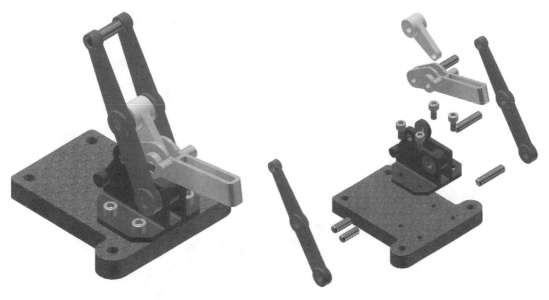

Exercise 18-4. Create a new project Exercise-18-04 with the path set to …Chapter 18\Exercises\Exercise-18-4. Run the AVI file Explosion-Ex-18-04.avi. Open the file EX04-Latch-Assembly.iam. Change the color styles and create an exploded view, without trails, similar to the AVI file. Create an AVI file similar to Explosion-Ex-18-04.avi.

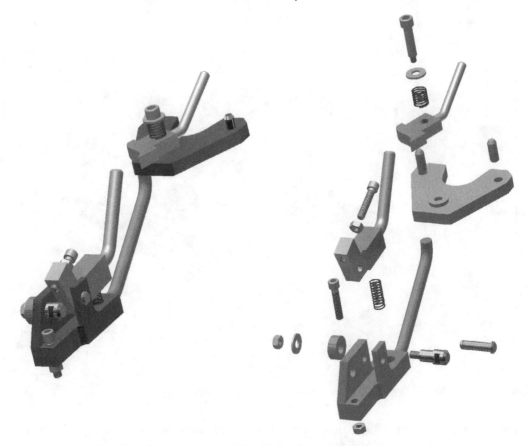

Exercise 18-5. Create a new project Exercise-18-05 with the path set to …Chapter 18\Exercises\Exercise-18-5. Run the AVI file Explosion-Ex-18-05.avi. Open the file EX05-VALET.iam. Create an exploded view, without trails, similar to the AVI file. Create an AVI file similar to Example-18-05.avi.

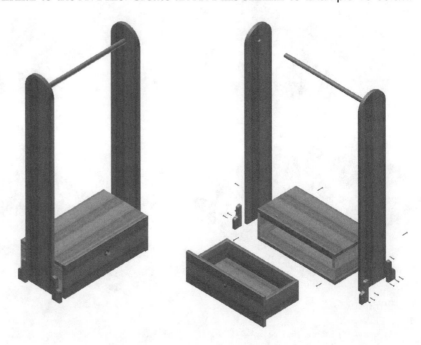

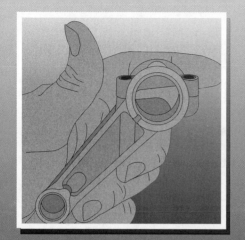

Sheet Metal Parts

Objectives

After completing this chapter, you will be able to:

* Define terms related to sheet metal parts.
* Create sheet metal styles.
* Add features to a sheet metal part with the **Face** tool.
* Specify relief settings.
* Override sheet metal styles.
* Create a flat pattern from a folded part.
* Create cutouts in sheet metal parts.
* Add features to a sheet metal part with the **Flange** tool.
* Create rounds and chamfers on the corners of sheet metal parts.
* Create hems on sheet metal parts.
* Create cutouts across bends.
* Create contour flanges.
* Create a folded sheet metal part from a developed pattern.
* Control the shape of the seam between two edges using the **Corner Seam** tool.
* Use punch tools to create holes or emboss sheet metal parts.

User's Files

*The following is a list of files that you will need to work through this chapter. These files can be found on the **User's Files** CD included with this text.*

Examples

Example-19-01.ipt	Example-19-10.ipt
Example-19-02.ipt	Example-19-11.ipt
Example-19-03.ipt	Example-19-12.ipt
Example-19-03.idw	Example-19-13.ipt
Example-19-04.ipt	Example-19-14.ipt
Example-19-05.ipt	Example-19-15.ipt
Example-19-06.ipt	Example-19-16.ipt
Example-19-07.ipt	Example-19-17.ipt
Example-19-08.ipt	
Example-19-09.ipt	

Practice

Practice-19-01.ipt
Practice-19-02.ipt

Exercises

Exercise-19-01.ipt
Exercise-19-02.ipt
Exercise-19-05-Pan.ipt
Exercise-19-06.iam
Exercise-19-07.ipt

The term *sheet metal part* generally refers to a part formed from relatively thin, flat sheet metal stock, which is typically between 0.01" and 0.18". In the United States the thickness of sheet metal is often specified by a gage number; the larger the number the thinner the material. For example, 12 gage cold rolled steel is 0.1046" thick. A table of the gages for several materials is listed in the appendix. There are many manufacturing processes for turning flat stock into parts. See **Figure 19-1.** These processes include spinning, sheet hydro-forming, and roll forming, but two of the most common are drawing and punching and bending.

Drawing uses matching pairs of dies in a press to stretch and flow the metal into complex curves. Solid models of drawn parts can be created in Inventor using sketches and the standard **Part Feature** tools such as the **Extrude** and **Revolve** tools.

Punching and *bending* uses tools in a press to cut the profile of the part, punch holes, form ribs and other small features, and then create bends. These bends are each about one axis, much like a piece of cardboard is bent to form a box. Solid models of these parts are created using the tools in the Inventor **Sheet Metal Features** menu. The basic steps in fabricating this type of part are:

1. Shear (cut) the sheet into a flat, rectangular shape called the *blank.* This is the basic shape from which the sheet metal part is produced.
2. Punch, drill, or broach various shaped holes, notches, and slots into the blank.
3. Bend the blank to form the part to its final shape.

Figure 19-1.
A—This Inventor part was created using the **Part Feature** tools. The actual part would be fabricated in a press with matching pairs of dies. B.— This Inventor part was created using the **Sheet Metal Features** tools. The actual part would be fabricated by punching the features into a metal blank and then performing four bends.

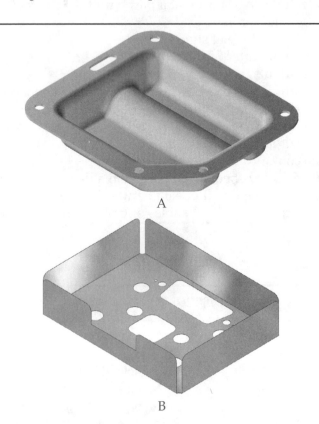

A

B

The problem is that the designer creates a part with a final shape that meets the design criteria. However, the dimensions of the flat pattern (the blank) must be calculated. These dimensions are called the *developed length* or layout dimensions. This is not simply a process of adding the dimensions to arrive at a total. The metal will actually dimensionally change as it is bent. The outside surface of the bend gets longer, the inside surface gets shorter, and the neutral axis moves away from the center of the bend.

Figure 19-2.
A—Two orthographic views of the folded part on the left and the flat pattern on the right. B—The completed part.

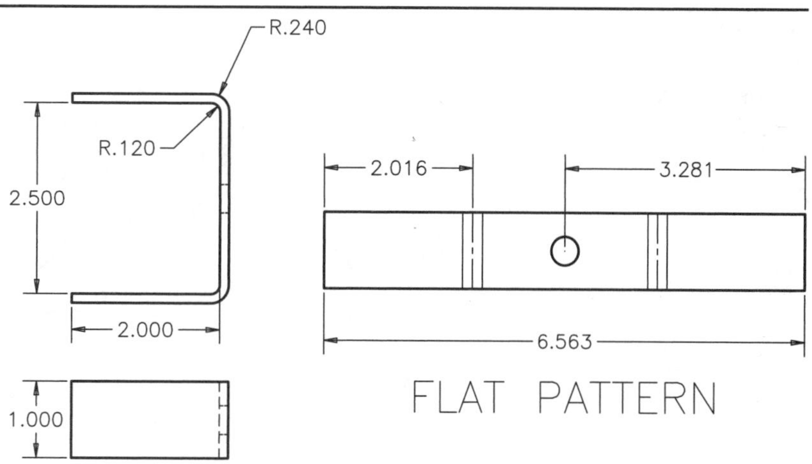

FLAT PATTERN

A

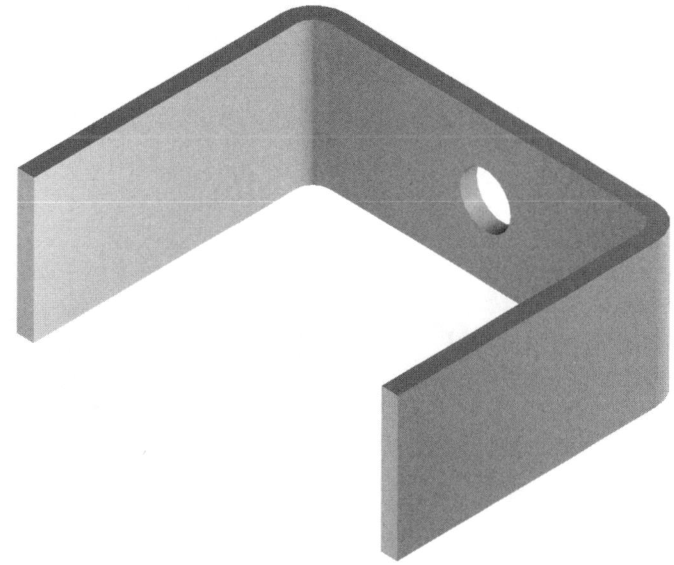

B

For example, the simple U-shaped bracket shown in **Figure 19-2** has to be 2″ deep and 2.5″ wide on the inside. The sheet metal is 11 gage (.12″) mild steel. The inside radius of the bend is .12″, which is equal to the metal thickness. The minimum recommended bend radius is usually the same as the material thickness. Using Inventor's sheet metal function, a flat pattern is automatically generated from the solid model and the length of the blank is calculated as 6.563″. The center of each bend is also calculated. The flat pattern can then be inserted into a drawing file and dimensioned.

Style Tool

In a standard part file, a profile is sketched and extruded a specified distance to give the part thickness. In a sheet metal part file, the thickness of the material is specified first using the **Sheet Metal Style** tool. The sketch is then extruded to the specified thickness.

Open Example-19-01.ipt, which is a sheet metal part file. See **Figure 19-3A.** You will create the finished sheet metal part shown in **Figure 19-3B.** First, you need to set a style for the part. The style for a sheet metal part includes information such as the material, material thickness, and several default values. To set the style, pick the **Sheet Metal Styles** button in the **Sheet Metal Features** panel. The **Sheet Metal Styles** dialog box is opened, **Figure 19-4.**

Figure 19-3.
A—The sketch for creating a face on a sheet metal part.
B—The completed part.

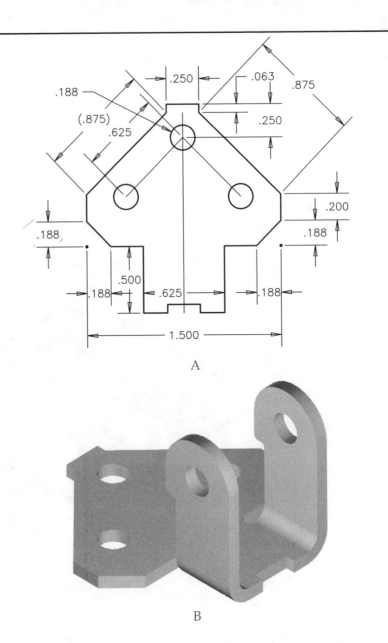

A

B

Figure 19-4.
The **Sheet Metal Styles** dialog box.

Name the style

Select a material

Enter the material thickness

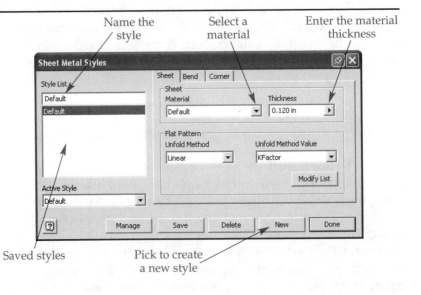

Saved styles

Pick to create a new style

The saved styles appear in the list at the left of the dialog box with the style Default. The active style appears at the bottom-left of the dialog box. To create a new style based on the current style, pick the **New** button. The new style is initially given the name Copy of *style*. To rename it, type a new name in the text box above the styles list. For this example, name the new style Example-19-01.

The type of material is set in the **Sheet** tab of the **Sheet Metal Styles** dialog box. The type of material is only important for the material shading color and the physical properties analysis. Pick the **Material** drop-down list in the **Sheet** area of the tab and select Steel, Mild. Also in this area, set the material thickness to .08″ by typing the value in the **Thickness** text box.

The other options in the **Sheet Metal Styles** dialog box are discussed later. For now, pick the **Save** button to save the new style. Then, select Example-19-01 from the **Active Style** drop-down list. Finally, pick the **Done** button to close the dialog box.

Face Tool

The **Face** tool is used to create sheet metal faces on the part. A *face* for a sheet metal part is a base feature. The base feature is generated from a sketch with holes and cutouts that is assigned the material thickness. As the extruded faces of the part are created, they will be joined with seams or bends to create the final sheet metal part. If you think of a folded sheet metal box that is open on one side, there are five faces to the part. The five faces are joined by four bends and four seams.

To create the first extruded face on the sheet metal part in this example, pick the **Face** button in the **Sheet Metal Features** panel. The **Face** dialog box is displayed, **Figure 19-5**. With the **Profile** button in the **Shape** tab on, select the profile. The area inside the circles should not be part of the profile. A preview of the extruded profile appears in green in the graphics window, just as it does when using the **Extrude** tool on a standard part. Picking the **Offset** button in the **Shape** tab changes the direction of the extrusion. Since this is the first face on the part, the direction is not important. Pick the **OK** button to create the face, **Figure 19-6**.

Figure 19-5.
The **Face** dialog box.

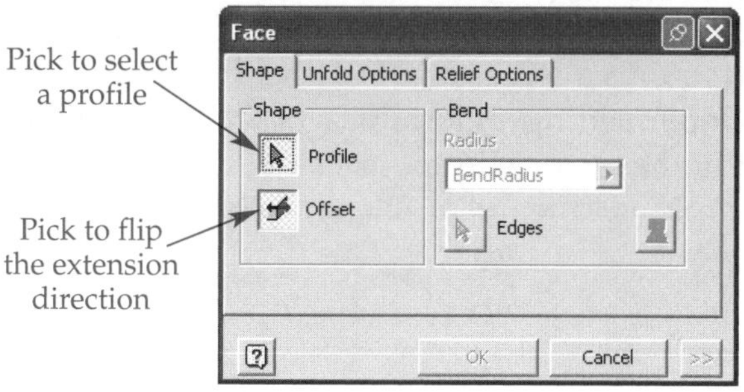

Pick to select a profile

Pick to flip the extension direction

Adding a Second Face

Now, you will add another extruded face to the sheet metal part. The new face will be at 90° to the face you just created. Open the Example-19-02.ipt where a sketch of the new face has already been created. The side face shown in red was selected as the sketch plane. A vertical construction line is used in the sketch to align the arc with the base part.

Figure 19-6.
The face is created based on the sketch in **Figure 19-3A.**

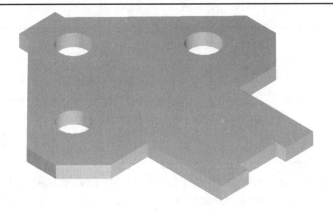

Specifying a bend radius

Before creating the face, you need to add a bend radius specification to the style. Open the **Sheet Metal Styles** dialog box and select the **Bend** tab. See **Figure 19-7.** The entry in the **Radius** text box is the inside radius of the bend. It is a good idea to express the radius as a function of the material thickness. In this way, if the material thickness is changed, the bend radius is automatically updated. Type Thickness in the **Radius** text box, if it is not already displayed. In other applications, you may type a formula here.

Specifying a relief

A *relief* is a notch punched in the blank that allows the material to bend without tearing. There are two choices for the shape of the relief—**Round** and **Square**. A round relief has a circular end or radiused corners. A square relief has square corners. You can also select **None** so that no relief is created.

If the relief is close to the edge of a part, a small "prong" or leftover piece of material may be created. This unwanted leftover material is called a *remnant.* The minimum allowable remnant size is set in the **Minimum Remnant** text box in the **Bend** tab. This is often expressed as a formula, as is the case with the default. If the remnant is less than this value, the width of the relief is increased to eliminate it.

The **Relief Width** setting determines the width of the relief. As with the bend radius, the relief width is often related to the material thickness. This is because thicker material requires a wider relief to prevent distortion.

The **Relief Depth** setting determines how far the relief extends into the part beyond the start of the bend. Minimum value is half of the material thickness.

For this example, set the **Relief Shape** to Round. Leave all other relief settings at their defaults.

Figure 19-7.
Creating a sheet metal style.

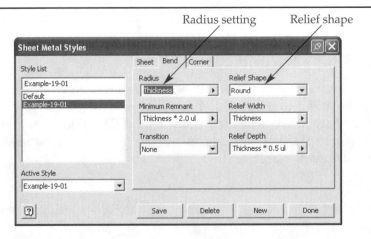

Saving the style

Once you have made changes to the style, it must be saved. Pick the **Save** button in the **Sheet Metal Styles** dialog box to save the style. Then, pick the **Done** button to close the dialog box.

Creating the face

Open the **Face** dialog box. Then, select the profile, **Figure 19-8.** The offset is important here as it determines the width of the part. The extruded face should extend out from the existing extruded face, not into the existing face. If needed, pick the **Offset** button in the **Shape** area to reverse the extrusion direction. You can enter a new bend radius in the **Radius** text box of the **Shape** tab to override the value set in the style, if needed. Using the settings in the **Relief Options** tab, you can override the relief settings for the style. Overriding the style is discussed later. For now, pick the **OK** button to accept the default settings and create the new face.

Notice that a bend is automatically applied between the new face and the existing face. See **Figure 19-9.** The bend is generated based on the settings in the style and any overrides. Also, notice how the relief is created in relation to the bend.

Figure 19-8.
The preview of the new face. The offset is important and should appear as shown here.

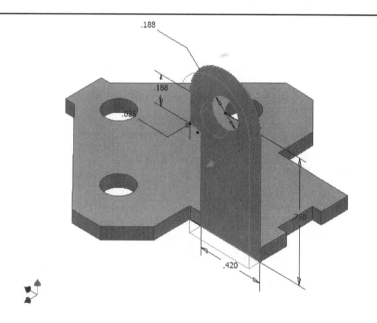

Adding a Third Face

Now, you need to add a third face that is the mirror image of the face you just created. The **Mirror Feature** tool will work in this situation, but here you will practice the sketch technique. You need to create a new 2D sketch and project the geometry from the second extruded face onto the sketch plane. Then, you can create the symmetrical feature. Start by selecting the opposite face on the first extruded face as the sketch plane. This face is colored purple in the file. Then, use the **Project Geometry** tool to project the geometry from the second extruded face. If you pick within the area of the surface, as shown in **Figure 19-10,** the outline is projected. Notice how there is a gap between the projected geometry and the existing extruded face. This is OK because the bend will fill in this area. Finish the sketch. Then, open the **Face** dialog box and select the profile. Again, the extrusion should extend out from the existing extruded face. Now, pick the **Edges** button in the **Bend** area of the **Shape** tab. Pick the top edge of the existing extruded face, as indicated in **Figure 19-10.** Finally, pick the **OK** button in the dialog box to create the new face. Notice how Inventor fills in the bend and creates the relief.

Figure 19-9.
A—The new face is created. The bend is automatically applied. B—The top view clearly shows the relief that was created.

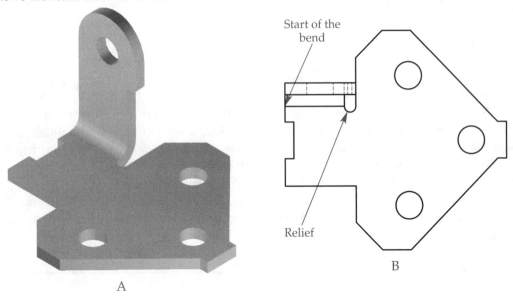

Figure 19-10.
Creating a third face by projecting the second face.

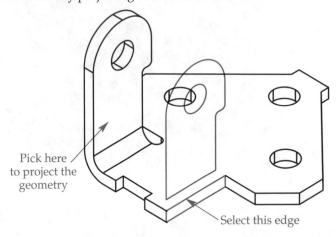

Overriding Styles

The settings for the style in the **Sheet Metal Styles** dialog box control the entire part. However, some of these settings can be overridden when creating features. For example, some of the settings in the **Bend** tab that control the relief can be set when creating a new face. The bend radius can also be changed for an individual bend. However, some settings, such as material thickness, cannot be overridden. Settings that can be overridden are:

- **Radius**
- **Relief Shape**
- **Minimum Remnant**
- **Relief Width**
- **Relief Depth**

For example, you can set the default choice in the **Bend** tab of the **Sheet Metal Styles** dialog box to Round, as in the previous example. Then, for each individual bend, you can choose Default, None, Round, or Straight in the **Relief Shape** drop-down list in the **Relief Options** tab of the **Face** dialog box. This override only applies to the bend on the face you are currently creating. To use the style setting, choose Default.

PROFESSIONAL TIP

If you want a unique relief for a particular bend, construct the reliefs as sketches and cut them from the part using the **Extrude** tool. Make sure the sketched relief is deep enough so that it goes past the start of the bend. Then, when creating the face, override the style by selecting None in the **Relief Shape** drop-down list in the **Relief Options** tab.

Creating a Flat Pattern

A *flat pattern,* or *developed blank,* results when the sheet metal part is unfolded into a flat sheet. Inventor can automatically develop and unfold a sheet metal part into a flat pattern. This pattern is generated in a new window and can be inserted into a drawing layout. Once in the layout, the flat pattern can be dimensioned. The **Flat Pattern** tool is used to create a flat pattern.

Open Example-19-03.ipt. You will create a flat pattern of this folded sheet metal part. Pick the **Flat Pattern** button in the **Sheet Metal Features** panel. A new graphics window is opened and the flat pattern is generated. The new window is called Flat Pattern: Example-19-03.ipt. Using the **Windows** pull-down menu, you can switch between the pattern and the part, or tile the windows to see both. See **Figure 19-11.**

Figure 19-11.
The folded part and the flat pattern are displayed in different graphics windows.

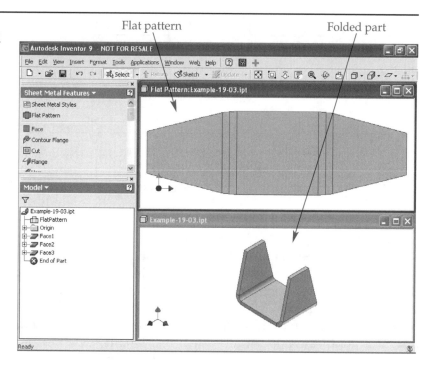

Now, open Example-19-03.idw, which is a drawing set up with two orthographic views of the folded part. Pick the **Base View...** button in the **Drawing Views Panel**. In the **Component** tab of the **Drawing View** dialog box, select **Flat Pattern** from the **Sheet Metal View:** drop-down list. In the **Scale** text box, enter 2 for the scale. Then, move the view above the two orthographic views and pick to place the view and close the dialog box. Finally, using the **General Dimension** tool in the **Drawing Annotation Panel**, place an overall dimension on the drawing, as shown in **Figure 19-12**. Notice how only one dimension is required on the flat pattern. This is because the two fold lines are perpendicular to the dimensioned length. If the part had another fold not perpendicular to this length, the height would need to be dimensioned as well.

Figure 19-12.
A drawing shows the dimensions of the folded part and the overall dimension of the flat pattern.

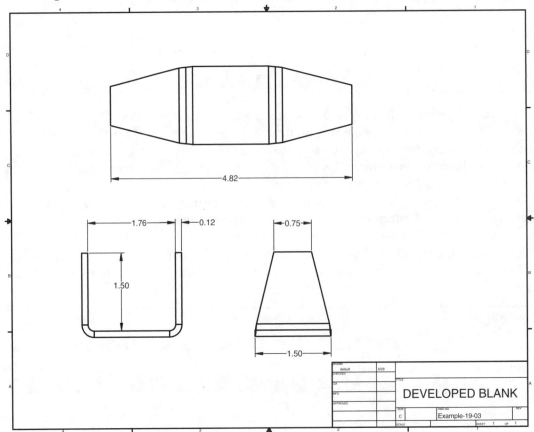

Cut Tool

Now, you will add a hole to the folded sheet metal part and see how the flat pattern is affected. In the part file (Example-19-03.ipt), start a new 2D sketch on the right-hand vertical surface. See **Figure 19-13**. Sketch a circle and dimension as shown in the figure. Finish the sketch and pick the **Cut** button in the **Sheet Metal Features** panel. The **Cut** dialog box is opened, **Figure 19-14**. The **Cut** tool is equivalent to the **Cut** option of the **Extrude** tool used to create part features. With the **Profile** button in the **Shape** area on, select the interior of the circle as the profile. Then, in the **Extents** area of the dialog box, pick **All** in the upper drop-down list. Finally, pick the **OK** button to create the hole. Notice how the hole passes through both upright extruded faces on the part.

Return to the drawing layout. Notice how the flat pattern view is automatically updated to reflect the addition of the hole. Dimension the hole as shown in **Figure 19-15**.

Figure 19-13.
Sketching a circle to be used with the **Cut** tool.

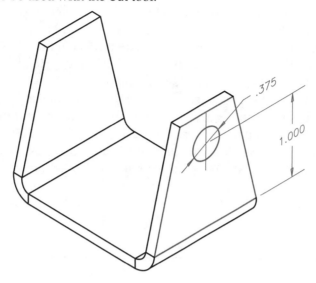

Figure 19-14.
The **Cut** dialog box.

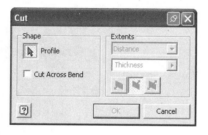

Figure 19-15.
The drawing of the part is automatically updated when the hole is cut in the part. Add the dimensions related to the hole.

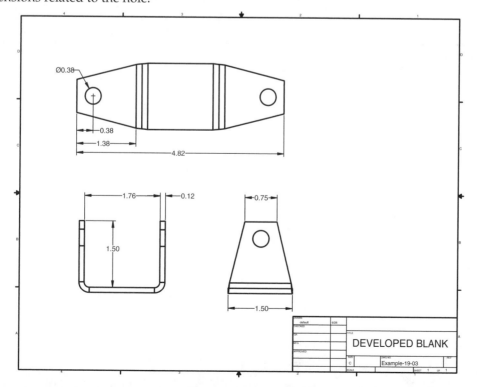

Projecting a Flat Pattern into a Sketch

You can project a tracing of the flat pattern into a sketch. Open Example-19-04.ipt. Start a new 2D sketch. Select the top surface as the sketch plane. Refer to **Figure 19-16.** Then, pick the arrow next to the **Project Geometry** button in the **2D Sketch Panel**. From the flyout, select the **Project Flat Pattern** button. Pick the vertical face. The outline of the unfolded part is projected onto the sketch plane.

Now, you can use the projected flat pattern to create a hole. Construct and dimension the circle shown in **Figure 19-17.** Finish the sketch, and pick the **Cut** button in the **Sheet Metal Features** panel to open the **Cut** dialog box. Select the interior of the circle as the profile. Then, check the **Cut Across Bend** check box. This will project the circle onto the surface that was projected to create the sketch. If you do not check this check box, an error will result. Also, when the check box is checked, the **Extents** options are grayed out. You will examine these options later in this chapter. Finally, pick the **OK** button to create the hole. See **Figure 19-18.**

Figure 19-16.
Projecting a flat pattern into a sketch.

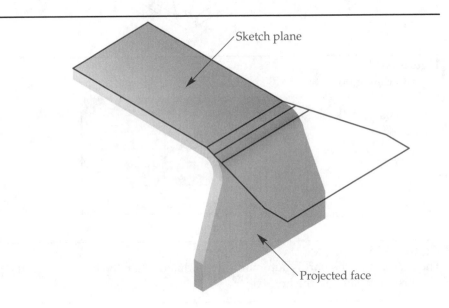

Sketch plane

Projected face

Figure 19-17.
Sketching a circle to use with the **Sketch** tool.

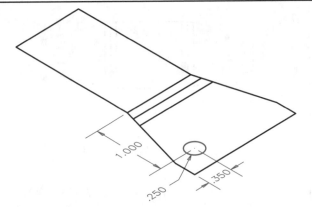

1.000

.250

.350

Figure 19-18.
The hole is created
using the **Cut**
Across Bend option
of the **Cut** tool.

PRACTICE 19-1

❑ Open the file Practice-19-01.ipt.
❑ Using the techniques just covered, construct the part, flat pattern, and drawing shown below.
❑ The style should reflect the material as 6061 aluminum and the thickness of 1 mm.

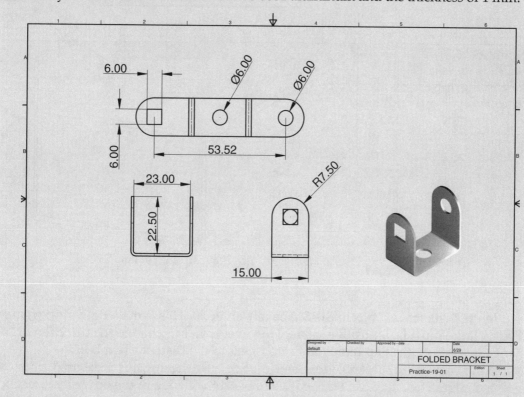

Flanges

A *flange* is a rectangular face on a sheet metal part that is at an angle to an existing face. The **Flange** tool is used to add a flange to a sheet metal part. The advantage of this tool is you can easily specify the bend angle. The entire edge or a portion of an edge can be flanged with or without reliefs. However, the disadvantage of the **Flange** tool is that you can only create a rectangular face.

Open Example-19-05.ipt, which is a sheet metal part with a base face already created. Then, pick the **Flange** button in the **Sheet Metal Features** panel. The **Flange** dialog box is displayed, **Figure 19-19.** With the **Select Edge** button in the **Shape** area of the **Shape** tab on, pick Edge A shown in **Figure 19-20.** A preview of the flange appears as a green wireframe in the graphics window.

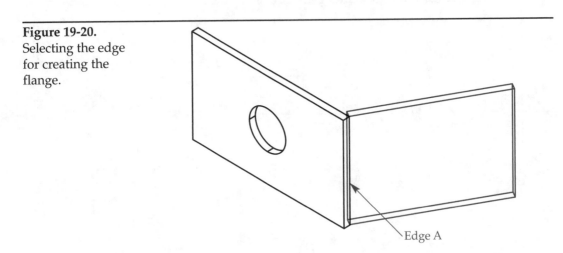

Figure 19-19.
The **Flange** dialog box.

Length of flange

Bend angle

Pick to expand the dialog box

Figure 19-20.
Selecting the edge for creating the flange.

Edge A

In the **Distance** text box in the **Shape** tab, enter 20. This is the length of the flange. The distance must be a positive value. However, you can change the direction of the flange by picking the **Flip Direction** button next to the **Distance** text box.

Also, enter 60 in the **Angle** text box. Notice how the angle is measured from the selected edge. The angle can be negative to bend the flange in the opposite direction. An **Angle** value of 90 creates a right angle flange. Realistically, a flange would not be created with an **Angle** value of 0. Therefore, values less than 3 result in an error.

The **Flip Offset** button in the **Shape** area of the **Shape** tab determines if the inside or outside edge of the flange is inline with the selected edge. In effect, this determines if the width of the existing extruded face remains the same or is increased.

Once all settings have been made, pick the **OK** button to apply the flange and close the dialog box. You can now modify the flange by sketching on one of its faces and using the **Face** tool. You can also create cuts by putting a sketch plane on the original extruded face and projecting the flange. You can edit the flange feature by right-clicking on its name in the **Browser** and selecting **Edit Feature** from the pop-up menu. The **Flange** dialog box is displayed. Make the necessary changes and pick **OK** to update the feature.

More on Flanges

The maximum allowable **Angle** value in the **Flange** dialog box is 180°. However, for angles much larger than 90° the results can be unpredictable. Open Example-19-06.ipt. The original extruded face is 9" long. A flange was applied to one end using a 1" bend radius, 10" length, and 90° angle. The overall length of the part is 10" by 10". Each flat surface is 9" long. The 1" bend radius adds 1" to each length measurement. See **Figure 19-21**.

Edit the flange feature by right-clicking on Flange1 in the **Browser** and selecting **Edit Feature** from the pop-up menu. Change the **Angle** value to 145° and pick the **OK** button to update the feature. However, notice that the length of each flat surface is now 6.828". Verify this using the **Measure Distance** tool located in the **Tools** pull-down menu. Edit the flange again and increase the angle to 168.5°. The length of each flat surface is reduced to .069" and you have a very small part. Any further increase in the angle will cause the bend to fail as the start of the bend will be off of the existing part. For bend angles larger than 135°, use the **Hem** tool that is covered later in this chapter.

Figure 19-21.
When creating a flange, the angle of the bend dramatically affects the shape and size of the part.

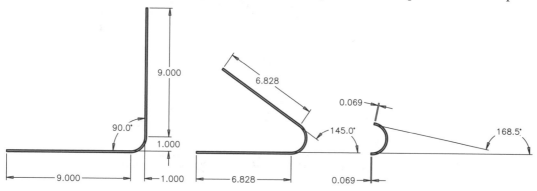

Applying a Flange to a Portion of an Edge

The flanges you have worked with so far have extended across the edge selected in the operation. However, you can create a flange that is only applied to a portion of the edge. Open Example-19-07.ipt. This file contains a sheet metal face; the material is 3 mm thick aluminum. Open the **Flange** dialog box and pick Edge A indicated in **Figure 19-22**. Enter 25 in the **Distance** text box and 90 in the **Angle** text box.

Next, expand the dialog box by picking the **>>** button, **Figure 19-23**. There are three choices in the **Extents** drop-down list—**Edge**, **Width**, and **Offset**. The **Edge** option creates the flange across the entire edge, which is the default option. The **Width** option allows you to specify a portion of the edge on which the flange is created. The **Offset** option allows you to create a flange offset a distance between two points. Select the **Width** option. You must select a starting point. With the **Select Start Point** button on, pick the point indicated in **Figure 19-22**. Finally, enter 5 in the **Offset** text box, 10 in the **Width** text box, and pick the **OK** button to create the flange. Notice that a relief is generated based on the settings in the sheet metal style.

Figure 19-22.
A flange can be
applied to a portion
of the selected edge.

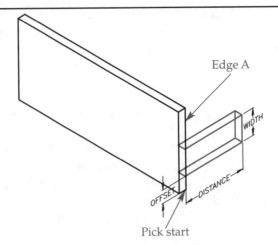

Figure 19-23.
The expanded
Flange dialog box.

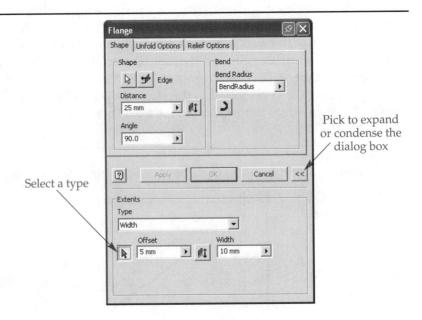

Rounds and Chamfers

You can apply rounds and chamfers to sheet metal parts. Fillets, which are internal arcs, are generally created on sheet metal parts by the bending and folding operations. The bend radius is the fillet radius. Rounds and chamfers are applied to the corners of sheet metal parts. Corners are parallel to the thickness of the material. The tools for placing rounds and chamfers on sheet metal parts are **Corner Round** and **Corner Chamfer**. The tools work in much the same way as the **Fillet** and **Chamfer** tools found in the **Part Features** panel for standard parts.

PROFESSIONAL TIP  You may need to add a round to create a safer edge or add a chamfer for welding purposes. These are applied using the **Fillet** and **Chamfer** tools listed in the **Part Feature** panel.

Figure 19-24.
Corner rounds are
added to the part.

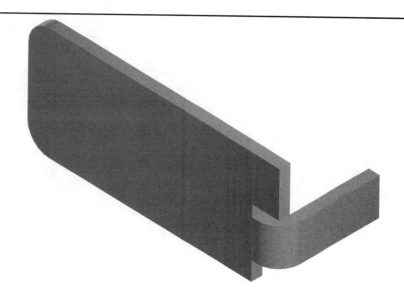

Using the **Corner Round** tool, place a 6 mm round on the two corners opposite the flange you added in the last section. See **Figure 19-24.** Notice as you select the corners that you cannot select the edges.

Hems

A *hem* is a short distance of metal at the edge of a sheet metal part that is folded back onto itself. Hems are often added to strengthen a long span of unfolded sheet metal. Open Example-19-08.ipt and refer to **Figure 19-25.** This part has a single hem placed on the right-hand edge. The face to which the hem was applied was 4". Right-click on Hem1 in the **Browser** and select **Edit Feature** from the pop-up menu. The **Hem** dialog box is displayed, **Figure 19-26.** Like other features, the dialog box used to edit the feature is the same as the one used to create the feature.

There are four types of hems available in the **Type** drop-down list in the **Shape** tab. The type is currently set to Single. This type of hem is defined with a gap and length. The **Gap** setting is currently twice the material thickness. The **Length** setting is currently four times the thickness. The overall length of the part will remain unchanged, which is 4" in this case. The angle for a single hem is fixed at 180°. If you expand the dialog box, the three choices available for the width of the hem are the same as those offered with the flange.

Figure 19-25.
A hem is added to
the right edge of
this part.

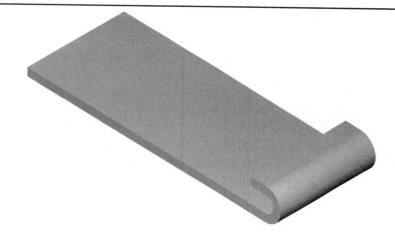

Figure 19-26.
The **Hem** dialog box, shown expanded.

Select the type of hem →

Hem

Shape | Unfold Options | Relief Options

Type
⯈ Single ▼

Shape
↖ ∃I Edge

Gap
Thickness * 2 ul ▶

→ Define the hem

Length
Thickness * 4.0 ul ▶

? | OK | Cancel | <<

→ Pick to expand or condense the dialog box

Extents
Type
Edge ▼

Select a type →

Select Teardrop in the **Type** drop-down list. The settings for defining the hem change to **Radius** and **Angle**. Also, notice the preview in the graphics window. The range of angles Inventor will accept for the **Angle** is from 181° to 359°.

Select Rolled in the **Type** drop-down list. The settings for defining the hem are **Radius** and **Angle**. The range of acceptable values for the angle is from 3° to 359°. Notice the difference between the previous teardrop hem and the rolled hem. With the same settings, the hem appears differently. See **Figure 19-27**.

Now, select **Double** in the **Type** drop-down list. The settings for defining the hem are **Gap** and **Length**. For a double hem, the **Length** value must be greater than the **Gap** value plus twice the material thickness. Also, notice the preview of the hem. This type of hem is common for strengthening an edge of the part.

Figure 19-27.
These two hems were created with the same radius and angle settings. Notice the difference between the two.

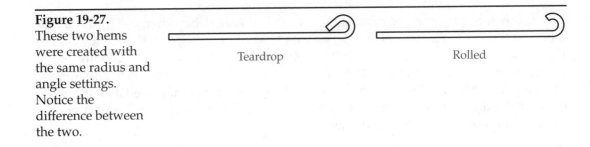

Teardrop   Rolled

Holes and Cuts across Bends

Holes and slots that are punched on or near the bend line in the flat pattern change shape when the part is bent (folded). If you create a sketch on a folded face of the part and cut a hole using the sketch, the hole will not be distorted. However, these features should display in their true, distorted shape in the folded part. To do this, the folded faces need to be projected onto a sketch plane, as you did earlier in this chapter. Open Example-19-09.ipt and refer to **Figure 19-28**. The left-hand, rectangular face was created as a flange and the right-hand, trapezoidal face was created as a face, both with a bend radius of .5".

Figure 19-28.
The folded part before cuts across bends are added.

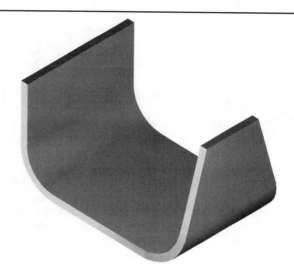

First, you will add a circular cutout through the bend between Face1 and Flange1. Face1 is the original face. Start a new 2D sketch on the top face of Face1. Refer to **Figure 19-29.** Next, pick the **Project Flat Pattern** button in the **2D Sketch Panel**. Pick the flat portion of the inside face of Flange1 to project it onto the sketch plane. You must pick the inside face because the sketch plane is on the inside face of Face1. If the sketch plane was on the outside face of Face1, you would need to pick the outside face of Flange1. At this point, you should either create a sliced graphics display or change to a wireframe view. Construct a .75″ diameter circle centered at the midpoint of the bend centerline. Finish the sketch and open the **Cut** dialog box. Select the interior of the circle as the profile. In the **Cut** dialog box, check **Cut Across Bend** check box. When you check this check box, all of the options in the **Extents** area are grayed out. Pick the **OK** button to create the circular cutout. Notice how the circle is distorted as the cutout is created. Then, use the **Flat Pattern** tool to see the developed pattern. See **Figure 19-30.**

Figure 19-29.
The flat pattern is projected into the sketch and a circle is created.

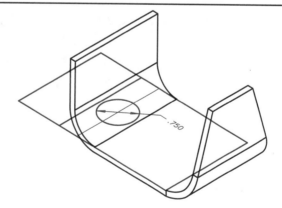

Now, you will add a slot cutout to the other bend. Rotate the view to see the inside of Face2, which is the trapezoidal face. Start a new 2D sketch on the top of Face1. Using the **Project Flat Pattern** tool, project the inside face of Face2. Remember, select the flat part of the face, not the bend. Sketch the slot as shown in **Figure 19-31.** Then, finish the sketch and, using the **Cut** tool, create the slot. The distortion may not be as apparent in this feature, but it is there. Finally, right-click on FlatPattern in the **Browser** and select **Open Window** in the pop-up menu to see the developed pattern, **Figure 19-32.**

Figure 19-30.
A—The circle is distorted as the hole is created. This is what actually happens when a bend is created through a hole. B—The hole is correctly shown in its true shape and size in the flat pattern.

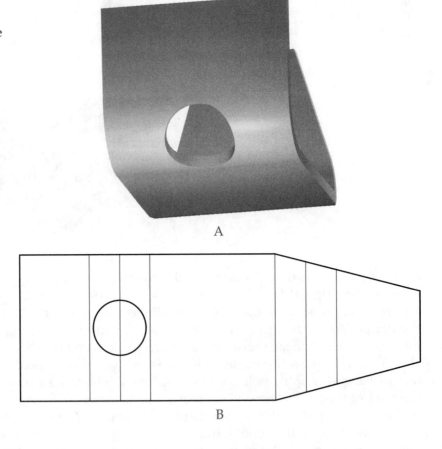

A

B

Figure 19-31.
In a second sketch, the slot is drawn.

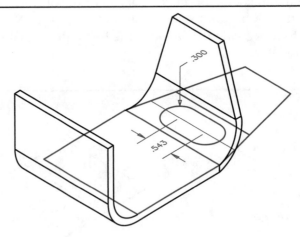

.300

.543

PROFESSIONAL TIP

One of the important things to note about projecting holes across bends is that you must have a flat portion of a face or flange to project onto the sketch plane.

Learning Inventor

Figure 19-32.
A—The slot is distorted as it is created. B—In the flat pattern, the slot is shown in its true shape and size.

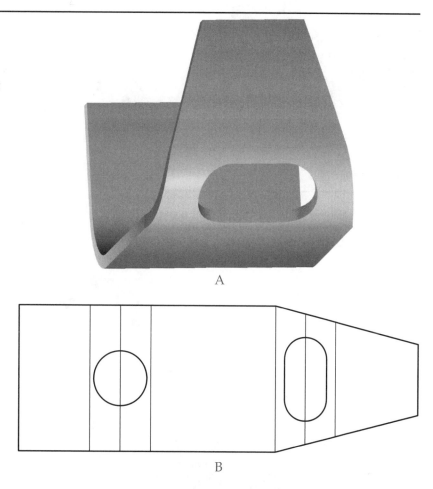

A

B

Contour Flanges

If a part has a uniform cross section, it can be created from an open sketch with the **Contour Flange** tool. The sketch can contain lines, arcs, and splines, but it cannot have any sharp internal corners. A valid sketch and the resulting part are shown in **Figure 19-33.** Notice that a 2 mm fillet has been applied to eliminate the square corner. Also, tangent constraints are required between the vertical lines and the large arc.

Creating a Contour Flange

Open the file Example-19-10.ipt, which contains the sketch shown in **Figure 19-33.** Pick the **Contour Flange** button in the **Sheet Metal Features** panel to display the **Contour Flange** dialog box. See **Figure 19-34.** With the **Profile** button on in the **Shape** tab, pick any part of the open sketch. Since there are no sharp corners, the entire sketch is selected. The **Flip Offset** button determines which side of the profile the material thickness is applied.

Since this is the first feature to be created, the **Contour Flange** dialog box is expanded and **Distance** is automatically set in the **Type** drop-down list in the **Extents** area. The drop-down list is disabled, so you cannot select a different option. In the **Distance** text box, type 15. Then, pick the **OK** button to create the contour flange. Once the flange is created, you can cut slots and add rounds and chamfers to the part, as shown in **Figure 19-35.**

Figure 19-33.
A—The open sketch that will be used to create a contour flange. B—The resulting contour flange.

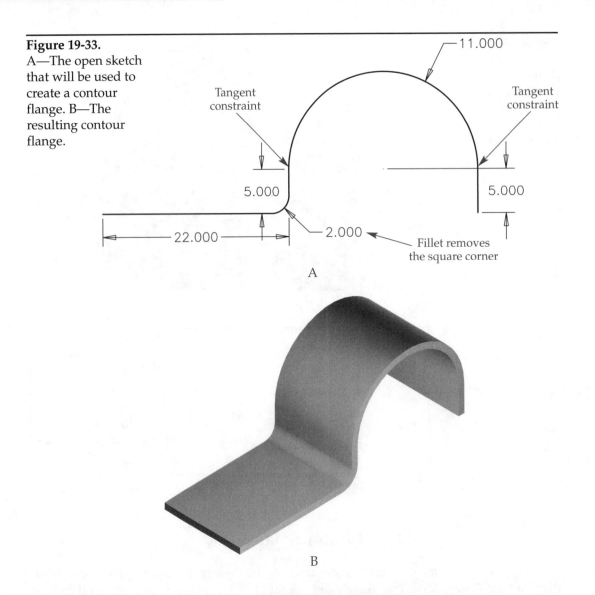

A

B

Figure 19-34.
The **Contour Flange** dialog box, shown expanded.

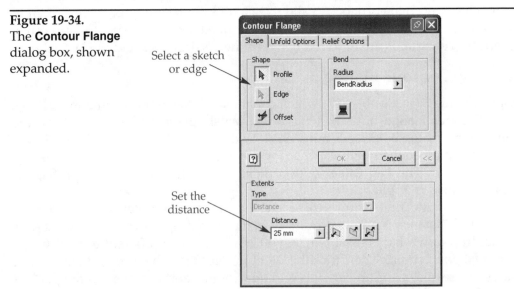

Figure 19-35.
Two corner chamfers and a cutout (slot) are added to the contour flange. The rounded end was created using the **Cut** tool, not the **Corner Round** tool.

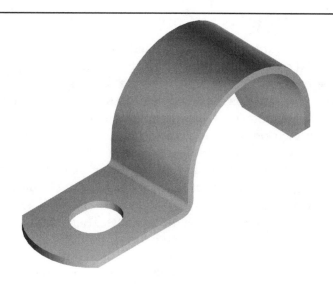

Extending a Contour Flange from an Edge

A contour flange can be created from the edge of an existing sheet metal part based on the sketched profile. Open the **Contour Flange** dialog box and first select the profile. Then, pick the **Edge** button in the **Shape** tab and select an edge from which the contour flange should extend. Expand the dialog box. The same options are available in the **Extents** area as for the **Flange** tool.

Open the file Example-19-11.ipt. Open the **Contour Flange** dialog box and select the sketch as the profile. Then, with the **Edge** button on, select the bottom edge of the part. Now, expand the dialog box and select **Width** from the **Type** drop-down list in the **Extents** area. With the **Select Start Point** button on, pick the left endpoint of the edge as the start. Then, type .5 in the **Offset** text box and 1.5 in the **Width** text box. Notice the preview in the graphics window. The top surface of the contour flange should be flush with the top surface of the existing part. See **Figure 19-36.** If not, pick the **Offset** button in the **Shape** area of the **Shape** tab. Finally, pick the **OK** button to create the contour flange. Notice the reliefs that are created based on the sheet metal style.

Figure 19-36.
Adding a contour flange to a portion of an existing edge.

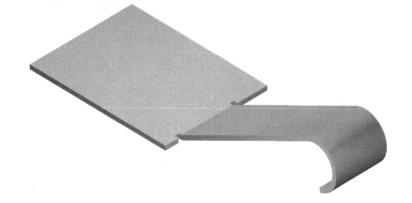

Complex Patterns as Contour Flanges

Contour flanges can also be used to create round, elliptical, or complex-shaped sheet metal tubes or ducts for which flat patterns are required. The sketch, of course, must be open and the trick is to have a small straight line at one end. Open the file Example-19-12.ipt. This file contains three sketches, two of which you will use to create a round sheet metal duct with an angled end.

Open the **Contour Flange** dialog box and select the circular sketch as the profile. If you zoom in on the circle, you can see that this is not a circle, rather an arc and a straight line with a small gap between its ends. If you zoom in close, you will see the small straight line on the left side of the gap. The **Contour Flange** tool does not work with closed shapes. However, the gap can be made small enough to be within the allowable tolerances for sheet metal ducts. Then, enter 12 in the **Distance** text box and pick the **OK** button to create the contour flange. The part is now a round duct with two square ends. See **Figure 19-37.**

Pick the **Cut** button in the **Sheet Metal Features** panel. The trapezoidal sketch should be automatically selected as the profile. If not, select it. Set the extents to **All** and pick the midplane button. Then, pick **OK** to create the cut. One end of the duct is now angled. See **Figure 19-38.**

Figure 19-37.
A revolution is added to the small flange to create a round duct with square ends.

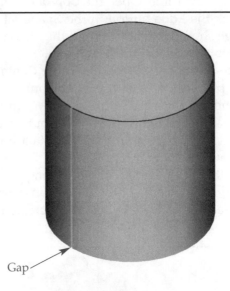

Gap

Figure 19-38.
The angled end is created using the **Cut** tool.

Figure 19-39.
The flat pattern for
the duct with an
angled end.

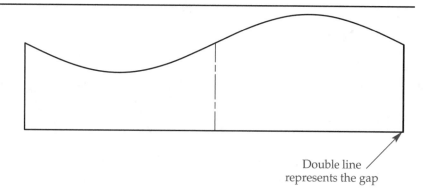

Double line
represents the gap

Now, use the **Flat Pattern** tool to create the developed pattern, **Figure 19-39.** In order for this operation to work correctly, you must select the part in the graphics window before picking the **Flat Pattern** button in the **Sheet Metal Features** panel.

Cylindrical and conical ducts can also be created with the **Revolve** tool using an angle of revolution of just less than 360°. Open the file Example-19-13.ipt. This part has a long, narrow face extruded with the **Face** tool. It is named Face1 in the **Browser**. There is also an unconsumed sketch that was created by projecting Face1 onto a sketch plane. A centerline was drawn in the sketch to use as the axis of revolution.

Pick the **Revolve** button in the **Part Features** panel to display the **Revolve** dialog box. If this panel is not displayed, pick the arrow next to the current panel name and select **Part Features** from the drop-down list. The sketch should be automatically selected as the profile. Also, since the axis is a centerline linetype, it is automatically selected as the axis. In the **Extents** drop-down list in the dialog box, select Angle. Then, enter 358 in the text box. Finally, pick **OK** to create the revolution. See **Figure 19-40.** Now, display the **Sheet Metal Features** panel. Then, pick the **Flat Pattern** button to create the developed pattern. See **Figure 19-41.**

You can use the **Project Flat Pattern** tool in a sketch if the small flat face is selected as the sketch plane. The revolved part can be projected. However, a sketched circle will not generate a hole in the revolved part using the sheet metal **Cut** tool.

Figure 19-40.
A conical duct
created with the
Revolve tool.

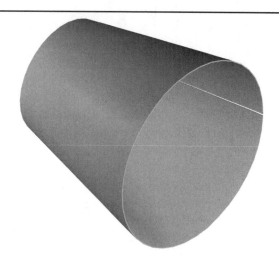

Figure 19-41.
The flat pattern for
the conical duct
shown in **Figure 19-40.**

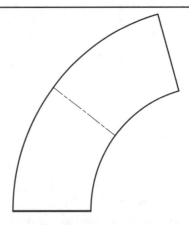

Fold Tool

The process you have used to create sheet metal parts so far is to create the folded part and then produce the unfolded, developed pattern. However, you can start with the developed pattern and then create the folded part. First, sketch the developed blank (flat pattern) and then use the **Face** tool to create a flat part. Next, sketch bend lines on the part using any linetype; however, the lines must be straight. Finally, use the **Fold** tool to create the folded part.

Open Example-19-14.ipt. There are two unconsumed sketches in this part. Each sketch is of a line that will be used as a bend line. Refer to **Figure 19-42.** A sketch is consumed as the **Fold** tool is used. Therefore, since there are to be two folds on the part, two sketches are necessary.

Figure 19-42.
A—This part was
created from the flat
pattern drawing.
The two bend lines
are shown here in
color. B—The part
after the two bends
are created.

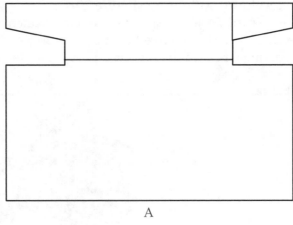

A

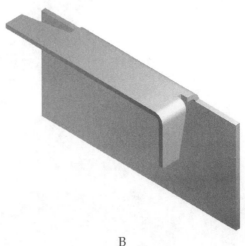

B

Figure 19-43.
The **Fold** dialog box.

Pick to select a bend line

Choose a location

Set the direction

Set the angle

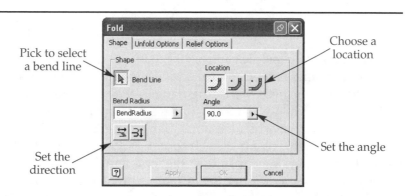

Pick the **Fold** button in the **Sheet Metal Features** panel to open the **Fold** dialog box, **Figure 19-43.** With the **Bend Line** button in the **Shape** tab on, pick the short vertical line in Sketch4 as the bend line. Refer to Figure 19-42A. Only one bend line can be selected. Once you select a bend line, you cannot select a different line without deselecting the first line. If you need to do so, pick the **Bend Line** button, hold down the [Shift] key and pick the first line again. Then, select the new line.

Now look at the preview in the graphics window. See **Figure 19-44.** The bend line is displayed in cyan. The area of the bend is outlined by a red box, as is the area of the relief. There are also two green arrows displayed, one straight and one curved. The straight arrow indicates which side of the line is going to be bent. The portion of the part to which the arrow points will be bent about the line. The curved arrow shows the direction of the bend. In the **Shape** tab of the dialog box, there are two buttons below the **Bend Radius** text box. The **Flip Side** button is used to change the direction of the straight arrow. The **Flip Direction** button is used to change the direction of the curved arrow. Using these two buttons, there are four possible bends. Pick the buttons as needed so the arrows point as shown in Figure 19-44.

There are three **Location** buttons in the **Shape** tab—**Centerline of Bend**, **Start of Bend**, and **End of Bend**. These buttons are used to determine how the bend is applied in relation to the bend line sketch. Pick the **Centerline of Bend** button so the bend is applied equally on each side of the bend line.

The bend can be from 0° and 180°. To set the bend angle, type a value in the **Angle** text box in the **Shape** tab. If you enter 180°, the result is similar to that produced by the **Hem** tool. For this example, set the angle to 90°.

When all settings have been made, pick the **OK** button. The tab on the part is folded 90° about the bend line. A relief is added based on the settings in the sheet metal style. Create another fold using the horizontal line shown in Figure 19-42A as the bend line. The top of the part should bend 90° down. This time, pick the **Start of Bend** button. Notice how the red box representing the bend shifts. The final part with the two folds is shown in Figure 19-42B.

Figure 19-44.
The preview arrows show which side of the bend will be folded and the direction. The red lines represent the extent of the bend and relief.

Bend line

Preview arrows

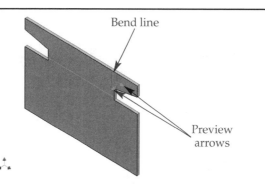

The **Corner Seam** tool controls the shape of the seam between the two edges. The gap and the relief can be controlled with the tool. Open Example-19-15.ipt. The sheet metal box was formed by creating flanges on the two long edges of the face and then on the two short edges. The long flanges have no relief as they were the full length of the edge. However, when the short flanges were created, reliefs were applied. This relief forms the hole or gap at each corner, as shown in **Figure 19-45.** The seam created by the flange operations results in the long flange overlapping the short flange.

You can use the **Corner Seam** tool to modify the seams between the short and long flanges. Pick the **Corner Seam** button in the **Sheet Metal Features** panel to open the **Corner Seam** dialog box, **Figure 19-46.** First, you must select the two edges. With the **Edges** button in the **Shape** tab on, pick two edges of one of the seams. The selection order is important in determining the overlap. For this example, pick an edge on the short flange first and then an edge on the adjacent long flange. Pick either inner or outer edges, but do not pick one of each.

After selecting the edges, you can set the type of seam. There are three choices. See **Figure 19-47.** The **No Overlap** button will open the seam. The end of each face (flange) will stop at the inner edge of the adjacent face (flange). The **Overlap** button will lengthen the first selected face to overlap the second face. The gap is applied between the extension of the first face and the second face. The **Reverse Overlap** will lengthen the second selected face to overlap the first face. In this case, there is no change because as the second choice (long side) already overlaps the first choice (short face). For this example, pick the **Overlap** button.

Figure 19-45.
The **Corner Seam** tool can be used to control the seam between two edges.

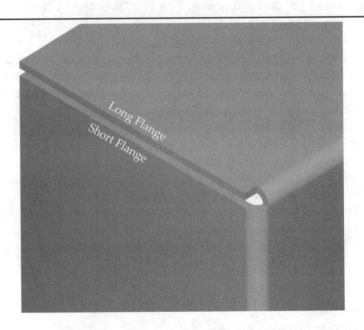

Figure 19-46.
The **Corner Seam** dialog box.

Pick to select edges

Choose a seam option

Set the gap

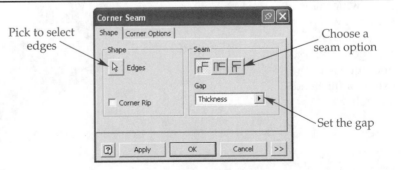

Figure 19-47.
There are three options for the type of seam. A—Reverse Overlap. B—No Overlap. C—Overlap.
The order in Inventor is No Overlap, Overlap, and Reverse Overlap.

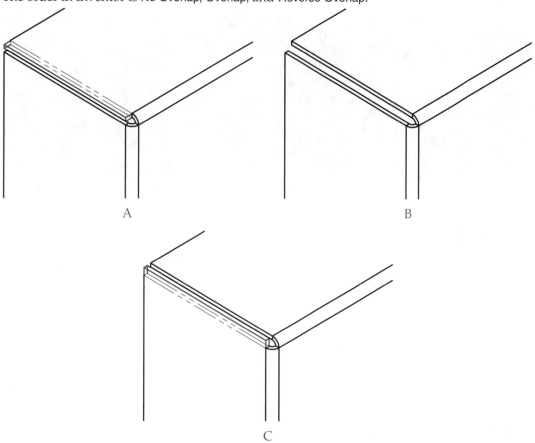

A

B

C

The shape and size of the relief at the corner is controlled by the settings in the **Corner Options** tab. There are five options for the shape of the relief. These are selected in the **Relief Shape** drop-down list in the **Corner Options** tab. The **Default** option creates the relief based on the settings in the sheet metal style. The **Trim to Bend** option produces no corner relief. The other three options are **Round, Square**, and **Tear** and are illustrated in **Figure 19-48.**

Once all settings have been made, pick the **OK** button to create the corner seam. You can edit the feature as you would any other feature.

Punch Tools

Punching is the process of creating holes in sheet metal. Stamping is similar, but the tool deforms or embosses the metal instead of cutting through it. Inventor has a catalog of standard punch tools that can be used to punch or stamp sheet metal parts. Nine of the punch tools provided with Inventor create holes and two emboss shapes into the part. **Figure 19-49** shows the shapes of the available punches. They all can be sized and positioned on the face of the part.

Open Example-19-16.ipt, which contains a simple sheet metal part. In order to use the **Punch Tool**, there must be at least one unconsumed sketch. Depending on which tool is being used, the sketch may or may not need to contain any geometry. Start a new sketch on the large rectangular face. Then, place a hole center near the center of the face and finish the sketch. There should now be an unconsumed sketch in the **Browser**. Now you can use the **Punch Tool**.

Figure 19-48.
You can control the shape of the relief created with the **Corner Seam** tool. A—Round.
B—Square. C—Tear.

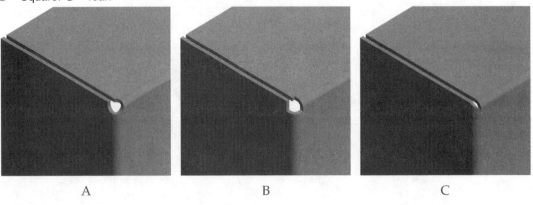

A   B   C

Figure 19-49.
The eleven punch tools that come with Inventor. Tools A through I cut, while tools J and K
emboss. A—Curved slot. B—D-sub connector 2. C—D-sub connector 3. D—D-sub connector
4. E—D-sub connector 5. F—D-sub connector. G—Keyhole. H—Keyway. I—Obround.
J—Round emboss. K—Square emboss.

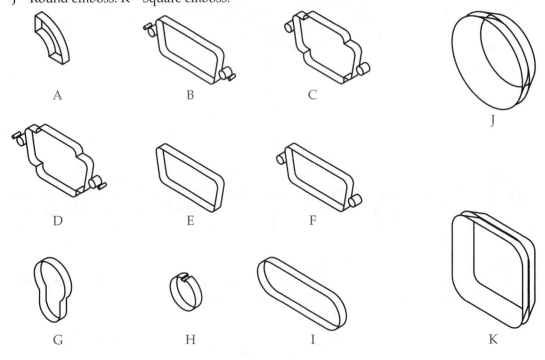

Pick the **Punch Tool** button in the **Sheet Metal Features** panel. The **Punch Tool**
dialog box is displayed, **Figure 19-50.** This dialog box works much like a wizard. At
the top of this first panel is the current path to the punch tools. By default, this is the
Inventor folder Catalog\Punches. To specify a different folder, pick the **Browse...**
button and navigate to the folder. In the list at the left are the punch tools in the
current folder. Select curved slot.ide in the list; a preview appears to the right. Then,
pick the **Next>** button to continue to the next panel.

The next panel is where you specify the center of the curved slot, **Figure 19-51.**
Depending on which punch tool you are using, you may need to select a face or other
geometry in this panel. Since there is a single hole center on the unconsumed sketch,
it is automatically selected as the center. To choose a different center, pick the **Center**
button and select the center in the graphics window. When the center is selected, pick
the **Next>** button to continue to the next panel.

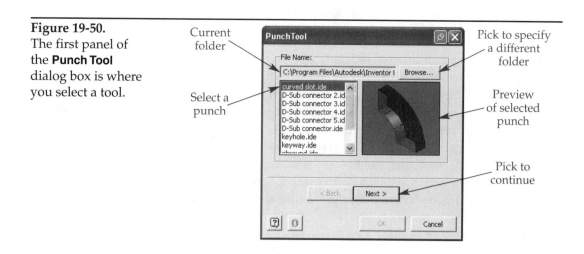

Figure 19-50.
The first panel of the **Punch Tool** dialog box is where you select a tool.

Current folder

Select a punch

Pick to specify a different folder

Preview of selected punch

Pick to continue

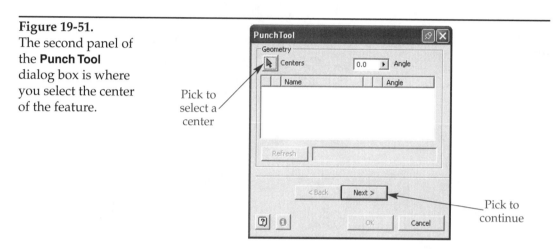

Figure 19-51.
The second panel of the **Punch Tool** dialog box is where you select the center of the feature.

Pick to select a center

Pick to continue

In the next panel, you specify the dimensions of the punch tool. The curved slot tool has three dimensions that can be specified, **Figure 19-52.** To change a dimension, pick on the value to display a text box. Then, type a new value and press [Enter]. Finally, pick the **OK** button to create the punched hole. Notice in the **Browser** that the punched feature is called **iFeature**x. There is another tool in the **Sheet Metal Features** panel called **Insert iFeature**, which can be used to manually insert a "punch" into the part.

Right-click on the punch feature name in the **Browser** and select **Edit Sketch** from the pop-up menu. Using dimensions, locate the hole center 1″ from the top edge and .5″ from the left edge. Then, finish the sketch. If needed, you can change the dimensions of the punch tool by right-clicking on the feature name in the **Browser** and selecting **Edit Feature** or **Edit iFeature** from the pop-up menu.

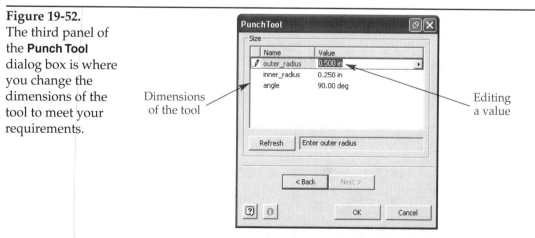

Figure 19-52.
The third panel of the **Punch Tool** dialog box is where you change the dimensions of the tool to meet your requirements.

Dimensions of the tool

Editing a value

The punch tools that cut can create holes across bend lines. Also, the **Fold** tool can create a bend through a hole created by a punch tool. However, the punch tools that emboss, or deform but do not cut, cannot be applied across bend lines. The **Fold** tool is also unable to create a bend across an embossed punch feature. Open Example-19-17.ipt. This part has two punch features, a slot created with the obround.ide punch and a boss created with the square emboss.ide punch. A bend line has been sketched across each of the punch features. Using the **Fold** tool, make a 90° bend about the line through the slot. Refer to **Figure 19-53.** Since the punch feature is a cut feature (hole), the bend is correctly applied. Now, use the **Fold** tool to make a 90° bend about the line through the boss. An error is not generated. However, since the punch feature is an embossed feature, the bend is not correctly applied and the result makes no sense.

Figure 19-53.
You can create bends across punched features that are cut. You cannot bend across embossed features.

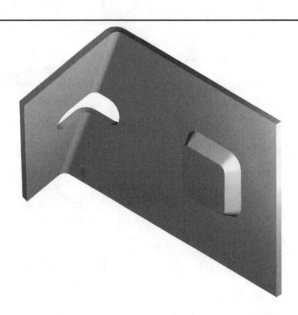

PROFESSIONAL TIP

The area affected by the punch tool does not need to be fully contained within the area of the part. A punch tool that cuts can be used to notch the edge of a sheet metal part. A punch tool that embosses can also be used on the edge of the part. However, the tool actually adds material to compensate for the area of the tool that is not within the boundary of the part.

❑ Open Practice-19-02.ipt.

❑ Using the **Punch Tool**, place a curved slot, obround slot, D-sub connector, and square emboss as shown below. Approximate the location of these features.

❑ Sketch a bend line through the center of the slot. Create a 45° bend about the line.

❑ Add two more bend lines to create corner tangs. Create 90° bends about those lines. Refer to the illustration below.

A

B

Chapter Test

Answer the following questions on a separate sheet of paper.

1. What is a *sheet metal part?*
2. Define *drawing.*
3. Define *bending.*
4. What is a *blank?*
5. What is the *developed length?*
6. Briefly describe a sheet metal style in Inventor.
7. What determines the thickness of a face created with the **Face** tool?
8. What is a *relief* and what purpose does it serve?
9. Define *remnant.*
10. List the five settings of a sheet metal style that can be overridden when creating a feature.
11. Define *developed blank.*
12. Which tool is used to create a developed blank?
13. What is the purpose of the **Cut** tool?
14. Using the **Project Flat Pattern** tool, you can project the flat pattern of a folded feature into a sketch. Why would you need to do this?
15. What is an advantage of using the **Flange** tool over using the **Face** tool?
16. Which tools are used to apply rounds and chamfers to the corners on sheet metal parts?
17. What is a *hem?*
18. Why is a hem created?
19. How many types of hems can be created?

20. From which type of geometry is a contour flange created?
21. What is the purpose of the **Fold** tool?
22. What does the **Corner Seam** tool do?
23. What are the five options for the relief shape when creating a corner seam?
24. What is the default location of the punch tools that come with Inventor?
25. Which type of punch tool can be used across a bend line, or through which a bend line can pass?

Chapter Exercises

Exercise 19-1. Open Exercise-19-01.ipt. Construct the face on the bracket as shown below. Use the **Face** tool to create the face. Add the rounds at the edge of the cutout using the **Corner Round** tool.

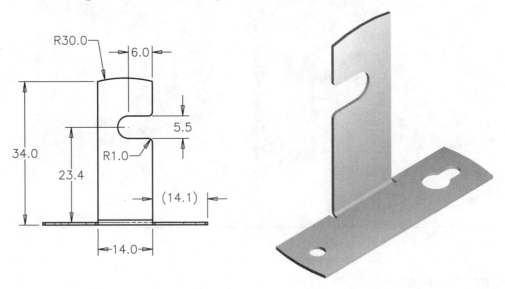

Exercise 19-2. Open Exercise-19-02.ipt. Construct a sheet metal shelf using the dimensions given below. The material is .06" thick mild steel. Use the **Flange** and the **Hem** tools as needed.

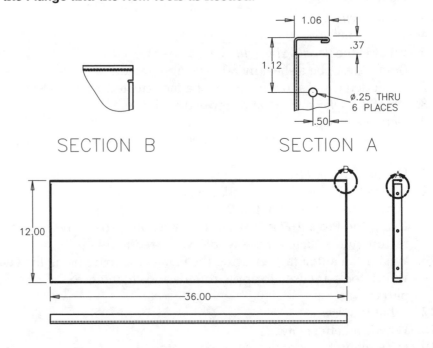

Exercise 19-3. Start a new English Sheet Metal (in).ipt file. Construct the hinged cover shown below from .15″ thick 6061 aluminum. Add a 1″ flange to the base face before constructing the double hem. Use a sketch and the **Cut** tool to create the notches in the hinge.

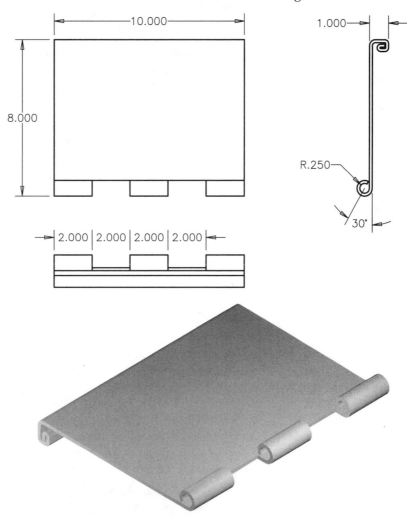

Exercise 19-4. Start a new English Sheet Metal (in).ipt file. Construct the part shown below. First, construct the flat pattern and then create the bends. The material is .125″ thick 6061 aluminum.

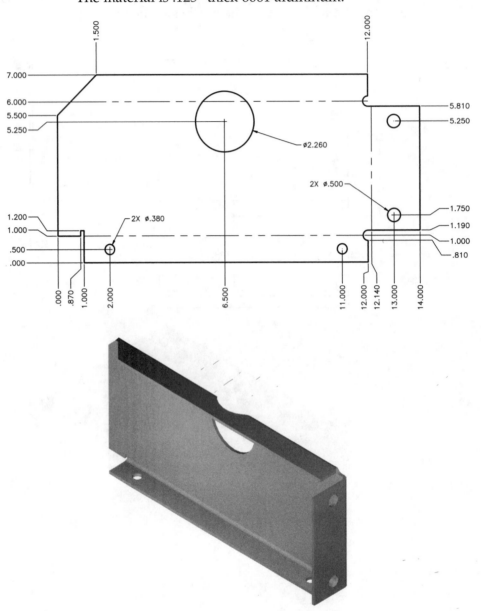

Exercise 19-5. Start a new English Sheet Metal (in).ipt file. Create a bracket shown below and save it as Exercise-19-05-Bracket.ipt. The material is .28" thick 6061 aluminum. Then, create a project and start a new assembly file. Build the assembly shown below in an exploded view. The four pressed-in fasteners are round-head, ribbed-neck bolts located in Inventor's ANSI Specialty Head Types library. The pan is created for you in the file Exercise-19-05-Pan.ipt.

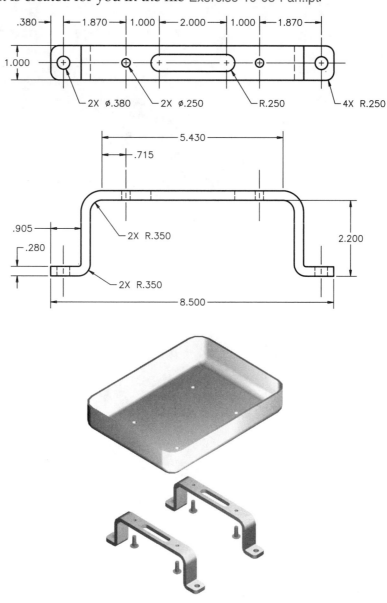

Exercise 19-6. Open the file Exercise-19-06.iam. This assembly contains two round ducts. Construct a cone to join the two ducts as a combination of a contour flange and a revolution, as shown below. Approximate the dimensions of the cone. Be sure to subtract the profile of the crossing duct from the cone.

Exercise 19-7. Open the file Exercise-19-07.ipt. Complete the part as shown below. Use the punch D-Sub Connector2.ide with its default settings to create the cutout as shown. Use geometry to create the other cutouts.

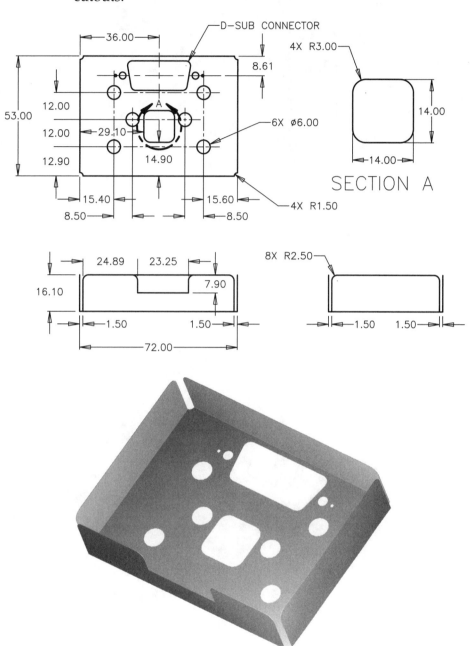

Exercise 19-8. Start a new metric Sheet Metal (mm).ipt file. Construct the bracket shown below from 0.8 mm mild steel. Approximate dimensions that are not given.

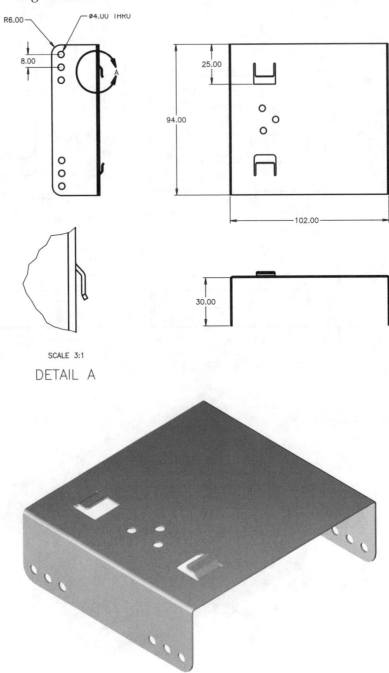

R6.00

Ø4.00 THRU

8.00

A

25.00

94.00

102.00

30.00

SCALE 3:1

DETAIL A

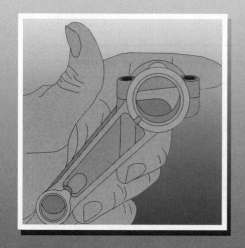

Using VBA

Objectives

After completing this chapter, you will be able to:

* Explain the basic concepts of macros and the automation of Inventor.
* Create and run macros.
* Load and save macros.

User's Files

*The following is a list of files that you will need to work through this chapter. These files can be found on the **User's Files** CD included with this text.*

Examples

Example-20-01.ivb
Example-20-02.ivb

This chapter introduces aspects of Inventor macro programming to those that are already familiar with the BASIC programming language. The programming language of Inventor macros is Visual Basic for Applications (VBA), which is a form of Visual Basic. Due to the extreme power of Inventor, only a full-featured macro language like VBA is capable of supporting the task of customizing the system. However, even VBA has limitations when compared to the other Application Programmer Interfaces (APIs) available for use with Inventor. VBA is the only API supplied with Inventor and the one discussed in this chapter.

Purpose of Macros

There are times when you will find yourself doing the same sequence of operations over and over again inside Inventor. This is where macros fit into the system. *Macros* provide an easy way to consolidate repetitive command sequences. They are programs you write and are unique to your application environment. Macro programming is a deep subject, and one that is very personal to you. The types of customization you may desire will be different from those of some other user.

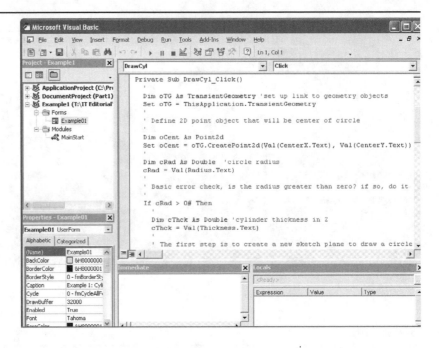

Figure 20-1.
The **Visual Basic** editor is a graphic interface to the BASIC programming language.

Introduction of VBA

VBA is a powerful programming environment where you define, save, and run macros. Contained within VBA are:
- The ability to define and edit dialog boxes for user input and reporting.
- A full-featured editor to enter and edit source code. See **Figure 20-1.**
- A visual interface to the BASIC programming language.

As the name implies, BASIC is pretty easy to learn and use. VBA is an extension of BASIC that includes the developer environment and numerous tools for interfacing to the Windows operating system. Because of these extensions, VBA is a deep programming language with many features. It will take some time to truly master VBA.

Working with Macros

The macro system inside of Inventor blends very well into the overall system. The real question boils down to how you want your macros to work into this blend. When you consider an Inventor macro, there are two basic paths you can take. See **Figure 20-2.** However, some macros will involve a combination of these paths.
- **Procedural flow.** This is when a series of commands is run given some variable input. In this type of programming, known as *procedural parameter programming*, there is a definite start followed immediately by input, processing, and output.
- **Event-driven flow.** This is when the macro is run as events take place in the system. Building event-driven macros involves an initialization or starting phase, and then the program waits to be notified when an event takes place. The remainder of the program is not activated until one of the related events occur.

For most macros, the advantage of the procedural approach is that the action and reactions involved are pretty straightforward. Inventor users who understand the design problem they face and can identify a frequently performed series of commands, can often quickly outline the macro sequence. For example, suppose the task is to create a block of steel with holes and pockets at locations based on parameters. The macro may start by displaying a dialog box (an event-driven object) in

Figure 20-2.
A comparison of macro approaches: procedural versus event driven.

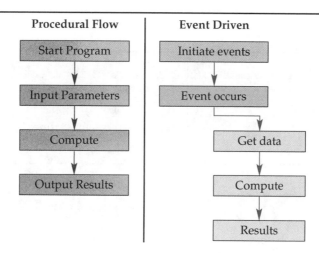

which to input the parameters, calculate points from the parameters, and then draw the part sketch. This is pretty much the same sequence as you would use when completing the part manually. Coding this in the VBA macro language requires some practice, but the results can be very productive.

Event-driven macros offer the ability to interface with the design behind the scene. While the user is modeling in Inventor, the macros can run and take advantage of the activity to accomplish their tasks. Almost all VBA programs are event driven as that is the method used to program dialog boxes. But, there are other events that take place where your macros might be able to obtain data and do extra work for the user. For example, a macro that controls access to various levels of document management might contain routines that keep an external database or spreadsheet up-to-date. Another example is to provide a running calculation as the design is changed or updated. And, you can use this tool to create custom animations of the parts that follow a complex set of rules.

Working with Objects

VBA is an object programming language. Note that VBA is not an object-*oriented* programming language because it lacks a few abilities, which are not explored in this text. It is important to realize that the VBA language can work with what are known as objects. An *object* is a container that holds properties, methods, and other objects. Other VBA terminology is defined in **Figure 20-3.**

The object is a key feature of VBA and why the language is well suited to Inventor. Inventor is exposed, or made available, to VBA in the form of a massive object library. You need to have a basic understanding of object manipulations to take advantage of the library.

An object contains methods (functions) and properties (data). An object may also contain events or functions that run when certain conditions occur in the system. Objects can even contain object references.

The entire Inventor application is treated as an object. Your VBA macros can direct the application object to perform various tasks, such as starting a new part design or plotting a series of designs. The key is to learn how the object tree is structured and then, once you have that under control, you know where to start looking when a question comes up.

The application object contains a collection of documents. *Collections* are groups of objects that may be manipulated as a single entity. For example, the document collection is an object that contains all of the open documents. When a macro is running, it has access to all the open documents, thus allowing data to be shared across multiple documents.

Figure 20-3.
Selected VBA terminology.

Object Programming Term	Definition
Object	A container holding properties, methods, and other objects.
Collection	A group of similar objects.
Method	A function or subroutine that works inside of an object.
Property	Data inside of an object

Object Tree

When objects are grouped into a chart showing the hierarchy, the result is an *object tree.* Using the analogy of a tree, the main trunk of the tree is the application object, in this case Inventor. The application object holds the main branches and is rooted into the ground. The largest branches stemming off of the main trunk of the tree are the documents. The branches stemming off of the document branches are the features within the document. Each new feature added to the document causes the document branch to grow a bit longer to hold the new feature. Some features may be inside of other features, such as a profile sketch inside of an extrusion, and are represented as another branch growing from a branch on the document branch. Each document branch holds smaller branches until you get to the leaves. In this analogy, the leaves represent data.

The key to success when learning an object programming system is to understand the context of a given function or variable value. In other words, you need to figure out where the object appears in the tree.

A drawing depicting the object tree of the API system in Inventor is provided on the Autodesk Web site at www.autodesk.com/developinventor. This object tree is very complex, but very well structured once you understand the concepts underlying the data storage inside Inventor.

VBA Development Environment

The VBA development environment is the **Visual Basic Editor**. This is a multiple document-window text editor and macro project management tool. Programs written for Inventor are created and maintained inside this environment. Open the **Visual Basic Editor** by selecting **Visual Basic Editor** from the **Macro** cascading menu in Inventor's **Tools** pull-down. Three of the windows in the **Visual Basic Editor** are where you will be working most of the time. See **Figure 20-4.** You can arrange these windows to fit your screen and preferred style of working.

Normally, you will start working in the **Projects** window. This window contains a list of the macros and forms that make up a project. It is where you can keep track of and maintain the various components in your macro.

The **Properties** window is located in the lower-left corner of **Figure 20-4.** While building dialog boxes using the forms editor, you will frequently use the **Properties** window to change details, such as the color or font of text.

The window you will spend most of the time working inside of is the **Code** window. This is the largest window shown in **Figure 20-4.** It is in this window where VBA source code is typed for the macros you will create.

Figure 20-4.
The **Project**, **Properties**, and **Code** windows in the **Visual Basic Editor**.

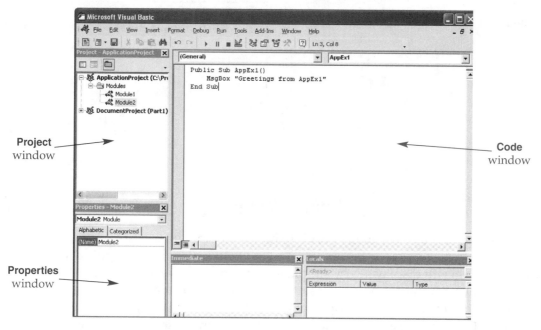

Project window

Properties window

Code window

Macros Are Stored in Projects

A VBA program may include dialog boxes as well as the code to accomplish a task. These elements are grouped together to create a project. Projects are stored in files with an extension of .ivb and can be associated with Inventor in one of three ways:

- **Application project.** This type of project loads every time Inventor is started. The project is stored in a standard location so that it is automatically available in all documents. An application project is best for macros that will be used in all documents.

- **Document project.** This type of project is loaded when a specific document is opened. A document project is best for macros that are applied only to a specific document. Macros of this nature are automatically moved with the document and available on other workstations when that document is open. A typical application for a macro associated with a document is for the creation of a specialized bill of materials or manufacturing notes.

- **User-defined project.** This type of project is loaded inside the VBA system itself from a user-specified project file name. A user-defined project is best for code that is undergoing revisions and testing.

The contents of a project may be viewed using the **Project** window. See **Figure 20-5.** A project may have several components. These components will vary based on what you used in your macro, but may include forms, modules, class modules, and Inventor objects.

- **Forms.** These are the definitions of dialog boxes and the code associated with the dialog box elements.

- **Modules.** These are the definitions of the macros. Sometimes, if no dialog boxes are used by the project, modules are the only place where code will be found.

- **Class modules.** These are definitions of functions and data for storage and manipulation as a class or object. Class modules can be used as libraries of common utility functions to handle your own specific processing requests.

- **Inventor objects.** These are the code and variable definitions associated directly with the document object.

Figure 20-5.
The **Project** window displays the contents of a project.

Simple Example

The best way to learn about VBA is to jump right in with a simple example. In this section, you will create your first macro. You will add a macro to the application project that when run will display a very simple message on the screen greeting the operator.

Start by opening the **Visual Basic Editor** from inside Inventor. Then, select the ApplicationProject entry in the **Project** window. If you are creating a document macro, select the DocumentProject entry. Next, select **Module** in the **Insert** pull-down menu. A new entry named Module2 should appear inside the Modules folder in the **Project** window. Double-click the entry for Module2 in the **Project** window to display the **Code** window. It should be empty. See **Figure 20-6.** Now, type the following code in the **Code** window.

```
Public Sub AppEx1 ()
        MsgBox "Greetings from AppEx1"
End Sub
```

To test the macro, first make sure the blinking cursor is within the block of code inside the **Code** window. Then, pick the **Run Macro** button on the toolbar, press function key [F5], or select **Run Macro** from the **Run** pull-down menu. The screen switches to Inventor and a small dialog box appears with the greetings from the VBA macro. See **Figure 20-7.** The **MsgBox** function displays the message box.

Figure 20-6.
Starting a new project.

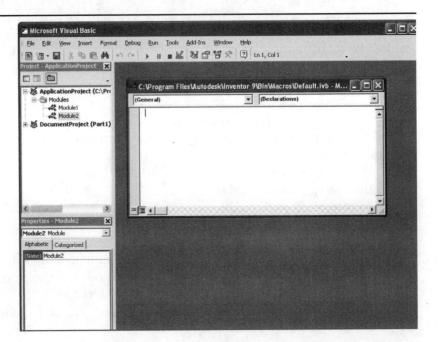

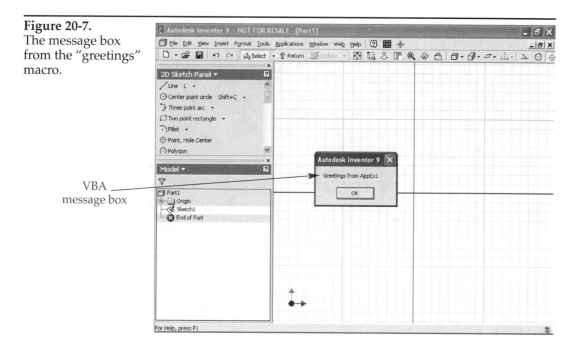

Figure 20-7.
The message box from the "greetings" macro.

VBA message box

Congratulations! You have just completed your first macro. This one does not do very much, of course. However, you are now on your way to making the computer do what you tell it to and in the way you want it done.

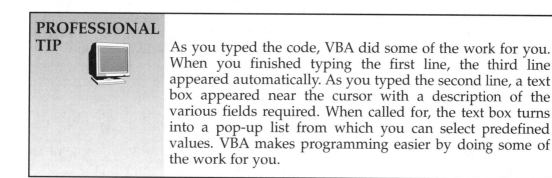

PROFESSIONAL TIP

As you typed the code, VBA did some of the work for you. When you finished typing the first line, the third line appeared automatically. As you typed the second line, a text box appeared near the cursor with a description of the various fields required. When called for, the text box turns into a pop-up list from which you can select predefined values. VBA makes programming easier by doing some of the work for you.

Saving a Macro

It is always a good idea to save your macro before any testing or after typing in several lines of code. To save a project, pick the **Save** button on the toolbar or select **Save** from the **File** pull-down menu. The default project for applications is stored in the Inventor \Macros folder. A document project is saved inside the document file itself.

Dialog Based Input

When creating a macro based on a procedure, the first action almost always involves input from the operator. VBA provides an easy-to-use tool for programming dialog boxes that will satisfy most, if not all, of the input requirements for most macros. *Dialog boxes* are rectangular areas of the screen that contain smaller rectangular objects, known as *controls.* There are various types of controls that you will recognize from your previous experience with Inventor and Windows. A partial

Figure 20-8.
A partial list of the VBA controls used with forms (dialog boxes).

Control	Purpose
TextBox	User input of text using the keyboard.
Label	Text appearing in the dialog box. This is normally a prompt string or description of the input requested.
ListBox	A list of text options.
ComboBox	User input of text or a selection from a list of options that can pop up when an arrow is selected.
CheckBox	A toggle box that can be checked or unchecked by the user.
OptionButton	Normally presented as a series of options where the operator can select any one option.
CommandButton	A button that, when pushed, will cause the program to perform some activity.
Image	A graphic button that can be used to activate some activity or just display graphic information describing the input required in the remainder of the dialog box.

listing of controls is provided in **Figure 20-8.** There are many more available than those shown. However, you will find the ones listed in the illustration to be the most common.

Dialog box programming is event-driven programming, even if the macro is procedural. You start by initializing the dialog box. The remainder of the program is connected via the various controls in the dialog box. Selecting a control or changing it in some way is an event, which initiates the code that you created to be run when the control is manipulated.

In-Depth Example

In this section, you will examine a macro that is a "real world" example of an Inventor application. The purpose of this example is to acquaint you with the various elements that go into building a macro for Inventor by "walking through" the code. This macro set demonstrates the following items.

- Dialog box with buttons and data
- Creating a work plane and sketch
- Creating a circle from parameter data
- Creating a profile and extruding it

The macro we will be exploring involves a dialog box. The dialog box allows the user to enter parameters that will be used to generate some new geometry in the part document.

The complete macro is provided on the *User's Files* CD in the file Example-20-01.ivb. To load the project, start the **Visual Basic Editor** from inside Inventor. Then, select **Load Project...** from the **File** pull-down menu and locate the IVB file on the CD. There should be two **Code** windows and a **Form** window displayed on the right-hand side of the **Visual Basic Editor**. See **Figure 20-9.**

Form

Select the title bar of the **Form** window. The window is placed on top of the others and the forms **Toolbox** is displayed. The form is the dialog box displayed by the macro. The **Cylinder Draw** form, or dialog box, contains five text boxes for input of the

Figure 20-9.
The Example1 project loaded into the **Visual Basic Editor**.

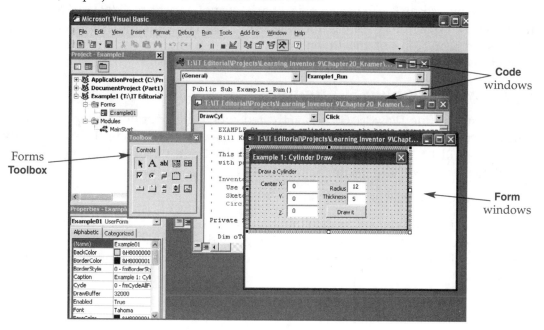

center point coordinates, radius, and thickness parameters. There are six different text labels that indicate to the user what information is needed and what action the macro performs (**Draw a Cylinder**). The form also contains a button to activate the drawing sequence (**Draw it**).

Programming an interface of this nature is very easy. Start by defining the dialog box and its contents using the tools found in the forms **Toolbox**. To add a control, pick on the button in the **Toolbox**. Then, pick and drag on the form to place and size the control. You can resize and relocate the controls using the pointing device. First, pick the control on the form so it is highlighted ("grips" appear). Then, use the standard Windows resizing and relocation methods. You can also use the **Properties** window to adjust the properties for the controls. This will change the appearance of the controls and the way in which the controls are referenced inside of your program (the control's "name").

In this example, the labels were inserted on the blank form and then the default caption and text justification properties for the labels were changed. The text boxes were inserted, then their names were changed so that they could be referenced in the actual program code with meaningful names. Default values were also established for the text boxes. Finally, the controls were all positioned so they line up (for the sake of appearance) and that completed the user input section of the coding.

Code

For this application, all the code is written to service the **Draw it** button. This button is the "go" button for the program. When the program is run, the user inputs all of the data required to draw the cylinder and then picks the button to tell the macro to do the work.

To draw the cylinder, there are several steps that must be taken. These are essentially the same steps you would take when drawing the cylinder manually in Inventor.

1. Establish a work plane and sketch.
2. Draw a circle object in the sketch.
3. Finish the sketch to create a profile.
4. Extrude the profile.

When performing these operations in VBA, you use object variables to keep track of the various things you would normally keep track of on the screen and through interactive mouse clicks. To use an object variable, you must first "declare" a variable name with a type from the Inventor object options. The value of that object is then established using one of the Inventor functions. Inventor functions are normally related to other objects, so a complex tree of objects may result.

DrawCyl_Click Step Subroutine

The subroutine DrawCyl_Click is activated when the operator picks the **Draw it** button in the dialog box. This button is named DrawCyl in the code. This is the only place you will be seeing VBA code. Activate the **Code** window that displays the complete macro code and locate the following line of code:

```
Private Sub DrawCyl_Click()
```

This subroutine, or **Sub**, is a private function inside of the dialog box module. When you double-click on the button while defining the dialog box, the **Code** window displayed and the start and end of the **Sub** are automatically created in the code area of the module.

Now, look at the next part of the code:

```
Dim oTG As TransientGeometry
Set oTG = ThisApplication.TransientGeometry
```

The Visual Basic language is extended through objects and object libraries. Even though VBA is running inside of Inventor, you still need to establish links to the various items your application will be using. The **TransientGeometry** object contains the tools that allow a program to convert data from basic numbers into points and other geometric items. From this point forward in the program we can use the *oTG* object (variable) reference to access those tools.

The first tool needed in this program is the one to create a 2D point object. The 2D point object will be used in the creation of the circle object.

```
Dim oCent As Point2d
Set oCent = oTG.CreatePoint2d(Val(CenterX.Text), Val(CenterY.Text))
```

Taking the values from the control objects located in the dialog box that were named *CenterX* and *CenterY*, the **Val** function converts the text into double-precision numbers. These values are passed to the **CreatePoint2d** method in the transient geometry object, resulting in a 2D point object. From this point on, the center point of the circle is simply referenced as an object.

Obtaining the radius of the circle and the thickness of the cylinder is just as simple:

```
Dim cRad As Double
cRad = Val(Radius.Text)
Dim cThck As Double
cThck = Val(Thickness.Text)
```

The text boxes with the name properties *Radius* and *Thickness* contain the values needed. Using control names that mean something makes the code all that much more readable. Now, the code can check the radius value to see if a nonzero value was supplied. A radius that is zero or less than zero does not make any sense. Refer to the full code in the **Visual Basic Editor** for the error checking code.

Once references to the data are obtained, the next step is to establish a working plane and sketch. Both of these features are found inside of the component definition of the part.

```
Dim oCompDef As PartComponentDefinition
Set oCompDef = ThisApplication.ActiveDocument.ComponentDefinition
```

The component definition is where the part containers can be found. From an Inventor operations point of view, a *part container* is the upper level of the features that make up a part. The new plane and sketch space that will be added to the part document will be placed in the component definition collections.

The next step is to determine a center point Z value.

```
Dim vOff As Variant
vOff = Val(CenterZ.Text)
```

The center point Z value is used to establish the work plane. This value is the offset from the XY base plane when the new plane is created.

Next, begin to establish the work plane.

```
Dim wPlane As WorkPlane
Dim oPlane As WorkPlane
Set wPlane = oCompDef.WorkPlanes.Item(3)
```

The XY base plane is the third element in the **WorkPlanes** collection inside of the component definitions of the current part. This work plane exists in all normal part documents along with the base planes for the other axis pairs. The above code dimensions work plane object *wPlane* as the XY base plane.

Now, create the new work plane definition relative to the XY plane with the offset supplied. The offset is the center point Z of the cylinder to be created.

```
Set oPlane = oCompDef.WorkPlanes.AddByPlaneAndOffset(wPlane, vOff)
```

There are other "add" functions you can use to create a new work plane. Refer to the online documentation for more information.

By default, work planes are created as visible. The following code makes the work plane invisible to the operator.

```
oPlane.Visible = False
```

Once an object relationship is established, you can manipulate it with ease.

A work plane allows us to quickly define a new sketch for the part using the **Add** method.

```
Dim oSketch As PlanarSketch
Set oSketch = oCompDef.Sketches.Add(oPlane)
```

Sketches and all other feature components are found in collections. The **Add** method adds new members using a series of default settings and parameter types. There are several "add" methods to facilitate the creation of new objects.

When working with a part document, you need to switch to the edit sketch mode before adding new graphics.

```
oSketch.Edit
```

This method opens the newly created sketch in edit mode.

The **Circle** object is added to the sketch using the **AddByCenterRadius** method.

```
Dim oCirc As SketchCircle
Set oCirc = oSketch.SketchCircles.AddByCenterRadius(oCent, cRad)
```

The circle is created and a references to it is obtained (*oCirc*).

Only one circle is being added to the sketch, so the sketch can now be finished.

```
oSketch.ExitEdit
```

When starting in sketch mode, it is not required that you start with **Edit** and end with **ExitEdit**. These statements are used when starting in another mode and switching to sketch mode inside your program.

The next step is to convert the closed objects in the sketch (the circle) into a profile.

```
Dim oProf As Profile
Set oProf = oSketch.Profiles.AddForSolid
```

Once converted into a proper profile reference, the geometry can be extruded to form the desired 3D shape, which is a cylinder in this case.

The thickness of the cylinder is applied when the extrusion feature is created. Using the profile just created from the sketch, the extrusion feature is added to the component definitions collection set.

```
Dim oExtr As ExtrudeFeature
Set oExtr = _
    oCompDef.Features.ExtrudeFeatures.AddByDistanceExtent( _
        oProf, cThck, kNegativeExtentDirection, kJoinOperation)
```

Notice that there are flags to indicate the direction of the extrusion (**kNegativeExtenDirection**) and what to do when encountering other solid objects (**kJoinOperation**).

After the extrusion is completed, the macro is finished. A new work plane, which is invisible, and extrusion are now part of the features in the part document. The extrusion is based on a sketch that contains a single circle.

Selection Based Input

Many times your program will need to obtain information from the design via a user selection. For example, your program might need to calculate the area of a face that the user selects. Using VBA in the Inventor, select input is based on an object known as the *interaction manager.* The interaction manager is the parent object for the various interfaces of Inventor, including the mouse, keyboard, and a special object selection driver.

Using selection-based input is simple. An object is created to link to an instance of the interaction manager. That object is also used to link to an instance of the **Select** object. The properties of the two objects are set for the specific type of objects desired and the event is started. At this point, your program runs a loop calling the function **DoEvents**. When objects are selected, they are placed in a collection inside the **Select** object. Each time an object is selected, an event occurs that results in a function running. This, in turn, allows you to perform a variety of tasks related to the selected object.

Using a class module, objects can be defined that contain events. The class description for the input handling functions starts by declaring object references with events enabled. The following code normally appears near the top of the module, not inside any of the function definitions.

```
Private WithEvents oIntEv As InteractionEvents
Private WithEvents oSelEv As SelectEvents
Private bNotDoneYet As Boolean
```

The third line of code is a declaration of a Boolean flag. It is used in the module to indicate that selection is complete. The variable is defined at the top of the listing in the general declarations so that it will be available to functions and subroutines inside of the module.

The event handler will call functions related to the interaction and select events as they occur in the process of the run.

```
Set oIntEv = ThisApplication.EventManager.CreateInteractionEvents
```

The event manager is referenced inside of the application object to obtain a link to the interaction events. Interaction implies that your program is interacting with the operator.

There are two flags to set up before starting:

```
oIntEv.SelectionActive = True
bNotDoneYet = True
```

The first flag is associated with the interaction events manager. It informs the event manager that the event involves a selection. Thus we want the selection features enabled. The second flag is the indicator that selection is not completed.

The selection event object is then linked using the interaction object just created.

```
Set oSelEv = oIntEv.SelectEvents
```

Each of these objects provide you with various control options of the input system. The interaction object controls the overall process, while the selection event object is used for the specific selection activity and as a place to store the selected object references.

PROFESSIONAL TIP  Basic user input functions can be placed in a class module stored in the application project. This class module can always be imported into a different project and changed to suit the needs at the moment. Most of the time, you can use this class module to achieve what you need quickly, since it will be specialized for accessing the type of parts and components you typically use.

Filters in Selection

When selecting objects inside of a program, even with user support, you want to control exactly what is being selected. This is accomplished using *filters.* Inventor supports a variety of selection filters that enable you to specify valid selections. The type of feature required will depend on the application and the reason behind the selection.

For examples of the features discussed in this section, refer to the macro provided on the *User's Files* CD in the file Example-20-02.ivb. This macro contains similar code to the examples shown in this section. To load the project, start the **Visual Basic Editor** from inside Inventor. Then, select **Load Project...** from the **File** pull-down menu and locate the IVB file on the CD. There should be two **Code** windows and a **Form** window displayed on the right-hand side of the **Visual Basic Editor**.

The first step is to establish the filter to apply to the selection. Only those objects matching the filter(s) supplied in this step can be selected by the user. As many filter options can be defined as needed by making repeated calls to the **AddSelectionFilter** method. A complete list of the available filter settings is available in the online documentation for the **Visual Basic Editor**.

```
oSelEv.AddSelectionFilter kPartEdgeFilter
```

The name *kPartEdgeFilter* is part of an enumeration for the select event object. *Enum* is short for enumeration and is used in VBA programming to mean a list of coded entries. Each code is associated with a name designed to improve the readability of the code. Inventor VBA has many enums for various codes that you will use. They are all detailed in the online help. When an object like *oSelEv* is created in the declaration area of the module, it is also added to the list of objects in the module.

Once the filters are set up, you can start the interaction event by simply calling the start method. Control will return to the program once the interaction event has started.

```
oIntEv.Start
Do While bNotDoneYet
    Do Events
Loop
```

The **Do While** loop will repeat so long as the variable *bNotDoneYet* remains true. The value of *bNotDoneYet* is controlled in a different function. **Do Events** is a call from the macro to the VBA system manager to let the system manager know that it should run any pending events.

The use of the **WithEvents** causes a series of event names to be imported for you to define. These event names start with the name of the object and then the specific event name. When the event happens, the event handling function is called by Inventor and can take whatever action is desired.

The following code sets the *bNotDoneYet* flag to false when something is selected. This is how the event manager connects back to your program. The event manager looks for named subroutines to call whenever a specific event occurs.

```
Private Sub oSelEv_OnSelect (_
        ByVal JustSelectedEntities As ObjectEnumerator, _
        ByVal SelectionDevice As SelectionDeviceEnum, _
        ByVal ModelPosition As Point, _
        ByVal ViewPosition As Point2D,_
        ByVal View As View)
    bNotDoneYet = False
End Sub
```

Functions such as the **OnSelect** callback will have parameters that point to information related to the cause of the callback. **OnSelect** starts with five parameters. They include an object enumerator (collection of object references from Inventor) of the objects just selected, a code indicating the selection mechanism used (from the browser list or on screen), the point where the object was selected relative to the model, the point relative to the screen coordinate system, and the view object. The view object is the actual view in which the object was selected.

In the majority of cases, you will be ignoring the data sent to the **OnSelect** subroutine since it will be the selected features themselves you are interested in obtaining. In those cases, the function will be simple, like the one shown where the flag *bNotDoneYet* is set to indicate the completion of the selection process. That means you are allowing only one object to be selected.

Selected objects do not need to be saved directly. They are included in the object reference for the select event. Back in the main program, after the **While** loop, the program can iterate the collection of selected objects. In this simple example, only one object will be selected at a time.

Getting Data from Objects

Most of the data that your application needs can be found as object properties or the result of methods. The key to success here is that you must know exactly which object you have in hand. That can be accomplished using selection filters. Using selection filters you can control most of the user input options so that you do get what you expect.

The following code, extracted from Example2, shows how simple it is to use the selection class described earlier to obtain user input and then access associated data. The first step is to hide the dialog box:

```
Example02.Hide
```

If you do not hide the dialog box, it will stay on the screen making object selection a bit more difficult in some cases. At the very least it is confusing to the user who is not sure about what to do next.

Using the custom class defined for object selection (and found in the *clsSelect* class module) a new instance is defined and a face object requested for selection.

```
Dim oSelect As New clsSelect
Dim oFace As Face
Set oFace = oSelect.Pick(kPartFaceFilter)
```

When the function **Pick** is completed, the result is a reference to the object that is selected, if any object is selected. That value needs to be tested:

```
If Not oFace Is Nothing Then
```

Testing to see if the result of the **Pick** function is valid allows our function to proceed without any further concerns. The question now is what can be learned from a face object.

The face object information can be viewed using the programming help in Inventor's online documentation. Most of the data available is concerned with connecting the face to other geometry in the part. At first glance, there does not appear to be much there for analysis purposes. But looking another level deeper, at the **SurfaceEvaluator** object, you might find what you need for a particular calculation. For example, the following code retrieves the area of the face object.

```
Dim myArea As Double
myArea = oFace.Evaluator.Area
```

When finished accessing the selection, the dialog box can be displayed again using the **Show** method:

```
Example02.Show
```

Each of the features has different properties that may be of interest in your application. The dialog box in Example2 allows you to select a face or an edge. See **Figure 20-10.** When the face is selected, the area of the selection is displayed in the dialog box. A label control in the dialog box named Answer is used to show the result.

Note that the dialog box in Example2 contains a toggle that allows the user to display the result in inches instead of the default centimeters. A global variable named dUnitDiv is initially set to 1.0, indicating a divisor of one (centimeters). When the inches flag is selected, the value is set to 2.54, the conversion divisor of centimeters to inches.

```
myArea = myArea / (dUnitDiv ^ 2)
```

If the value is to be output in centimeters, the divisor will be one and the result of the division will be nothing changed. Otherwise, the value is converted to square inches.

Chapter 20 Using VBA

Figure 20-10.
The dialog box that
is displayed when
the Example2 macro
is run in Inventor.

The best way to convert numeric data into string data is the **Format** function in VBA.

```
S$ = Format(myArea, "#.###")
```

The variable S$ is a string variable that will hold the conversion of the area value into a text string. The string #.### specifies that the converted value is to have digits on each side of the decimal and a precision of three places to the right of the decimal.

The following code checks the unit divisor and adds the units to the variable S$.

```
If dUnitDiv = 1# Then
        S$ = S$ & " cm^2"
    Else
        S$ = S$ & " in^2"
End If
```

That value will either be square centimeters (cm^2) or square inches (in^2). The value 1# is the way VBA denotes an integer that is a double-precision value.

By having the code adjust the caption property for the label named Answer, the user can see the result in the dialog box.

```
Answer.Caption = "Face area = " & S$
```

The caption is a string variable and will be updated as soon as the assignment operation is completed.

Obtaining Details from Edges

To obtain more specific data from an object, you need to investigate and use the various properties and methods contained in the object definitions. In this section, the portion of the Example2 code related to the dialog box button **Pick an edge** is discussed. When the user picks the button, they can then select an edge on the part. The details about the edge are displayed in the label area of the dialog box named Answer.

To view the function, load the Example-20-02.ivb project in the **Visual Basic Editor** and make the **Form** window active. Next, double-click on the button labeled **Pick an edge** in the **Form** window. The **Code** window is displayed with the code for the click selection callback.

When the Pick subroutine in Example2 is called looking for an edge object, it returns a reference to an object. However, all the program knows at that time is that the object is an edge of some kind. The edge could be a line, arc, B-spline, or circle. The Example2 macro processes arcs, lines, and circles. B-spline edges are detected, but no numeric information is extracted for report.

Starting inside of the callback function for the edge details, the program sets up an edge object reference to be set by the **Pick** function from the custom selection class. This is the same as the earlier discussion, but **Pick** is used with **kPartEdgeFilter** inside of **kPartFaceFilter**. After getting the edge data from the **Pick** function, the data are tested to see if nothing was returned.

```
Dim oEdge As Edge
Set oEdge = oSelect.Pick(kPartEdgeFilter)
If Not oEdge Is Nothing Then
```

If the edge selection was a success, then the value of *oEdge* is not empty and can be used.

The variable *oEdge* is a generic edge definition and does not have the type of details to be reported. For those data items, the application must reference another level in the **Geometry** object.

```
Dim oGeo As Object
Set oGeo = oEdge.Geometry
```

The geometry object reference is a direct reference to the specific types of geometry available. You can use the **TypeOf** function to test *oGeo* for a match against the known types. **TypeOf** *object* **Is** *object class name* will return **True** if the object in question matches the type. Otherwise, the statement will return **False**.

```
If TypeOf oGeo Is Arc3d Then
```

Once you know for certain that the object is of a particular class, you can set it to a local variable declared of that type.

```
Dim myArc As Arc3d
Set myArc = oGeo
```

After the set statement, the variable *myArc* can be used to reference the properties and methods for the arc object. See **Figure 20-11.** For example, the arc length can be quickly computed by multiplying the sweep angle and the radius:

```
Dim ArcLength As Double
ArcLength = myArc.SweepAngle * myArc.Radius
```

It is important to note that angles in VBA are *always* in radians.

Figure 20-11.
The methods and properties of the **Arc3d** object.

Method or Property	Data
GetArcData	Returns all arc data as parameters.
Center	Point object.
EndPoint	Point object.
Evaluator	Link to CurveEvaluator object.
Normal	Vector object.
Radius	Double real number.
ReferenceVector	Vector object.
StartPoint	Point object.
SweepAngle	Double real number; included angle in radians.

One of the most common places to customize Inventor is in regards to the batch, or group, processing of documents. You can access the documents collection from the application object level. In manipulating this collection, you can open and close documents, and work with the open documents. Each of the document types can be manipulated in this manner. The type of document will determine what you can do.

The following function loops through the documents that are currently open and displays the open files in the **Immediate Window** in the **Visual Basic Editor**. If this window is not displayed in the **Visual Basic Editor**, it can be displayed by pressing [Ctrl]+[G].

```
Public Sub ListOpenFiles()
    Dim oDoc As Document
    For Each oDoc In ThisApplication.Documents
        S$ = oDoc.FullFileName
        Select Case oDoc.DocumentType
            Case kAssemblyDocumentObject
                Debug.Print "Assembly - " & S$
            Case kDesignElementDocumentObject
                Debug.Print "Design element - " & S$
            Case kDrawingDocumentObject
                Debug.Print "Drawing - " & S$
            Case kPartDocumentObject
                Debug.Print "Part - " & S$
            Case kPresentationDocumentObject
                Debug.Print "Presentation - " & S$
        End Select
    Next oDoc
End Sub
```

Every drawing in Inventor has a set of properties associated with it. These properties include details for tracking the documents, such as the edit dates, names, and titles. There may also be design-related information included inside of a document. All of this is stored using a **Property** object. Each property contains a name and a value.

To illustrate the access of properties inside a document, look at the following code. It changes the value of the company property to the string A New Company for all open documents. To achieve the task, the program must dig several levels down from the application level through a series of collections. The short coding required to accomplish this task is a testament to the power of VBA and object handling.

```
Public Sub myCompany()
    Dim oDoc As Document
    Dim oPS As PropertySet
    Dim oPr As Property
    '
    For Each oDoc In ThisApplication.Documents
        For Each oPS In oDoc.PropertySets
            For Each oPr In oPS
                If oPr.Name = "Company" Then
                    oPr.Value = "A New Company"
                End If
            Next oPr
        Next oPS
    Next oDoc
End Sub
```

The property being sought is named Company. It is found inside the collection of properties known as a **PropertySet**. The **PropertySet** is contained in a collection inside of the document known as the **PropertySets**. At the top level, the document is housed in a collection of documents in the application object.

Learning More about Objects

The main problem with objects is learning the "who" and "where." Inventor macro programmers can expect to spend a lot of time searching for the "perfect object" for any given situation. There are many different objects and sometimes the obvious choice does not provide what you are seeking.

So how do you learn more about the various objects available to you? The best place to start is the online documentation. Inventor is supplied with a rather extensive online library of references pertaining to customization. To access this information, select **Programming Help** from Inventor's **Help** pull-down menu. See **Figure 20-12.** Some of the concepts behind writing Inventor macros are explained along with a list of almost all of the various objects, methods, properties, and predefined constants. It is these lists to which you will find yourself referring most often.

Another place to find out more about objects is the **Object Browser** in the **Visual Basic Editor**. This is displayed by selecting **Object Browser** from the **View** pull-down menu in the **Visual Basic Editor**. See **Figure 20-13.** When you select a class on the left side of the **Object Browser,** its members are displayed on the right side. When you select a member, its description is displayed at the bottom of the **Object Browser**. When looking at a specific item, you can press the [F1] function key to display the online documentation about that particular detail.

A third place to learn about objects is to take advantage of the "break point" system in VBA and explore the data in the immediate and locals windows. The locals window will display the current values of variables in your macro and you can explore their details searching for the undocumented aspects of VBA.

Figure 20-12.
Inventor's online documentation contains programming help.

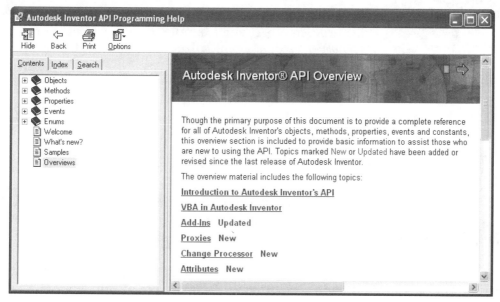

Figure 20-13.
Using the **Object Browser** to obtain information about objects.

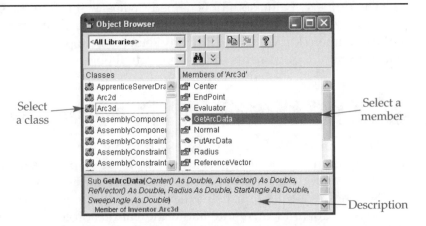

Select a class

Select a member

Description

Where Next?

VBA is the perfect introduction to the customization of Inventor. The object programming approach is well serviced in the VBA environment for most applications. However, it may not be long before you decide you need more. There are some forms of customization that are simply best done using the additional tools of the API.

If you want to leverage your knowledge of VBA, then the next step is to go into Visual Basic (VB). VBA is supplied inside Inventor, while VB is a product from Microsoft. Everything you have learned in VBA can be applied to VB. However, your projects will not migrate with ease. Migration will have to be done one module at a time. Export each module to disk inside VBA and then import them into a VB project to accomplish the migration.

VB is different from VBA from a programmer's perspective. In VBA, you start with connections to the Inventor objects being quickly available. In VB, you will have to make those connections. Another difference is that VB is more powerful than VBA when it comes to working with the operator system and data. It is also faster because it is compiled.

PROFESSIONAL TIP

Define global variables of the same name as the system variables used in VBA. For example, create a variable named *ThisApplication* that is initialized to point to the Inventor application object.

The other programming tool for Inventor customization is Microsoft Visual C++. Using the Inventor API library and associated definitions, you can create programs in C++ that perform all the same functions as one in VB or VBA. The main advantage is that in C++ you have better control over what is going on inside of the system. Applications will run faster because of the way C++ can talk to the Inventor interfaces. In fact, it will run many times faster than the same program written in VBA. But, the expense of a C++ program is that the program is more difficult to create. C++ coding involves more lines of code in most cases and a solid understanding of what is going on inside the computer.

But the quest for speed must be tempered with the quest for automation. Spending several days writing a C++ program compared to a couple hours to write a VBA program means that the application must be run many times in order to achieve the payback. There are other incentives to using C++ and the Inventor API, but most of these are for third-party developers who must support a large number of users.

Chapter Test

Answer the following questions on a separate sheet of paper.

1. Where are the three places that macros can be stored?
2. What do *objects* contain?
3. What do VBA *projects* contain?
4. Which two statements are used to dimension and set variables?
5. What is a *form* in VBA?
6. What VBA element do dialog boxes contain?
7. Which control is used to display read-only text?
8. On which language is VBA based?
9. What is a list of names associated with code numbers called in VBA?
10. Which function displays a text message to the user?
11. What are the two types of programming paths?
12. Which object is used to convert numbers to points?
13. What is the function called when an object is selected and you have started a select event object named *oSel?*
14. Where is the area of a face object stored?
15. In VBA, which unit of measure is used to store angles inside of objects?

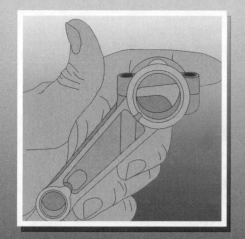

Appendix

Common Shapes of Metals

Shapes	Length	How Measured	How Purchased
Sheet less than 1/4″ thick	Up to 144″	Thickness × width, widths to 72″	Weight, foot, or piece
Plate more than 1/4″ thick	Up to 20′	Thickness × width	Weight, foot, or piece
Band	Up to 20′	Thickness × width	Weight or piece
Rod	12′ to 20′	Diameter	Weight, foot, or piece
Square	12′ to 20′	Width	Weight, foot, or piece
Flats	Hot rolled 20′-22′ Cold finished	Thickness × width	Weight, foot, or piece
Hexagon	12′ to 20′	Distance across flats	Weight, foot, or piece
Octagon	12′ to 20′	Distance across flats	Weight, foot, or piece
Angle	Up to 40′	Leg length × leg length × thickness of legs	Weight, foot, or piece
Channel	Up to 60′	Depth × web thickness × flange width	Weight, foot, or piece
I-beam	Up to 60′	Height × web thickness × flange width	Weight, foot, or piece

INCHES		MILLI-METERS	INCHES		MILLI-METERS
FRACTIONS	DECIMALS		FRACTIONS	DECIMALS	
	.00394	.1	15/32	.46875	11.9063
	.00787	.2		.47244	12.00
	.01181	.3	31/64	.484375	12.3031
1/64	.015625	.3969	1/2	.5000	12.70
	.01575	.4		.51181	13.00
	.01969	.5	33/64	.515625	13.0969
	.02362	.6	17/32	.53125	13.4938
	.02756	.7	35/64	.546875	13.8907
1/32	.03125	.7938		.55118	14.00
	.0315	.8	9/16	.5625	14.2875
	.03543	.9	37/64	.578125	14.6844
	.03937	1.00		.59055	15.00
3/64	.046875	1.1906	19/32	.59375	15.0813
1/16	.0625	1.5875	39/64	.609375	15.4782
5/64	.078125	1.9844	5/8	.625	15.875
	.07874	2.00		.62992	16.00
3/32	.09375	2.3813	41/64	.640625	16.2719
7/64	.109375	2.7781	21/32	.65625	16.6688
	.11811	3.00		.66929	17.00
1/8	.125	3.175	43/64	.671875	17.0657
9/64	.140625	3.5719	11/16	.6875	17.4625
5/32	.15625	3.9688	45/64	.703125	17.8594
	.15748	4.00		.70866	18.00
11/64	.171875	4.3656	23/32	.71875	18.2563
3/16	.1875	4.7625	47/64	.734375	18.6532
	.19685	5.00		.74803	19.00
13/64	.203125	5.1594	3/4	.7500	19.05
7/32	.21875	5.5563	49/64	.765625	19.4469
15/64	.234375	5.9531	25/32	.78125	19.8438
	.23622	6.00		.7874	20.00
1/4	.2500	6.35	51/64	.796875	20.2407
17/64	.265625	6.7469	13/16	.8125	20.6375
	.27559	7.00		.82677	21.00
9/32	.28125	7.1438	53/64	.828125	21.0344
19/64	.296875	7.5406	27/32	.84375	21.4313
5/16	.3125	7.9375	55/64	.859375	21.8282
	.31496	8.00		.86614	22.00
21/64	.328125	8.3344	7/8	.875	22.225
11/32	.34375	8.7313	57/64	.890625	22.6219
	.35433	9.00		.90551	23.00
23/64	.359375	9.1281	29/32	.90625	23.0188
3/8	.375	9.525	59/64	.921875	23.4157
25/64	.390625	9.9219	15/16	.9375	23.8125
	.3937	10.00		.94488	24.00
13/32	.40625	10.3188	61/64	.953125	24.2094
27/64	.421875	10.7156	31/32	.96875	24.6063
	.43307	11.00		.98425	25.00
7/16	.4375	11.1125	63/64	.984375	25.0032
29/64	.453125	11.5094	1	1.0000	25.40

Solutions to Triangles

	$A + B + C = 180°$ $S = \dfrac{a+b+c}{2}$	Right (right triangle: A, b, 90° C, c, a, B)	Oblique (oblique triangle: A, c, b, a, B, C)

Have	Want	Formulas for Right	Formulas for Oblique
abc	A	$\tan A = a/b$	$1/2A = \sqrt{(s-b)(s-c)/bc}$
	B	$90° - A$ or $\cos B = a/c$	$\sin 1/2B = \sqrt{(s-a)(s-c)/a \times c}$
	C	$90°$	$\sin 1/2C = \sqrt{(s-a)(s-b)/a \times b}$
	Area	$a \times b/2$	$\sqrt{s \times (s-a)(s-b)(s-c)}$
aAC	B	$90° - A$	$180° - (A + C)$
	b	$a \cot A$	$a \sin B/\sin A$
	c	$a/\sin A$	$a \sin C/\sin A$
	Area	$(a^2 \cot A)/2$	$a^2 \sin B \sin C/2 \sin A$
acC	A	$\sin A = a - c$	$\sin A = a \sin C/c$
	B	$90° - A$ or $\cos B = a/c$	$180° - (A + C)$
	b	$\sqrt{c^2 - a^2}$	$c \sin B/\sin C$
	Area	$1/2a \sqrt{c^2 - a^2}$	$1/2 ac \sin B$
abC	A	$\tan A = a/b$	$\tan A = a \sin C/b - a \cos C$
	B	$90° - A$ or $\tan B = b/a$	$180° - (A + C)$
	c	$\sqrt{a^2 - b^2}$	$\sqrt{a^2 + b^2 - 2ab \cos C}$
	Area	$a \times b/2$	$1/2ab \sin C$

Area Equivalents

1	area = radius^2 × 3.1416 or diameter^2 × .7854
1	circumference = diameter × 3.1416 or diameter ÷ .3183
2	when the area of a circle & square are equal, D = S × 1.128
2	when the area of a circle & square are equal, S = D × .8862
3	side of inscribed square – diameter × .7071
3	diameter of circumscribing circle = S × 1.1412
4	surface area of a sphere = diameter × circumference
4	volume of a sphere = diameter^3 × .5236

Equivalents

Fahrenheit and Celcius
$°F = (1.8 \times °C) + 32$ $°C = (°F - 32) \div 1.8$

Weight
1 gram = .03527 oz (av.)   1 oz = 28.35 grams   1 kilogram = 2.2046 pounds   1 pound = .04536 kilograms   1 metric ton = 2,204.6 pounds   1 ton (2000) lbs in U.S. = 907.2 kg.

Volume
1 U.S. quart = 0.946 liters   1 U.S. gallon = 3.785 liters   1 liter = 1.0567 U.S. quarts   1 liter = .264 U.S. gallons

Rutland Tool and Supply Co., Inc.

Length Conversions

multiply	by	to obtain
Inches	25.4	Millimeters
Feet	304.8	Millimeters
Inches	2.54	Centimeters
Feet	30.48	Centimeters
Millimeters	.03937008	Inches
Centimeters	.3937008	Inches
Meters	39.37008	Inches
Millimeters	.003280840	Feet
Centimeters	.03280840	Feet
Inches	.0254	Meters

Square Area Conversions

multiply	by	to obtain
Millimeters	.00001076391	Feet
Millimeters	.00155003	Inches
Centimeters	.1550003	Inches
Centimeters	.001076391	Feet
Inches	645.16	Millimeters
Inches	6.4516	Centimeters
Inches	.00064516	Meters
Feet	.09290304	Meters
Feet	929.0304	Centimeters
Feet	92,903.04	Millimeters

Rutland Tool and Supply Co., Inc.

Physical Properties of Metals

Metal	Symbol	Density	Specific Heat	Melting Point*		Lbs. per Cubic Inch
				C°	F°	
Aluminum (cast)	Al	2.56	.2185	658	1217	.0924
Aluminum (rolled)	Al	2.71	–	658	1217	.0978
Antimony	Sb	6.71	.051	630	1166	.2424
Bismuth	Bi	9.80	.031	271	520	.3540
Boron	B	2.30	.3091	2300	4172	.0831
Brass	–	8.51	.094	–	–	.3075
Cadmium	Cd	8.60	.057	321	610	.3107
Calcium	Ca	1.57	.170	810	1490	.0567
Chromium	Cr	6.80	.120	1510	2750	.2457
Cobalt	Co	8.50	.110	1490	2714	.3071
Copper	Cu	8.89	.094	1083	1982	.3212
Columbium	Cb	8.57	–	1950	3542	.3096
Gold	Au	19.32	.032	1063	1945	.6979
Iridium	Ir	22.42	.033	2300	4170	.8099
Iron	Fe	7.86	.110	1520	2768	.2634
Iron (cast)	Fe	7.218	.1298	1375	2507	.2605
Iron (wrought)	Fe	7.70	.1138	1500-1600	2732-2912	.2779
Lead	Pb	11.37	.031	327	621	.4108
Lithium	Li	.057	.941	186	367	.0213
Magnesium	Mg	1.74	.250	651	1204	.0629
Manganese	Mn	8.00	.120	1225	2237	.2890
Mercury	Hg	13.59	.032	−39	−38	.4909
Molybdenum	Mo	10.2	.0647	2620	47.48	.368
Monel metal	–	8.87	.127	1360	2480	.320
Nickel	Ni	8.80	.130	1452	2646	.319
Phosphorus	P	1.82	.177	43	111.4	.0657
Platinum	Pt	21.50	.033	1755	3191	.7767
Potassium	K	0.87	.170	62	144	.0314
Selenium	Se	4.81	.084	220	428	.174
Silicon	Si	2.40	.1762	1427	2600	.087
Silver	Ag	10.53	.056	961	1761	.3805
Sodium	Na	0.97	.290	97	207	.0350
Steel	–	7.858	.1175	1330-1378	2372-2532	.2839
Strontium	Sr	2.54	.074	769	1416	.0918
Tantalum	Ta	10.80	–	2850	5160	.3902
Tin	Sn	7.29	.056	232	450	.2634
Titanium	Ti	5.3	.130	1900	3450	.1915
Tungsten	W	19.10	.033	3000	5432	.6900
Uranium	U	18.70	–	1132	2070	.6755
Vanadium	V	5.50	–	1730	3146	.1987
Zinc	Zn	7.19	.094	419	786	.2598

*Circular of the Bureau of Standards No.35, Department of Commerce and Labor

Density of Specific Engineering Materials

PLASTICS

Material	Density (g/cc)
Polyethylene, High-Density	0.935-0.960
Polyethylene, Low-Density	0.910-0.925
Polystyrene, High-Impact	1.05-1.15
ABS	1.05-1.19
Acetal (Delrin)	1.41
Nylon, 66	1.14
Nylon 66, 40% glass	1.44-1.47
Polycarbonate	1.20
Polycarbonate, 30% glass	1.43
PPO	1.04-1.07
PPO, 30% glass	1.28
Polyester teraphthalate, 30% glass	1.70
Polypropylene	0.90
Polypropylene, 40% Talc	1.26

THERMOSET PLASTICS

Material	Density (g/cc)
Carbon fiber,epoxy	1.80
Fiberglass, epoxy	1.90
Fiberglass, polyester	1.90
Phenolics	1.60

STEELS

Material	Density (g/cc)
Stainless, T304	7.90
Low-Carbon, 1020	7.87
Low-Carbon, cold rolled	7.87
High Strength low alloy (High-Carbon)	

ALUMINUM

Material	Density (g/cc)
1100	2.71
6061, T6	2.70
7075, T6	2.81
A360	2.68
A380	2.76
B390	2.71

TITANIUM

Material	Density (g/cc)
Grade 2	4.51
Grade 3	4.50
Grade 4	4.51

ZINC

Material	Density (g/cc)
Zamak 2	6.60
Zamak 5	6.70
ZA8	6.30
ZA12	6.04
ZA27	5.02

BRASS

Material	Density (g/cc)
Brass 70/30	8.52
Brass 60/40	8.38
Phosphor bronze	8.92

MAGNESIUM

Material	Density (g/cc)
AZ91D	1.81
AMGOB	3.58

CAST IRON

Material	Density (g/cc)
Gray	7.15-7.89
Nodular	7.30

MONEL

Material	Density (g/cc)
Nickel copper	8.80

SYMBOL FOR:	ANSI Y14.5M	ASME Y14.5M	ISO
STRAIGHTNESS	—	—	—
FLATNESS	▱	▱	▱
CIRCULARITY	○	○	○
DYLINDRICITY	⌭	⌭	⌭
PROFILE OF A LINE	⌒	⌒	⌒
PROFILE OF A SURFACE	⌓	⌓	⌓
ALL AROUND	←⊖	←⊖	←⊖ (proposed)
ANGULARITY	∠	∠	∠
PERPENDICULARITY	⊥	⊥	⊥
PARALLELISM	//	//	//
POSITION	⊕	⊕	⊕
CONCENTRICITY	◎	◎	◎
SYMMETRY	NONE	⩶	⩶
CIRCULAR RUNOUT	*↗	*↗	↗
TOTAL RUNOUT	*↗↗	*↗↗	↗↗
AT MAXIMUM MATERIAL CONDITION	Ⓜ	Ⓜ	Ⓜ
AT LEAST MATERIAL CONDITION	Ⓛ	Ⓛ	Ⓛ
REGARDLESS OF FEATURE SIZE	Ⓢ	NONE	NONE
PROJECTED TOLERANCE ZONE	Ⓟ	Ⓟ	Ⓟ
TANGENT PLANE	Ⓣ	Ⓣ	NONE
FREE STATE	Ⓕ	Ⓕ	Ⓕ
DIAMETER	∅	∅	∅
BASIC DIMENSION	50	50	50
REFERENCE DIMENSION	(50)	(50)	(50)
DATUM FEATURE	–A–	*⫟Ⓐ	*⫟ or *⫟Ⓐ
DIMENSION ORIGIN	⊕→	⊕→	⊕→
FEATURE CONTROL FRAME	⊕ ∅0.5Ⓜ A B C	⊕ ∅0.5Ⓜ A B C	⊕ ∅0.5ⓂA B C
CONICAL TAPER	▷	▷	▷
SLOPE	◁	◁	◁
COUNTERBORE\SPOTFACE	⊔	⊔	NONE
COUNTERSINK	⌵	⌵	NONE
DEPTH\DEEP	↧	↧	NONE
SQUARE	□	□	□
DIMENSION NOT TO SCALE	15	15	15
NUMBER OF TIMES\PLACES	8X	8X	8X
ARC LENGTH	⌒105	⌒105	⌒105
RADIUS	R	R	R
SPHERICAL RADIUS	SR	SR	SR
SPHERICAL DIAMETER	S∅	S∅	S∅
CONTROLLED RADIUS	NONE	CR	NONE
BETWEEN	NONE	*↔	NONE
STATISTICAL TOLERANCE	NONE	⟨ST⟩	NONE
DATUM TARGET	⊘∅6/A1	⊘∅6/A1 or ⊘A1—∅6	⊘∅6/A1 or ⊘A1—∅6
TARGET POINT	✕	✕	✕

* MAY BE FILLED OR NOT FILLED

Sheet Metal Gage Designation

Gage	Cold Rolled Steel USS Gage Rev.		Galvanized Steel USS Gage		Long Terne USS Gage		Stainless Steel USS Gage Lbs per Sq. Foot			Monel USS Gage	
	Thickness in Inches	Lbs. per Sq. Foot	Thickness in Inches	Lbs. per Sq. Foot	Thickness in Inches	Lbs. per Sq. Foot	Thickness in Inches	Chrome Alloy	Chrome Nickel	Thickness in Inches	Lbs. per Sq. Foot
32	0.0100		0.0130	0.563			0.0100	0.418	0.427		
31	0.0110		0.0140	0.594			0.0109	0.450	0.459		
30	0.0120	0.500	0.0157	0.656	0.0120	0.518	0.0125	0.515	0.525		
29	0.0135	0.563	0.0172	0.719	0.0140	0.581	0.0140	0.579	0.391		
28	0.0149	0.625	0.0187	0.781	0.0150	0.643	0.0156	0.643	0.656		
27	0.0164	0.689	0.0202	0.844	0.0170	0.706	0.0171	0.708	0.121		
26	0.0179	0.150	0.0217	0.906	0.0180	0.768	0.0187	0.772	0.787	0.0187	0.820
25	0.0209	0.875	0.0247	1.031	0.0210	0.893	0.0218	0.901	0.918	0.0218	0.960
24	0.0239	0.100	0.0276	1.156	0.0240	1.018	0.0250	1.030	1.050	0.0250	1.140
23	0.0269	1.125	0.0308	1.281	0.0270	1.143	0.0281	1.158	1.181	0.0281	1.286
22	0.0299	1.250	0.3360	1.406	0.0300	1.268	0.0312	1.287	1.312	0.0312	1.420
21	0.0329	1.375	0.0360	1.531	0.0330	1.393	0.0343	1.416	1.443	0.0343	1.562
20	0.0359	1.500	0.0396	1.656	0.0360	1.518	0.0375	1.545	1.575	0.0375	1.700
19	0.0419	1.730	0.0456	1.906	0.0420	1.768	0.0437	1.802	1.837	0.0437	1.975
18	0.0479	2.000	0.0516	2.156	0.0480	2.018	0.0500	2.060	2.100	0.0500	2.297
17	0.0538	2.250	0.0575	2.406	0.0540	2.268	0.0562	2.317	2.362	0.0562	2.572
16	0.0599	2.500	0.0630	2.650	0.0600	2.518	0.0625	2.575	2.625	0.0625	2.848
15	0.0673	2.812	0.0710	2.969	0.0680	2.831	0.0703	2.896	2.953	0.0703	3.216
14	0.0747	3.125	0.7850	3.281	0.0750	3.143	0.0781	3.218	3.281	0.0781	3.583
13	0.0897	3.750	0.9400	3.906	0.0900	3.768	0.9370	3.862	3.937	0.0937	4.272
12	0.1046	4.375	0.1084	4.531	0.1050	4.393	0.1093	4.506	4.393	0.1093	5.007
11	0.1196	5.000	0.1233	5.136	0.1200	5.018	0.1250	5.150	5.250	0.1250	5.742
10	0.1345	5.625	0.1382	5.781	0.1350	5.643	0.1406	5.793	5.906	0.1406	6.431
9	0.1494	6.250	0.1532	6.406			0.1562	6.437	6.562	0.1562	7.166
8	0.1644	6.875	0.1681	7.031			0.1718	7.081	7.218	0.1718	7.855
7	0.1793	7.500					0.1875	7.590	7.752	0.1975	8.590

Metal Sheet Materials Chart

Material (Sheet less than 1/4″ thick)	How Measured	How Purchased	Characteristics
Copper	Gage number (Brown & Sharpe and Amer. Std.)	24″ × 96″ sheet or 12″ or 18″ by lineal feet on roll	Pure metal
Brass	Gage number (Brown & Sharpe and Amer. Std.)	24″ × 76″ sheet or 12″ or 18″ by lineal feet on roll	Alloy of copper and zinc
Aluminum	Decimal	24″ × 72″ sheet or 12″ or 18″ by lineal feet on roll	Available as commercially pure metal or alloyed for strength, hardness, and ductility
Galvanized steel	Gage number (Amer. Std.)	24″ × 96″ sheet	Mild steel sheet with zinc plating, also available with zinc coating that is part of sheet
Black annealed steel sheet	Gage number (Amer. Std.)	24″ × 96″ sheet	Mild steel with oxide coating, hot-rolled
Cold-rolled steel sheet	Gage number (Amer. Std.)	24″ × 96″ sheet	Oxide removed and cold-rolled to final thickness
Tin plate	Gage number (Amer. Std.)	20″ × 28″ sheet 56 or 112 to pkg.	Mild steel with tin coating
Nickel silver	Gage number (Brown & Sharpe)	6″ or 12″ wide by lineal feet on roll	Copper 50%, zinc 30%, nickel 20%
Expanded	Gage number (Amer. Std.)	36″ × 96″ sheet	Metal is pierced and expanded (stretched) to diamond shape; also available rolled to thickness after it has been expanded
Perforated	Gage number (Amer. Std.)	30″ × 36″ sheet 36″ × 48″ sheet	Design is cut in sheet; many designs available

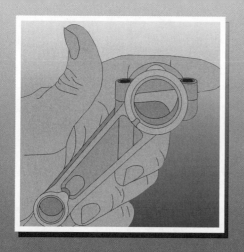

Index

Z

Zoom All button, 29
Zoom button, 29
Zoom Selected button, 30
Zoom Window button, 29